Energy metabolism in animals and man

Energy metabolism in animals and man

KENNETH BLAXTER Kt FRS
Former Director, Rowett Research Institute, Aberdeen, Scotland

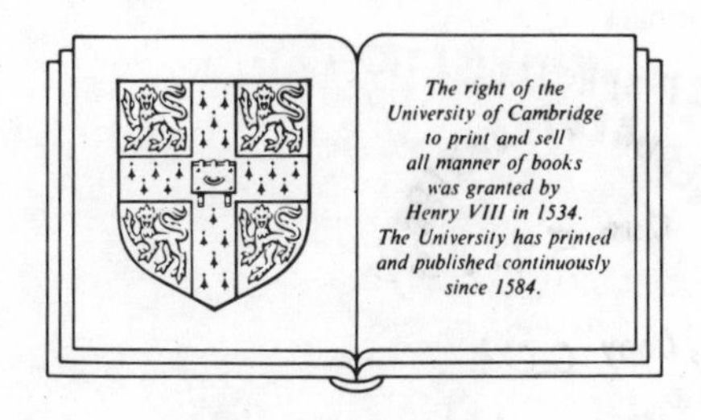

CAMBRIDGE UNIVERSITY PRESS
Cambridge
New York New Rochelle
Melbourne Sydney

Published by the Press Syndicate of the University of Cambridge
The Pitt Building, Trumpington Street, Cambridge CB2 1RP
32 East 57th Street, New York NY 10022, USA
10 Stamford Road, Oakleigh, Melbourne 3166, Australia

First published 1989

Printed in Great Britain at the University Press, Cambridge

British Library cataloguing in publication data

Blaxter, Sir Kenneth, 1919–
Energy metabolism in animals and man.
1. Animals. Energy metabolism 2. Man.
Energy metabolism
I. Title
591.1'33

Library of Congress cataloguing in publication data

Blaxter, K.L. (Kenneth Lyon), Sir.
Energy metabolism in animals and man / Kenneth Blaxter. p. cm.
Bibliography: p.
Includes index.
ISBN 0 521 36094 3. ISBN 0 521 36931 2 (paperback)
1. Energy metabolism. I. Title.
QP176.B53 1989
591.1'33–dc19

ISBN 0 521 36094 3 hard covers
ISBN 0 521 36931 2 paperback

SE

CONTENTS

ACKNOWLEDGEMENTS

My interest in the energetics of animals first arose in 1946, when, as a Commonwealth (Harkness) Fund Fellow, I worked with the late Professor H.H. Mitchell at the University of Illinois. On returning to Scotland to work, first at the Hannah Research Institute and later at the Rowett Research Institute, I was fortunate in having colleagues who had similar interests. I am grateful to these many colleagues and friends for their collaboration and help during the ensuing forty years in the exacting work which investigation of the energy exchanges of animals entails.

I am also indebted to many who have discussed various problems with me, read early drafts of the book and who have made helpful suggestions. Initial discussions with Dr J.A. McLean avoided overlap with his book with Dr G. Tobin 'Animal and Human Calorimetry'; we agreed that I would not deal with details of the construction of calorimeters and respiration chambers which they intended to and indeed did cover so admirably. Dr McLean also read some of the early drafts concerned with the physics of the heat exchange. Dr A.M. Prentice, Dr R.G. Whitehead and Professor J.S. Garrow all read the chapters dealing with the energy exchanges of man while Dr D.A.T. Southgate commented on the chapter dealing with the measurement of the energy value of human foods and diets. Dr L. Burkinshaw provided useful information on the newer, non-destructive methods for the determination of the composition of the body of man. Dr K.A. Nagy kindly provided information on field metabolic rates of different species in advance of their publication. I am particularly indebted to Dr A.J. Head of the National Physical Laboratory for his advice on the precision of measures of the enthalpy of combustion of organic compounds and for his searches of the chemical literature in attempts to find accurate combustion data. Professor W.D. Ollis was also helpful in providing advice on thermochemical matters.

Library staffs at the Rowett Research Institute, Aberdeen, the Institute of Animal Physiology, Babraham, the Food Research Institute, Norwich, and of the Department of Applied Biology, Cambridge gave freely of their time in tracing material for me and I am grateful to them.

Finally, the whole book has been read and commented on by Dr D.L. Ingram and by Professor D.G. Armstrong, both of whom are colleagues of long standing. They have made many useful suggestions and have corrected many of the infelicities of my written style. My thanks to them are considerable.

1
Introduction

1.1 The beginnings

At twelve minutes past eight on the morning of 3 February 1783, Antoine Lavoisier, working in close collaboration with Pierre Laplace, placed a guinea pig in the ice machine which Laplace had designed and which Lavoisier had built. This ice machine was the first animal calorimeter. It worked on the principle that the heat produced by the animal could be measured by the amount of ice melted. The experiment continued for over five hours and in the evening the same guinea pig was placed in the machine for a further ten hours. Holmes (1985) has written 'With these two experiments, carried out over a marathon twenty-four hour vigil, Lavoisier and Laplace began what Mendelsohn (1964) has called "the most important group of experiments in the history of metabolic-heat studies".' These experiments firmly linked the evolution of heat by animals with the consumption of oxygen and the formation of carbon dioxide.

In the period before the close of the eighteenth century, it was generally accepted that the source of animal heat was from an inborn fire producing an innate or vital heat. The innate heat was thought to arise in the left ventricle of the heart and the role of the lungs was to cool this internal fire through respiration. It had not been realised that physical and chemical processes were the source of this vital heat. Indeed the distinction between temperature and quantity of heat was not often made, and respired air was regarded simply as an elastic fluid, for its composition had not been elucidated. Priestley's discovery of oxygen (dephlogisticated air) and his experiments on the production of carbon dioxide (fixed air) by animals led to his theory that respiration and combustion were similar in that they both discharged phlogiston into the air (Priestley 1774). It was Lavoisier in his paper of 1777 who removed respiration from the confines of the phlogiston theory and established that during respiration the oxygen of air is diminished, the carbon dioxide content is increased and the nitrogen content remains unchanged (Lavoisier 1777). Parallel investigations were taking place in Scotland at this time, concentrating more on the heat produced rather than on the gaseous exchanges of animals. Adair Crawford measured the specific heat of oxygen and carbon dioxide and of arterial and venous blood. He concluded – and his conclusions were expressed in terms of the phlogiston theory – that animal heat derived from the air inspired, which in the lung was converted to carbon dioxide and received phlogiston. In quantitative terms he stated that the amount of air a man phlogisticated in a minute was the same as that altered by a burning candle (Crawford 1779). Thus before Lavoisier and Laplace made their critical experiments in February 1783, Lavoisier had made the major step of elucidating the process of respiration and Crawford had formulated ideas about the origin of animal heat in terms of the phlogiston theory. The studies with the guinea pig showed that heat produced could be related to the respiratory exchange in a quantitative way (Lavoisier & Laplace 1783).

Lavoisier, while sure that respiration was a slow form of combustion and fully analogous to the burning of charcoal or a candle, was uncertain about where the heat was generated, though he was convinced that carbon dioxide (fixed air) was produced in the lungs. The unifying hypothesis at the time was that carbon and hydrogen in the blood combined with vital air (oxygen plus caloric) in the lung and that the caloric (that is the heat) simultaneously united with the arterial blood to be circulated to the body where it was released (Seguin & Lavoisier 1793). The final proof that oxidation did not take place in the lung was not attained until 1847 when Magnus showed that arterial blood carried more oxygen and less carbon dioxide than did venous blood, and when von Helmholtz demonstrated in 1848 that isolated muscle produced heat (von Helmholtz 1848).

Lavoisier's contribution to the study of animal energetics was not limited to his elucidation of the relationship between respiration and the production of heat. His studies with Seguin on the metabolism of man, which involved the quantitative measurement of oxygen consumption and carbon dioxide production, uncovered hitherto unknown relationships. Some of these were listed in a letter which Lavoisier wrote to Joseph Black in November 1790 (Lavoisier 1790). Briefly, he showed that oxygen consumption is increased by the ingestion of food, by the performance of muscular work and by exposure to cold. He determined the minimal metabolism in the resting, post-absorptive state; he showed a proportionality between pulse frequency, respiratory frequency and metabolism; and, demonstrated that the volume of nitrogen expired is the same as that inspired. The earlier work with animals had also shown that, within a species, oxygen consumption is proportional to body size. These discoveries were major ones, and much of the material in this book can be traced back to these original observations made two centuries ago.

1.2 Animals and man

The number of animal species in which metabolism and energy exchange has been studied is very small when compared with the vast number of known species. For example, even the simple determination of resting metabolism has been undertaken for only about 100 out of approximately 25 000 known bird species. Detailed studies have been made with even fewer species. Most studies have been made with laboratory or domesticated animals, and of course, on man. The species coverage in this book is largely limited to mammals and birds. Despite the considerable amount of work undertaken with invertebrates they are not considered here, and scant attention is given to poikilothermic animals – fish, reptiles and amphibians – despite the knowledge that has accrued about their metabolism. Happily in recent years there has been an increased amount of investigative work undertaken with wild species of mammals and birds, and where possible, results obtained with these species have been included. Where wild species of animal are mentioned on the first occasion their latin name is given. The nomenclature used is that adopted by Bertin (1973). Nevertheless, the reader will discern a bias towards farm species and man; this simply reflects the greater amount of information about metabolism in these groups!

1.3 Units and symbols

The International System of Units (SI) (HMSO 1973) has been used throughout with one or two exceptions. Body temperatures and environmental temperatures are reported in the Centigrade rather than the Kelvin scale. Another exception relates to the use of the watt (W), the derived SI unit for power. Much of the study of energy exchange is concerned with the rates of energy expenditure per unit time, that is, with power. In nutritional studies in particular, to express power in terms of watts is inconvenient because a frequent calculation is to compute the mass of food needed each day by dividing the rate of energy expenditure per day by the specific energy of the food (energy/unit mass). To express the rate of expenditure as watts (i.e. J/s) would involve an awkward calculation involving the number of seconds in a day (86 400). It is more convenient to express the rate of energy expenditure as J/d and the specific energy of the food as J/kg. Although in many published studies power has been expressed as energy per minute or per hour, when such work has been quoted, the expressions have usually been replaced by rates per second or per day, depending on the context.

Symbols for variables conform as far as possible with those commonly used in respiratory physiology and thermal physiology and also agree with general practices adopted in the literature of bioenergetics. Where necessary variables are differentiated by the use of subscripts. Much of the subject of the book is concerned with rates – rate of heat production, $\mathrm{d}H/\mathrm{d}t$, rate of gain in weight $\mathrm{d}W/\mathrm{d}t$, etc. The Newtonian convention of placing a dot above the character is adopted to represent rates. Whenever a symbol is written in an equation it is immediately defined. It will be noted that multi-lettered symbols for quantities have not been used. With the advent of high-level computer languages, the use of multi-lettered symbols for quantities has become commonplace. Sometimes this aids understanding but equally can lead to cumbersome expressions when any algebraic manipulation of the quantities is involved.

1.3.1 *Calories and joules*

The agreed SI unit of energy is the joule (J), defined as $\mathrm{kg/(m^2s^2)}$. The older unit of energy, and one which traces back to the concept of 'caloric' of the late eighteenth century, is the calorie. This is broadly defined as the amount of heat required to increase the temperature of one gram of water by one degree Centigrade. This amount of heat is not a constant because the specific heat of water varies with temperature. At 0 °C its value is 4.218 J/g, and at 100 °C it is 4.216 J/g passing through a minimum of 4.178 J/g at a temperature of 34.5 °C. The calorie used in physiological studies is stated to be the '15 °C calorie', or, the heat required to increase the temperature of one gram of water from 14.5 to 15.5 °C. It differs appreciably from the Bunsen calorie which applies specifically to the ice calorimeter and a temperature of 0 °C or the steam tables calorie. The equivalent of the '15 ° calorie' in terms of joules is thus dependent on the precision of the measurement of the specific heat of water at this temperature. Because of this difficulty, the calorie used in physiological studies has been defined precisely as the heat required to increase the temperature of one gram water by one degree Centigrade when the specific heat of water is exactly 4.184 J/g. This definition of the calorie, or rather the thermochemical or Rossini calorie – 1 calorie = 4.184 joules – creates no problems in the conversion of data expressed in calories to joules since calorimeters are standardised using thermochemical grade benzoic acid. The heat of combustion of this standard is determined by national organisations in terms of a heat input which is measured in units of joules, by passing a known current through a known resistance for a known time. The primary standard on which thermochemical and biological energy measurements are based is thus one which is defined in joules.

Since the adoption of the International System in the early 1970s, biological scientists have adopted the joule as the unit of energy. The calorie as a unit of measure, however, persists in widespread use, notably in the USA and throughout dietetics. People concerned with their own diets continue to 'watch their calories'! No doubt with time this usage will disappear.

1.4 Presentation

Bioenergetics is a vast subject. It includes events taking place at the molecular and cell organelle level as well as those which relate to the energetics of individual animals and indeed of populations of animals. In dealing with this considerable range of material, the starting point has been a consideration of the energy exchanges of the individual animal. From this point departures have been made to deal in more detail with the underlying biochemical and biophysical processes involved. This approach necessarily results in a presentational structure which reflects the factors which affect the overall energy expenditures of animals and man. These include the amount of food ingested, the extent to which growth or lactation or pregnancy ensue, the amount and severity of physical exercise and the impact of the physical environment. Chapters 2–7 are devoted to the methods and approaches employed in the study of the metabolism of energy-yielding constituents and to the factors which affect metabolic rate. The biochemistry of oxidations and syntheses in the body are dealt with in a somewhat abridged form as are the physical processes concerned with heat loss. Chapters 8–11 consider the biology of minimal metabolism, muscular work, thermoregulation, reproduction and growth. Chapter 12 discusses the way in which the energy-yielding constituents of food elicit increases in metabolism, and Chapter 13 summarises the applications of animal bioenergetics. Throughout the book a comparative approach has been adopted, giving emphasis to inter-species similarities and drawing attention to the more interesting differences.

2
General principles

2.1 Bioenergetics and thermodynamics

Much of the study of the energy exchanges of animals and man can be undertaken within the framework of the first law of thermodynamics. The second law, concepts of entropy and of free energy, and other aspects of equilibrium thermodynamics are, however, relevant and so too are ideas of entropy generation and other concepts which have emerged from the study of non-equilibrium thermodynamics. Where pertinent in explaining biochemical and biophysical processes within the body they will be employed. However, most of what can be called the classical bioenergetics of whole organisms involves nothing more than a careful accounting in terms of the enthalpy of reaction of changes in the energy of the body and the basic concept of the first law of thermodynamics that energy is conserved.

2.1.1 *The discovery of the first law of thermodynamics*

One can perhaps find within the old caloric theory of heat the rudiments of the idea that energy is conserved. Caloric was regarded as an indestructible fluid which permeated the interstices of all matter, and in some of the writings of Lavoisier and Laplace there is an indication that they were approaching the same conclusion that was reached by Hess a half-century later. The caloric theory of heat was sufficient for the time; it explained the results of elementary calorimetry, accommodated the idea of specific heat and its variation with temperature and gave an explanation of latent heat. Indeed the second law of thermodynamics, which was formulated before the first law, by Sadi Carnot in 1824, is expressed in terms of the old caloric theory (Carnot 1824). Whatever the attraction and apparent utility of these views, they had to make way for a more general theory, and there is little doubt that the advances towards this end came more from physics and mechanics than they did from chemistry. The term 'energy' was first proposed by Thomas Young in 1807 as a synonym for the latin '*vis-viva*' of Newton, the latter being what we now understand as kinetic energy. The term only slowly gained acceptance. Ideas about the conservation of mass and of mechanical energy were certainly extant in the late eighteenth century but it was the difficulty of grasping the concept of energy which prevented a ready acceptance of Robert Mayer's formulation of the first law in 1842 (Mayer 1842). Before this, in 1838, Germain Henri Hess who occupied the Chair of Chemistry at the Chief Pedagogical Institute at St Petersburg, published his paper showing that the heat produced in a chemical reaction was always the same, regardless of whether the process went directly or proceeded through a number of intermediate steps. (Hess 1838). This finding, usually termed the law of constant heat summation, is a statement within the framework of the first law. (See Leicester 1951 for an account of Hess' life.) Mayer produced no new experimental evidence for his pronouncement of the law and it was the monumental work of James Prescott Joule showing the equivalence of work and heat that provided the basis for the explicit statement of the first law by Hermann von Helmholtz in 1847. Even here, it was apparent that Helmholtz thought in terms of the caloric theory and there was, additionally a semantic problem because the German '*kraft*' was used in the sense of 'force' as well as 'energy'. It was not until later in the century that the German word '*energie*' was employed.

The new 'dynamical theory of heat' was slowly accepted and it was then realised that it had great generality. Much controversy has followed its discovery, particularly regarding who should receive credit for priority (see for example Lloyd 1970). Kuhn (1955) quoted 12 simultaneous discoveries of the law, a conclusion which was criticised by Elkana (1974) who contended that the various discoveries were different aspects of the same thing. Elkana wrote

> Mayer discovered the indestructibility of forces of nature; Helmhotz discovered that the sum of the various kinds of force is a constant; Joule discovered the mutual convertibility of heat and 'mechanical power' as a result of which Thomson established the dynamical theory of heat on a firm foundation. Finally Clausius, Thomson and Rankine showed the equivalence of these various results and it becomes clear that one result is derivable from another, i.e. that all of them had actually discovered the same thing.

Perhaps Elkana should have included Hess in his list since his finding was just as seminal and important.

2.1.2 *Heat, work and enthalpy*

A minimal account of the first law is given here; for more complete accounts reference should be made to texts of thermochemistry (Nash 1970; McGlashan 1979).

Thermodynamics deals with systems, that is regions which have real or conceptual boundaries separating them from the rest of the universe. A system is defined by a small number of parameters – volume, quantity of chemical constituents (expressed in moles), pressure, temperature and electrical charge. Conceptually such an isolated system can be regarded as having rigid walls impermeable to heat, mass and charge transfer. Thus the internal energy, E, can be regarded as constant. The first law of thermodynamics states that the energy content of the system can be changed from an initial state, E_i, to a final state, E_f, by inputs of heat and work according to the equation:

$$E_f - E_i = \Delta E = q - w \quad (2.1)$$

where ΔE is the change in the internal energy of the system; q is the amount of heat absorbed by the system; w is the work done by the system, the negative sign indicating that work has been done on the system. The convention that the work in the first law is given a negative sign for work done on the system reflects the concern of investigators in the mid-nineteenth century with what could be achieved from engines in terms of a work output. The equation could be expressed in differential form, $dE = dq - dw$, but the alternative is used for the reason given by Lewis & Randall (1923); it is not reasonable to use dx for an infinitesimal value of x for it refers to an infinitesimal increment of x.

The work term can be expanded to distinguish between different types of work, notably mechanical work (force × distance), electrical work (electromotive work × charge transported), work of gaseous expansion (pressure × change in volume) and others. The first law can thus be written:

$$\Delta E = q - \Delta(PV) - w_{other} \quad (2.2)$$

where P = pressure and V = volume.

In biological systems, pressure is constant both within and without the system. A new quantity, $q_P = \Delta H$ = change in enthalpy, can be defined by placing P outside the operator when

$$\Delta E = \Delta H - P\Delta V - w_{other} \quad (2.3)$$

and, neglecting other work, the formulation of the law for most biological applications is thus:

$$\left.\begin{array}{lll} \Delta E & = \Delta H & - (P \times \Delta V) \\ \text{change in internal energy} & = \text{change in enthalpy} & - \text{(pressure × change in volume)} \end{array}\right\} \quad (2.4)$$

Enthalpy, like internal energy, and indeed the other thermodynamic functions, such as Gibb's free energy, and entropy, is a function of state. In transition from one state to another the change is a net one, independent of the path taken and determined solely by the initial and final values. Hess' law of constant heat summation, referred to earlier, is an expression of this property which is implicit in the first law. Hess' law is a very powerful tool in dealing wih the enthalpies of chemical reactions and with free energy and entropy changes.

2.1.3 *Enthalpies of reaction*

Thermochemical equations can be exemplified by a consideration of the oxidation of one mole of solid carbon as graphite and one mole of hydrogen gas by gaseous oxygen:

$$C(s) + O_2(g) = CO_2(g) : \Delta H = -393.513 \text{ kJ/mol} \quad (2.5)$$

$$H_2(g) + \tfrac{1}{2}O_2(g) = H_2O(l) : \Delta H = -285.84 \text{ kJ/mol} \quad (2.6)$$

The equations state whether the reactants and products are solids (s), liquids (l) or gases (g) and in these exothermic reactions ΔH is negative in accordance with the algebraic definition of the first law.

In equations (2.5) and (2.6) what is measured is a change of enthalpy not enthalpy itself. By choosing a reference or standard state and assigning zero enthalpy to the elements, enthalpies of formation of a compound from its elements can be formulated. The reference state chosen is atmospheric pressure (101.38 kPa, 760 mm Hg) and a temperature of 25 °C (298.15 K). In the formation of carbon dioxide and water in equations (2.5) and (2.6) above, the enthalpy of the reaction is clearly an enthalpy of formation, ΔH_F, since both compounds have been formed from their elements.

The great power of Hess' law stems from the fact that it shows that thermochemical equations can be multiplied, added or subtracted and thus used to calculate heats of reactions in cases where direct determinations are impossible. Provided one has a set of heats of formation of compounds, then the enthalpy of any reaction, whether exothermic or endothermic, can be calculated as the difference between the heats of formation of products and reactants:

$$\Delta H = n_p \sum \Delta H_{F(p)} - n_r \sum \Delta H_{F(r)} \quad (2.7)$$

where n_p is the number of moles of products with enthalpies of formation $\Delta H_{F(p)}$; and n_r is the number of moles of reactants with enthalpy of formation $\Delta H_{F(r)}$.

Enthalpies of formation are commonly computed from combustion data. The example below relates to the determination of the enthalpy of formation of D-glucose. The combustion of one mole of solid glucose by oxygen gives the following thermochemical equation:

$$C_6H_{12}O_6(s) + 6O_2(g) = 6CO_2(g) + 6H_2O(l) : \Delta H = -2803.1 \text{ kJ/mol} \quad (2.8)$$

By adding (with careful consideration of algebraic signs) six × equation (2.5) for the combustion of carbon and hence the heat of formations of carbon dioxide, and six × equation (2.6) for water, one can obtain the enthalpy of formation of glucose from its elements:

$$\begin{array}{llll} \text{product } CO_2(g) & + \text{product } H_2O(l) & - \Delta H_C & = \Delta H_F \\ (6 \times -393.51) & + (6 \times -285.84) & -(-2803.1) & = -1273 \text{ kJ/mol} \end{array} \quad (2.9)$$

Calculations of enthalpies of reaction from data on enthalpies of combustion are not always of high precision because they are obtained as differences between large numbers. Other methods can be employed which derive from the second law of thermodynamics, notably from the relation between the equilibrium constant and the reciprocal of absolute temperature (see Chapter 6). The slope of this relationship is a measure of ΔH. An example of this approach is the determination of the enthalpy of the reduction of nicotinamide adenine dinucleotide made by Burton (1974).

2.1.4 *Enthalpy of combustion and bomb calorimetry*

The three previous examples of thermochemical equations all relate to combustion in oxygen. The usual method of measurement is to burn the material in a heavily walled 'bomb' under a pressure of oxygen and to measure the heat loss. This procedure is termed bomb calorimetry and apparatus and techniques have been described by Sturtevant (1945) and McLean & Tobin (1987). The bomb maintains a constant volume but not a constant pressure. If the number of moles of gas produced is greater than the number consumed, then pressure rises. The measure of heat thus includes pressure–volume work and is not a measure of enthalpy. For example, in equation (2.8) six moles of carbon dioxide were produced and six moles of oxygen were consumed; the heat measurement was thus at constant pressure. In the combustion of palmitic acid, however, shown in equation (2.10), 23 moles of oxygen are consumed and only 16 moles of carbon dioxide are produced:

$$\begin{array}{l} CH_3(CH_2)_{14}COOH(s) + 23O_2 = 16CO_2(g) + 16H_2O(l): \\ \Delta H = -10\,014 \text{ kJ/mol} \end{array} \quad (2.10)$$

Using the gas laws, the first law of thermodynamics can be expressed:

$$\Delta H = RT\Delta n + \Delta E \quad (2.11)$$

where R is the gas constant (8.3144 J/mol), T is absolute temperature and Δn the difference between the number of moles of gaseous products and reactants. In equation (2.10) what was measured was ΔE (− 10 014 kJ/mol). The correction to obtain ΔH is $-7 \times 8.3144 \times 293.15 = 17.06$ kJ. The enthalpy of combustion is thus $-10\,014 - 17 = 10\,031$ kJ/mol. This correction is small. In most tabulations of enthalpies of combustion of organic molecules of known elemental composition, such corrections of observed heat have been made. Other corrections must also be applied to the crude data emanating from a bomb calorimeter experiment if results of high precision are required. These corrections, known as the Washburn corrections (see Sturtevant 1945), correct the combustion data to the standard state, and consider such points as the fact that some of the carbon dioxide will dissolve in the water in the bomb, that the oxygen is moist and other matters. The corrections are again very small. The Appendix lists the *enthalpies* of combustion of compounds of biological interest.

What is more pertinent, however, is whether combustion data and enthalpies of formation which have been determined for the standard state can be applied to biological systems. There are two considerations, firstly the temperature at which biological reactions take place and secondly the standard state of the products of reactions and in some instances the reactants as well. The effects of such departures from the standard state warrant examination, beginning with temperature effects.

2.1.5 *Effects of temperature on the enthalpy of reaction*

The standard conditions used in determinations of enthalpies of reaction are standard pressure and a temperature of 25 °C. It can be argued that in studies of the metabolism of warm-blooded animals the chosen temperature should be 37–39 °C. Enthalpies of reaction at different temperatures can be calculated using Kirchhoff's equation

$$\Delta H_2 = \Delta H_1 + \int_{T_1}^{T_2} C_P \mathrm{d}T \tag{2.12}$$

where C_P is heat capacity at constant pressure. Except when differences in temperatures T_2 and T_1, are large, heat capacity can be regarded as constant and thus Kirchhoff's equation reduces to:

$$\Delta H_2 = \Delta H_1 + \Delta C_P(T_2 - T_1) \tag{2.13}$$

The difference in heat capacities is again the difference between that of products and reactants. In the instance of the combustion of glucose the enthalpy of combustion at 293.15 K is 2803.1 kJ/mol. From the heat capacities of glucose, oxygen, carbon dioxide and water the difference between products and reactants is 225 J/K expressed per mole of glucose. At 39 °C the integral term in Kirchhoff's equation is 3.14 kJ/mol. The correction is about 0.1% of the value determined at 25 °C, and is sufficiently small that it is usually neglected in most calorimetric studies with animals and man. Calculation for other combustions of biological interest give similar proportional corrections for temperature effects.

2.1.6 *Protein oxidation*

The second question about the relevance of standard enthalpies of reaction relates to possible effects of departures from the standard state. Before dealing with this it is necessary to consider the oxidation of protein. In the animal body this is not complete; nitrogen is excreted largely as urea in mammals and largely as uric acid in birds. These are not in the standard state as solids but are in solution. In the computation of the enthalpy of the partial oxidation, allowance has to be made for the heat of solution of the nitrogenous end-product. The biological oxidation of alanine can be taken as an example. The thermochemical equation for the complete combustion of alanine is:

$$CH_3CHNH_2COOH(s) + 3\tfrac{3}{4}O_2(g) = 3CO_2(g) + 3\tfrac{1}{2}H_2O(l) + \tfrac{1}{2}N_2(g): \quad \Delta H = -1619.6 \text{ kJ/mol} \qquad (2.14)$$

The corresponding equation for the complete combustion of urea is:

$$CO(NH_2)_2(s) + 1\tfrac{1}{2}O_2(g) = CO_2(g) + 2H_2O(l) + N_2(g): \quad \Delta H = -631.8 \text{ kJ/mol} \qquad (2.15)$$

and the equation for the solution of urea to give a 0.1 molar concentration is:

$$H_2O(l) + CO(NH_2)(s) = CO(NH_2)_2(aq, 0.1 \text{ molar}): \quad \Delta H = -15.3 \text{ kJ/mol} \qquad (2.16)$$

The equation for the partial oxidation of alanine to yield urea in solution is obtained by subtracting 0.5 × equation (2.15) and 0.5 × equation (2.16) from equation (2.14) to give:

$$CH_3CHNH_2COOH(s) + 3O_2(g) = 2\tfrac{1}{2}CO_2(g) + 2\tfrac{1}{2}H_2O(l) + \tfrac{1}{2}CO(NH_2)_2 \text{ (aq 0.1 molar)}: \quad \Delta H = -1296.1 \text{ kJ/mol} \qquad (2.17)$$

Similar calculations can be made when the end-product of the oxidation is uric acid or ammonia. Heats of solution are given in compilations of chemical data and most of them date back to the International Critical Tables (1929). The enthalpy of solution of urea in the above example, however, relates to observations by Egan & Luff (1966).

To ignore the heat of solution of urea would result in an error of 0.6% in the estimate of the enthalpy of the biological oxidation of alanine. For a smaller amino acid such as glycine, which is richer in nitrogen than alanine, the error would be 1.2%. Thus small errors are incurred by assuming that the standard conditions employed in arriving at tabulated values of enthalpies of combustion can be applied to biological systems. Where nutrients are ingested in solution, or when metabolites are excreted in solution, some consideration should be given to the magnitude of the enthalpy terms involved, if only to reject the correction as too small to affect the outcome of the study.

2.2 Estimating heat from the gaseous exchange

If the amounts of the many organic compounds oxidised in the body are known the total heat produced by an animal can be calculated exactly by summating the enthalpies of their oxidation. Heat production can also be estimated from measurement of the oxygen consumed and the carbon dioxide and other end-products produced, that is, from the gaseous exchange and urinary excretion.

Equations (2.8), (2.10) and (2.17) describe the overall oxidation of glucose, palmitic acid and alanine, and, as a first approximation, these three compounds can be taken to be representative of the three major classes of organic compounds oxidised in the body – carbohydrate, lipid and protein. The three equations can be written as a series of simultaneous ones in which the molar enthalpies of reaction are regarded as determined by the molar amounts of oxygen consumed and the molar amounts of carbon dioxide produced and of nitrogen excreted:

Compound oxidised	$-\Delta H$ (kJ/mol) $= H_p$		Oxygen consumed (mol)		Carbon dioxide produced (mol)		Nitrogen excreted (mol)	
Glucose	2803	=	$6a$	+	$6b$	+	$0c$	(2.18)
Palmitic acid	10 039	=	$23a$	+	$16b$	+	$0c$	
Alanine	1296	=	$3a$	+	$2.5b$	+	$1c$	

Note that the heading of the second column is $-\Delta H$, making the entries in the face of the tabulation positive quantities. In biological applications it is the convention to ignore the way in which the first law was formulated and give the heat produced by an animal, that is the enthalpy of the thermochemical reactions within the body, a positive sign. Thus $-\Delta H$ of thermodynamics is equal to H_P in bioenergetics (see page 18). Equations (2.18) can be solved algebraically to give values of the coefficients, a, b, c, that is the heat produced in the body per mole of oxygen consumed, of carbon dioxide produced and of nitrogen excreted as urea. The coefficients can equally be expressed as heat per litre of the gases – since 1 mole of gas occupies 22.41 litres – and as grams of urinary nitrogen. For the above 'reference compounds' the values of the coefficients are:

$$\left.\begin{aligned} a &= 366.3 \text{ kJ/mol } O_2;\ 16.34 \text{ kJ/l } O_2 \\ b &= 100.8 \text{ kJ/mol } CO_2;\ 4.50 \text{ kJ/l } CO_2 \\ c &= -54.9 \text{ kJ/mol N};\ 3.92 \text{ kJ/g N} \end{aligned}\right\} \quad (2.19a)$$

The equation:

$$-\Delta H\ (\text{kJ/mol}) = 16.34V_{O_2} + 4.50V_{CO_2} - 3.92\text{N (grams)} \quad (2.19b)$$

where V_{O_2} is the volume of oxygen in litres and V_{CO_2} the volume of carbon dioxide in litres, results in exact prediction of the heat produced by an animal for any mixture of glucose, palmitic acid and alanine oxidised in the body.

The approach can be extended. In the instance of herbivores in which the carbohydrate of the diet is fermented anaerobically in the gut to give methane and lower steam-volatile fatty acids, and the latter are then oxidised to carbon dioxide and water, a further thermochemical equation can be added to the group with a new coefficient, *d*, representing the number of moles of methane formed. Similar methods can be used to accommodate other incomplete combustions such as those noted in ketosis. In ketosis, for example the thermochemical equation for the incomplete combustion of fat with the concomitant production of acetoacetic acid, β-hydroxybutyric acid and acetone is used and a further equation is introduced into the set in which the independent variable is the quantity of ketones excreted and the dependent one the heat of reaction. The introduction of further terms of this nature does not alter the original coefficients.

The constants in equation (2.19), however, are specific to the reference compounds chosen. If sucrose, oleic acid and glutamic acid had been taken to be representative of carbohydrate, lipid and protein respectively, the coefficients would be: $a = 355.7$, $b = 114.8$ and $c = -107.3$ kJ/mol. The difference between these coefficients and those for glucose, palmitic acid and alanine, illustrates that care must be taken about the choice of reference compounds. It is usual, in the instance of man and simple-stomached species, to take starch as the reference carbohydrate; in herbivores cellulose would be a more sensible choice, while in suckling animals lactose is preferred. For lipids it is usual not to take a fatty acid as the reference base but rather mixed triacylglycerols and particularly those which are close in composition to the lipids of the body. Amino acids are not used to represent protein; rather intact proteins are employed. The analytical data on which the choices of reference substances have been based have been discussed by Brockway (1987) and McLean & Tobin (1987). Table 2.1 summarises some of the factors which are or have been in use for man and animals, and here the additional term for methane excretion has been included. The low coefficient for protein in the factors for use with birds reflects their excretion of uric acid rather than urea.

The final column in Table 2.1 gives the heat production by an animal, computed from these different sets of factors; when its oxygen consumption is 500 l/d, its carbon dioxide production 400 l/d and its excretion of urinary

Table 2.1. *Coefficients for estimating heat production from oxygen consumption, carbon dioxide production, methane production and the excretion of nitrogen in the urine*

Authority	Oxygen consumption (kJ/l)	Carbon dioxide production (kJ/l)	Urinary nitrogen excretion (kJ/g)	Methane production (kJ/l)	Calculated heat (MJ)
Man and predominantly simple-stomached species					
Benedict[a]	16.20	4.94	−5.80	—	9.99
Weir (1949)	16.50	4.63	−9.08	—	9.96
Ben-Porat *et al.* (1983)	16.37	4.57	−13.98	—	9.80
Brockway (1987)	16.57	4.50	−5.90	—	10.00
Farm mammals, including ruminant species					
Hoffman (1958)	16.07	5.23	−5.26	−2.40	10.04
Brouwer (1965)	16.18	5.16	−5.93	−2.42	10.01
Birds					
Farrell (1974)	16.20	5.00	−1.20	—	10.08

Note:
[a] As modified by Lusk (1928) and by Cathcart & Cuthbertson (1931).
The heat was calculated for an oxygen consumption of 500 l, a carbon dioxide production of 400 l and a urinary nitrogen excretion of 15 g.

nitrogen 15 g/d. The variation between the values computed from the factors is small – excluding the value by Ben-Porat, Sideman & Bursztein (1983) it is about 1%. Generally it can be concluded that any set of factors, based on accurate heats of combustion of the dietary carbohydrates consumed and of the body lipids of the species concerned, leads to errors in estimating heat production which are unlikely to be greater than 1% (Blaxter 1962a, 1967). For virtually all purposes the factors of Brouwer (1965), which were arrived at by international consultation, suffice for both animals and man. In birds, however, the nitrogen coefficient of Farrell (1974) should be used.

2.2.1 *Synthesis of fat from carbohydrate*

The above analysis deals with the estimation of $-\Delta H$ arising from the oxidation of nutrients or bodily constituents from the respiratory exchange. It is as important to consider the estimation of the enthalpies of reaction when synthesis occurs. When glucose is converted to palmitic acid, each glucose molecule provides two two-carbon fragments, and the minimal formulation for the synthesis is:

$$4C_6H_{12}O_6(s)+O_2(g)=CH_3(CH_2)_{14}COOH(s)+8CO_2(g)+8H_2O(l); \quad \Delta H=-1173 \text{ kJ/mol} \qquad (2.20)$$

When the factors for oxygen (366.3 kJ/mole) and carbon dioxide (100.8 kJ/mole), derived from the oxidation of glucose and palmitic acid, are applied to this equation the estimated heat of reaction is precisely the same as that determined from heats of formation in equation (2.19). Thus the heat arising from synthesis of fatty acids from glucose can be determined using factors derived from their oxidation. Experimental proof of this was obtained by Bleibtreu (1901) and by Benedict & Lee (1937) who worked with geese which deposit fat readily and by Lusk (1915) working with dogs. Ben-Porat *et al.* (1983) and Acheson *et al.* (1984) have recently given additional support based on classical thermodynamic reasoning.

2.2.2 *Estimation of the heat of reaction from oxygen consumption alone*

An alternative way of estimating the enthalpy of combustion of the compounds oxidised by the body is to consider the oxygen consumption alone. The approach can be illustrated by the example already given in equation (2.19). When glucose is oxidised, as exemplified in equation (2.8), the ratio of carbon dioxide produced to oxygen consumed, a ratio called the respiratory quotient, is 1.00, and the enthalpy of combustion is −467.1 kJ/mol O_2 or −20.84 kJ/l O_2. When palmitic acid is oxidised, as in the equation (2.10), the respiratory quotient is 0.696 and the enthalpy of combustion per unit of oxygen consumed is −436.5 kJ/mol O_2 or −19.47 kJ/l. When alanine is partially oxidised the respiratory quotient is 0.833 and the enthalpy change is −432.0 kJ/mol O_2 or −19.28 kJ/l O_2. If the respiratory quotient is above 1.00, indicating that fat is being synthesised from carbohydrate, the enthalpy of reaction is also greater. Thus if the respiratory quotient is 1.3, the enthalpy change per unit of oxygen consumed is −507.4 kJ/mol O_2 or −22.63 kJ/l.

Employing the convention that $-\Delta H$ of thermodynamics is equal to H_P of bioenergetics, one dispenses with negative signs and regards the enthalpy of reaction per mole or per litre of oxygen consumed as a positive quantity, the 'heat equivalent of oxygen consumed'. It is obvious that in an animal or man the heat equivalent of oxygen consumed varies with the proportions of fat, protein and carbohydrate oxidised and whether net synthesis of fat takes place. It will also vary with the choice of the reference substances used to arrive at the basic thermodynamic equations. When the Brouwer respiratory exchange factors (which are those preferred for mammals) are employed, the range of heat equivalents, expressed as kJ/l O_2 consumed, are those given in Table 2.2. The range is considerable and clearly a choice has to be made. A common choice is 20.1 kJ/l O_2 which corresponds roughly to the average of the values for fat and carbohydrate. In fasting animals which commonly derive about 20–30% of their energy from body protein and the

Table 2.2. *The heat equivalent of oxygen and of carbon dioxide under different circumstances*

Substrate oxidised	Respiratory quotient	O_2 consumed (kJ/l)	CO_2 produced (kJ/l)
Lipid	0.711	19.7	27.8
Protein	0.809	19.2	23.8
Carbohydrate	1.000	21.2	21.2
Lipid synthesised	1.100	21.7	19.7
Lipid synthesised	1.300	22.7	17.5
Values normally chosen when only one gas determined	—	20.1	24.1

remainder from body fat a value of 19.7 would be more appropriate, and in animals which are laying down considerable amounts of body fat during times of dietary plenty, a value closer to 21.5 would lead to less error. Comparison of the numerical magnitude of the equivalents for different substrates might suggest that no major error would accrue by the adoption of the standard value of 20.1 kJ/l O_2. This might be true in some applications but in precise calorimetric work with animals and man it is not recommended.

2.3.3 *Estimation of the heat of reaction from carbon dioxide production alone*

The last column of Table 2.2 gives the heat of reaction per litre of carbon dioxide alone. The range of heat equivalent of carbon dioxide is considerably greater than it is for oxygen and the choice of a mean value is thus associated with a much greater potential error. This is of considerable importance because the technique of measuring the differential rates of loss of ^{18}O and ^{2}H radioisotopes when doubly-labelled water is given yields an estimate of the mean rate of carbon dioxide production over a long period of time (see Section 4.2.2). To convert such an estimate of carbon dioxide production to an estimate of heat entails estimating the heat equivalent of carbon dioxide.

If an animal or a man is neither losing chemical energy by combustion nor gaining it by synthesis of the organic constituents of its tissues, the sole source of heat is from the oxidation of its food. The determined or calculated respiratory quotient of the diet (or rather that part which is absorbed) can then be used to establish the heat equivalent of a litre of carbon dioxide. For a man consuming a western-type diet the respiratory quotient ranges from about 0.75 to about 0.95 and, using the Brouwer factors, the heat equivalent of a litre of carbon dioxide ranges from 26.6 to 22.0 kJ. The commonly accepted mean value of 24.1 kJ/l could lead to errors of about ±10%, even if the subject is in energy equilibrium; if the subject was fasting or receiving a diet which resulted in gain in body tissue, the error would be much greater.

2.3 Nomenclature in bioenergetics and thermochemistry

In thermochemistry the enthalpy of an exothermic reaction is given a negative sign and the operator, Δ, is employed to indicate that it is a difference between an initial and final state. It has already been noted that in dealing with the heat produced by chemical reactions in the body, it is more useful to employ a symbol with a positive sign (H_P). The convention employed in dealing with the work term in animal physiology is the same as that employed in the formulation of first law, namely, work (w), is defined as work done by the system – that is by the animal – on the surroundings. $H_P = -\Delta H$ is usually called the heat production of the animal; $H_P - w$ is the heat which arises in the body, for the work term is by definition energy which is transferred to the environment. This heat may, over short periods of time be stored in the body, as exemplified by a rise in its average temperature, but it is mostly dissipated by radiation, convection and conduction from the body surface or lost as latent heat by the evaporation of water from the skin and respiratory passages. These aspects of the biological energy exchange are discussed in Chapter 7. Heat storage is given the symbol, s, and the heat emitted the symbol H_L.

Unlike the relationships of thermodynamics which do not involve time, the bioenergetics of animals and man is largely concerned with time rates of energy. Thus the terms defined above are best expressed as rates, using the Newton convention, when:

$$\dot{H}_P - \dot{w} = \dot{H}_L + \dot{s} \tag{2.21}$$

where $\dot{H}_P$ = rate of heat production, i.e. the heat produced by the metabolism of food or bodily constituents; $\dot{w}$ = rate of work done by the animal on its environment; $\dot{H}_L$ = rate of heat loss; $\dot{s}$ = rate of storage of heat in the body.

This expression is important in understanding what is measured in energy metabolism experiments. When the gaseous exchange is measured in a respiration chamber and factors are applied to the oxygen consumed and the carbon dioxide, urinary nitrogen, and methane produced, the estimate is of the heat produced by oxidation of organic molecules, even if the subject is working or if his body temperature is not constant. When a subject is confined in a calorimeter which measures heat losses, the measurement of heat is equal to the heat production only if the heat content of the body is the same at the beginning and end of the experiment, and if all the heat arising from work is liberated in the calorimeter. If, for example, the work which the subject did was to lift 1000 one-kilogram weights from the floor of the calorimeter to a high shelf within it, the change in potential energy of the weights (mass × acceleration due to gravity × distance) would not be recorded. Such activities are unlikely but must be kept in mind. In practice, the longer the sojourn of a subject within a calorimeter, the smaller is the rate of heat storage since body temperature can vary only within narrow limits.

2.4 The formal framework for the study of the energy exchanges of animals and man

Over the years a series of concepts have been devised for the study of the overall energy exchanges of animals and man. These differ slightly from one discipline to another and in the account below alternative terms, which are mostly those employed in ecological energetics (see Brafield & Llewellyn 1982), are mentioned. Some of the terms are equivocal or not precisely defined or require definition when they are used. Examples are 'standard metabolism' or 'metabolic rate', and these are discussed in Chapter 8.

The *intake of energy as food* is the enthalpy of combustion of the dry matter of the food consumed, usually expressed as a daily rate. In studies with farm livestock it is sometimes called the *gross energy intake*. In ecological energetics it is usually termed the *ingestion rate*. Not all of this energy is available to the animal for some constituents of the diet are excreted undigested or partially digested in the faeces. The *faecal energy* is the enthalpy of combustion of the dry matter of the faeces, again usually expressed as a daily rate. In ecological energetics the terms *egesta rate* or *defaecation rate* have been employed to describe this term. Fermentations take place in the gut leading to the production of hydrogen and methane. Their enthalpy of combustion is referred to as a *rate of fermentative gas loss*, and lastly there is a loss of energy to the body through the excretion of organic compounds in urine, measured by the heat of combustion of the dry solids of the urine.

From these definitions, two other quantities can be defined; *apparently digested energy* and *metabolisable energy*. Apparently digested energy is the difference between the rate of intake of dietary energy, $\dot{I}$, and the rate of loss of energy in the faeces, $\dot{F}$. Metabolisable energy, $\dot{M}_E$, is the difference between the rate of intake of energy and the sum of the rates of loss of energy in the faeces, $\dot{F}$, in urine, $\dot{U}$, and as combustible gas, $\dot{G}$.

$$\dot{M}_E = \dot{I} - (\dot{F} + \dot{U} + \dot{G}) \tag{2.22}$$

Metabolisable energy, which is probably synonymous with the *assimilated energy* of ecology, is often regarded as a measure of the energy which is available to the animal for meeting the energy demands of maintaining the integrity of its body, muscular activity, growth, reproduction and lactation. This is but an approximation. To be an unbiased estimate of the enthalpy of combustion of the organic molecules which are absorbed from the gut and subsequently metabolised by the tissues proper, it would have to be reduced by the heat arising from the fermentations in the gut, the heats of hydrolysis of polysaccharides, lipids and proteins and the enthalpies of solution there, of the products of hydrolysis. These heat terms are included in the measured heat production of the animal. Even so, metabolisable energy is a useful term.

In a non-lactating mammal or a non-egglaying bird the rate of intake of metabolisable energy less the rate of heat production of the animal, represents the rate at which energy is retained in the body as the enthalpy of its tissues, $\dot{R}$. In a lactating animal there can be loss or gain of body tissue as well as milk secretion. The rate of milk energy secretion is the weight secreted per unit time × the specific enthalpy of combustion of its constituents and is given the symbol, $\dot{L}$. An analogous approach is used for the production of eggs where the rate of egg production represents the enthalpy of combustion of the egg × the rate of egg production. The nomenclature used in ecological energetics is different. Heat production is referred to as respiration and the difference between metabolisable energy and *respiration* is referred to as *production*. A distinction is then made between reproduction (which includes egg-laying and lactation) and growth. In growing animals and birds or in adults which are not reproducing the general expression is:

$$\dot{M}_E - \dot{H}_P = \dot{R} \tag{2.23}$$

When no food is given, equation (2.22) states that heat production is equal to $-\dot{R}$, that is the source of heat is the catabolism of body tissues. When, over a long period of time – usually several days – energy retention is zero, $\dot{M}_E$ is equal to heat production and this state is called *energy equilibrium* or '*maintenance*' and the dietary energy which supports maintenance is called the M_E *requirement for maintenance.*

As the $\dot{M}_E$ intake is increased from zero upwards, $\dot{H}_P$ also increases; the heat production at maintenance levels of nutrition is greater than it is during starvation. This increase in heat production associated with the ingestion of food is given several names. In human physiology it is termed the *specific dynamic effect* of food, in domestic animals it is usually termed the *heat increment* of food, and latterly the terms *thermogenic effect* of food or *dietary thermogenesis* have been used to deal with both animals and man. The neutral term, *heat increment* is the preferred one because it is explicit; the term *diet induced thermogenesis* is not so. It has been used to describe a particular type of increase in heat production associated with hyperphagia as well as with the general phenomenon (see Section 12.3). The heat increment is normally expressed as a proportion of the metabolisable energy of the food consumed. In a growing or adult animal unit increment in metabolisable energy intake thus results in a proportion, k, being retained in the body and a proportion, $1-k$, being lost as heat. The proportional retention, k, is termed *the efficiency of utilisation of metabolisable energy*. In a lactating animal the same relationship applies; the proportion, k, then represents the sum of the energy of milk and any change in the energy retained by the animal, and $1-k$ is the increment of heat associated with the ingestion of the additional unit of metabolisable energy by the animal.

Fig. 2.1. Components of the energy exchange and the relation between both energy retention and heat production and metabolisable energy intake.

Food – (Faeces + Urine + Combustible gas) – Heat produced = Retention

$$\dot{I} - (\dot{F} + \dot{U} + \dot{G}) - \dot{H}_P = \dot{R}$$

Metabolisable Energy $\dot{M}_E$

Comprising
Δ 'Fat'
Δ 'Protein'
Δ CHO

Heat loss + Work done + Heat stored

$$\dot{H}_L + \dot{w} + \dot{s} = \dot{H}_P$$

Convection $\dot{C}$
Conduction $\dot{K}$
Evaporation $\dot{E}$
$\dot{R}_I$ Infrared radiation
$\dot{R}_S$ Solar radiation

Retention
+
0
–
Slope k
Maintenance ($\dot{R} = 0$)
0
Food

Heat production
Slope $(1 - k)$
Minimal heat production
0
Food

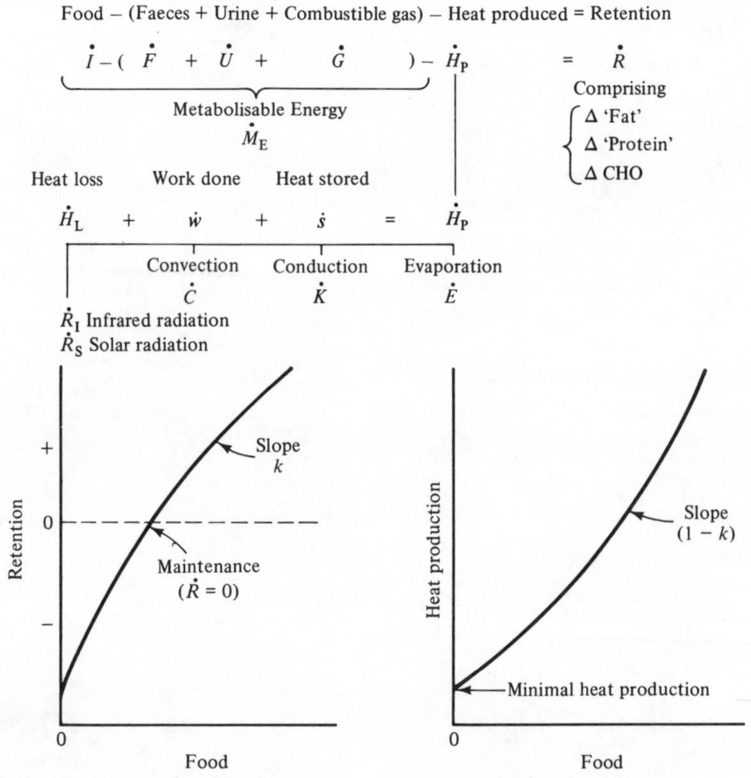

Obviously, if the relationship between retention and metabolisable energy intake is not linear, then the efficiency of utilisation is not constant but varies with the amount of food consumed. The factors which affect the thermogenic effect of food – or its converse, the efficiency of utilisation of the metabolisable energy of food – are dealt with in Chapter 12. Figure 2.1 shows in diagrammatic form the terms which are used.

3

Components of the energy budget: metabolisable energy

3.1 Energy budget and energy balance

Many studies of the bioenergetics of whole animals, including man, are concerned with what is called *energy budgetting*, that is, a careful accounting of the energy consumed in food, losses of energy from the body in excreta, heat produced by metabolism and the retention in or secretion from the body of energy represented by organic compounds. With man this budgetting is usually termed the *energy balance*. Such concern is obvious in nutritional studies; estimates are needed of the amount of food energy required to support particular levels of growth or secretion or to maintain weight under a variety of environmental and other circumstances. It is equally so in ecological studies when the transfer of energy from one trophic level to another is of interest.

The terms in an energy budget or energy balance can be expressed by combining equations (2.21) and (2.22):

$$\left.\begin{array}{llllll} \text{Intake of energy} & - \text{Energy of faeces} & - \text{Energy of urine} & - \text{Energy of combustible gas} & - \text{Heat produced} & = \text{Energy retained or secreted} \\ \dot{I} & -\dot{F} & -\dot{U} & -\dot{G} & -\dot{H}_{\mathrm{P}} & =\dot{R} \end{array}\right\} \quad (3.1)$$

or

$$\dot{M}_{\mathrm{E}}-\dot{H}_{\mathrm{P}}=\dot{R} \quad (3.2)$$

Additionally, if heat loss ($\dot{H}_{\mathrm{L}}$) is measured rather than heat production, terms for work done on the environment ($\dot{w}$) and heat storage ($\dot{s}$) have to be included in equation (3.2):

$$\dot{M}_{\mathrm{E}}-\dot{H}_{\mathrm{L}}+\dot{w}+\dot{s}=\dot{R} \quad (3.3)$$

Implicit in these relationships is the principle of the first law of thermodynamics, namely that energy is conserved. The experimental demonstration that the principle applies to animals and man is discussed in Section 5.4. Given that the first law applies – and there is nothing to suggest it does not –

the equations can be rearranged so as to estimate one component from knowledge of the others. Thus in a calorimetric experiment under laboratory conditions the retention of energy in the body can be estimated, not directly, but as a difference between measurements of the intake of energy as food, and the sum of energy excretions and the heat produced. Similarly heat production can be estimated from measurements of the metabolisable energy consumed and measurements of the energy retained in the body obtained by the use of direct or indirect approaches. Again, heat storage is usually determined as the difference between heat production and heat loss when no work is done on the environment. Clearly all such estimations made by difference accumulate any errors attached to the primary measurements on which they are based (see for example Schiemann, 1958 and Blaxter, 1967). Realisation of this accounts for the almost obsessive concern of those working in the field with matters related to the precision of their measurements. In studies with man the term *energy expenditure* is used as an alternative to *heat production*. The former term might seem to contradict the first law of thermodynamics because energy cannot really be expended – in the sense of being destroyed. It is retained here since its use is commonplace.

The components of an energy budget are expressed as rates per unit time. The time dimension employed depends on the context. The day, although it is a convenient base, is not always a satisfactory one since during the course of a day energy gain or loss from the body can vary appreciably. Food consumption is periodic and so too is the amount of muscular activity. Additionally, circadian rhythms of metabolism occur irrespective of the food consumed or the amount of muscular exercise. Bodily retention of energy thus varies considerably during the course of the day, and in animals and people free to consume as much food as they wish and to move about at their own volition, there is also a variation in retention from day to day. The physiological processes involved in short-term changes in the energy content of the body are of obvious interest. So to are the longer-term adjustments which result in constancy of body weight in adults, or indeed the slow accretion of fatty tissue in obesity. Observation of energy retention on a single day may be of little relevance in many situations. For example, in man there is evidence that the voluntary intake of food energy lags behind energy expenditure in exercise by about two days and that even 14 days of continuous observation reveal apparent retentions of energy in the body which are impossible to regard as sustainable, implying the existence of even longer-term metabolic integrations and adjustments of intake or expenditure or both (Garrow 1978).

3.2 Methods of determining energy budgets

The three components of an energy balance or budget – the intake of metabolisable energy, the production of heat and the retention or secretion of energy – are discussed in this and in the two following chapters. A distinction is made between what may be termed laboratory methods and field methods. Laboratory experiments are distinguished, not so much by the sophistication of the approaches, although this is usually a feature of them, but by the fact that dietary conditions and the environment are precisely controlled. The variation in intake and expenditure of energy characteristic of free-living animals is minimised and the three terms can be attributed to any imposed nutritional or environmental factors. It can be questioned whether such laboratory conditions have relevance to the normal life of the animal or man concerned. In housed domesticated livestock the conditions imposed in laboratory type calorimetric experiments are not too dissimilar to those under which they are kept, and some of the metabolic suites devoted to studies of human metabolism attempt to simulate the normal environment of modern sedentary man. With wild species of mammals and birds there is some doubt about the validity of extrapolating from laboratory to natural conditions and it is only by careful comparison and integration of laboratory and field work that credence is gained for such studies.

3.3 Measurement of metabolisable energy intake in the laboratory

In the laboratory, metabolisable energy can be measured with precision. The amounts of food offered and refused can be weighed accurately and so too can the amounts of excreta voided. The determination of their heats of combustion is equally precise, indeed the analytical error attached to a determination of metabolisable energy is usually considerably less than $\pm 0.5\%$. There are however possibilities of systematic error due to failure to take into account the time lag between ingestion of food and the excretion of its residues in the faeces. This time lag is best expressed as a mean retention time of food in the digestive tract (Blaxter, Graham & Wainman 1956; Warner 1981). Considering a single meal, the first appearance of its residues in the faeces is at about one-third of the mean retention time and the last appearance is at about four times the mean value, there being species differences in these two ratios. Table 3.1 gives some retention times for a number of species largely drawn from Warner's (1981) compilation. They are greatest in large herbivorous mammals – 40–60 h – and least in insectivora – about 2 h. There is a general relation between body size and retention time within groups of animals eating similar diets. Demment &

Table 3.1. *The mean retention time of food in the digestive tract of different species*

Species	Mean retention time (h)
Water shrew (*Neomys fodiens*)	2
Mink (*Mustela vison*)	4
Field vole (*Microtus agrestis*)	7
Bank vole (*Clethrionomys glareolus*)	11
Cat (*Felis cattus*)	13
Rabbit (*Oryctolagus cuniculis*)	15
Dog (*Canis familiaris*)	23
Rat (*Rattus norvegicus*)	28
Horse (*Equus caballus*)	29
Elephant (*Elephus maximus*)	33
Pig (*Sus scrofa*)	43
Goat (*Capra hircus*)	43
Man (*Homo sapiens*)	46
Sheep (*Ovis aries*)	47
Ox (*Bos taurus*)	60

Note:
Largely from Warner (1981).

van Soest (1985) have suggested that the lower limit to ruminant animal size is that at which the size-related retention time of food is equal to the time needed to digest coarse plant material. Man has a mean retention time of 46 h and this is highly variable (Cummings 1978). In those species in which measurements have been made, increasing the amount of food given increases the rate of passage, and, in omnivora, so too does an increase in fibre content. The faeces excreted in any 24 h period thus reflect only part, if any, of the current day's food consumption and parts of previous days' intakes. Adequate preliminary periods in which the amount of food and its composition is kept constant avoid this potential error.

Even when these precautions are taken, errors can arise because of the periodicity of faecal elimination. Periods of collection have to be sufficiently long to ensure that the mean 24 h estimate of the amount excreted is reliable. Errors of this sort can occur readily in man. For example, van Es *et al.* (1982) collected faeces for four days continuously, yet at low levels of intake the frequency of defaecation was such that the results were clearly in error. The sloth appears to hold the record for infrequent bowel emptying; it apparently defaecates once per week (Montgomery & Sunquist 1978).

3.4 Estimation of metabolisable energy in free-ranging animals

In free-ranging animals, whether wild or domesticated, the estimation of the intake of metabolisable energy is difficult, largely because of the problem of measuring the weight of food consumed. With wild animals much information is available on the species and parts of plants which are seen to be consumed, on time spent foraging, and on the items ingested as evidenced by the identification of food residues in gut contents. This information is largely qualitative and cannot easily be expressed in terms of daily rates of food consumption.

With grazing domesticated livestock much ingenuity has been displayed in providing estimates of these rates (Leaver 1982). A method which has been particularly useful, has been to estimate the faecal output of organic matter by giving animals a constant measured amount of chromium sesquioxide (which is inert in the digestive tract) and determining the ratio of chromium sesquioxide to organic matter in the faeces. If the apparent digestibility of the organic matter of the herbage is known, intake can then be estimated from the faecal excretion of organic matter. A linear relation has been found between nitrogen concentration in faecal organic matter and its apparent digestibility, varying somewhat with the nature of the herbage ingested. Provided that a reliable relation can be assumed, the method provides estimates of organic matter intake which are reasonably accurate. The intake multiplied by the metabolisability of the diet provides an estimate of the intake of metabolisable energy. Another approach to estimating the intake of herbage makes use of the fact that most plants contain hydrocarbons with n-alkanes containing odd-numbered carbon chains, with C_{19}–C_{32} chains predominating. The long-chain hydrocarbons are virtually indigestible (Mayes & Lamb 1983). Dosing with an indigestible marker to estimate the output of faeces and utilising the n-alkanes as an internal marker to determine digestibility enables an estimate to be made of metabolisable energy intake. Such methods are probably associated with errors at least ten times greater than those found in laboratory experiments and of course are prone to systematic errors arising from variation in the amount and nature of the diet ingested each day.

3.5 The metabolisable energy intake of man

In man there have been few investigations in which the excreta are collected and their heats of combustion determined – no doubt because such tasks are not pleasant ones. Most values quoted for the metabolisable energy intake of man consist of measures of the amounts of the foods or energy-yielding nutrients consumed, multiplied by factors which represent

Table 3.2. *Estimates of the intake of metabolisable energy (expressed as MJ/d ± S.D.) by civil servants made using different methods of dietary survey*

	Duplicate diet	24-hour recall	7-day weighed diet	28-day purchase	Food frequency interview
Women (9)	7.5 ± 1.6	7.8 ± 1.4	8.8 ± 1.4	10.1 ± 2.5	9.8 ± 2.5
Men (21)	8.9 ± 2.0	9.3 ± 2.1	11.3 ± 2.3	11.5 ± 3.2	13.0 ± 3.0
All subjects (30)	8.5 ± 2.0	8.9 ± 2.0	10.6 ± 2.4	11.1 ± 2.9	12.0 ± 3.2

Note:
From the results of Bull & Wheeler (1986).

the metabolisable energy provided by each. The precision of such estimates of the metabolisable energy intake (often simply called the *energy intake*, or in lay terms the *calorie intake*) depends on the accuracy of both the estimates of what is consumed and of the factors.

Methods for determination of the amount of food consumed range from preparation and subsequent analysis of duplicate meals with correction for any food left on the plate – a method which, properly conducted, is accurate – to retrospective listing of what has been eaten and an estimate of the size of portions (Marr 1971; Graham 1982). Retrospective methods are open to bias; there is for example evidence that there is under-recording of consumption by those with smaller than average intakes and over-recording by those with larger ones (van Staveren & Deurenberg 1984). Table 3.2 gives an indication of the precision with which the habitual dietary intake of metabolisable energy can be estimated using alternative methods under conditions of normal family life (Bull & Wheeler 1986). Bull & Wheeler regard the seven-day weighed intake as the most accurate estimate, but there is no way in which its absolute precision can be tested. Passmore, who has had considerable experience in this field, has written

> There is an immense literature on the reliability of dietary surveys and it is very difficult to assess the accuracy of the results in any one individual. In my opinion the error in assessing the calorie intake is unlikely to be less than ± 10% and often much greater. (Passmore 1967)

In normal life the daily intake of food is not constant. Items of diet change and so too do amounts. Although she fully informed her subjects about the purpose of a trial, Dr Dauncey found variation in daily intake to be two-fold during a three-week period of observation (Dr Dauncey, personal communication). The same problems discussed under laboratory determinations of the metabolisable energy intake apply, and attempts are made to obviate them by using long periods of observation. The methods all involve close cooperation of the subject, a matter not easily obtained over long periods of time.

3.5.1 *Metabolisable energy factors for human diets*

There is little doubt that a major source of error in estimating the metabolisable energy intake of man during normal life relates to the estimation of the amounts of the different items of the diet which are consumed. The application of factors to these amounts to estimate the metabolisable energy of the whole diet is usually considered to be a lesser source of error. However these same factors are often used to arrive at estimates of the intake of metabolisable energy in laboratory experiments in which the intake is carefully measured; their derivation and precision is thus of some importance.

The metabolisable energy factors for human diets derive from pioneering work carried out by Rubner (1885) in German and by Atwater (Atwater & Bryant 1900; Atwater 1902, 1910) in the USA. Atwater conducted about 50 experiments with three men, and, combining his results with earlier published ones, arrived at what have been termed the *Atwater factors* of 4, 9 and 4 kilocalories (kcal) of metabolisable energy per gram of protein, fat and carbohydrate, respectively. The factors were given in whole numbers and applied to the protein estimated from its nitrogen content, fat determined by Soxhlet extraction and carbohydrate determined by difference, taking into account the water and ash contents of the diet or food concerned. Carbohydrate so determined includes indigestible substances such as lignin, cellulose, pentosans, and pectic material as well as highly digestible ones such as starch and simple sugars. The nitrogen content of 'proteins' – which included non-protein nitrogenous material – was recognised to vary with the source as judged by older analyses. Atwater originally stated that different factors should be used for different items of the diet; it was only in his later work that he suggested the simplified ones given above. An abridged version of Atwater's original factors, on which the classic '4,9,4' factors were based, is presented in Table 3.3. The values have been calculated in terms of kJ/g from a summary of Atwater's data (Maynard 1944).

In the USA the 'standard' values for the metabolisable energy of individual foods are those listed in Merrill & Watt (1955). In these cognisance is taken of Atwater's original finding that the metabolisable energy of fat and protein varies from food to food and in the computation, carbohydrate is determined by difference. In the UK the standard tables are those of Paul & Southgate (1978). In these cognisance is taken of the variation of the nitrogen content of proteins (varying from 15.7% for milk proteins to 18.9% for nut proteins) and, as a major departure from older procedures,

Table 3.3. *The original extended version of Atwater's factors for estimating the metabolisable energy of human foods and diets from their chemical composiiton, expressed as kJ/g nutrient*

Food source	Protein	Fat	Carbohydrate
Animal products	17.8	37.7	15.9
Animal and vegetable fats		37.7	
Cereals, 70% extraction	15.5		17.1
Cereals, 85% extraction	15.1		15.9
Pulses and nuts	13.4	34.9	16.7
Vegetables	12.1		16.7
Fruits	13.2		15.1
Sugar (sucrose)			16.1
The rounded factors by different authorities are:			
Atwater '4,9,4'	16.7	37.7	16.7
Rubner '4.1,9.3,4.1'	17.1	38.9	17.1
Paul & Southgate[a]	17	37	16

Note:
[a] Available carbohydrate, not carbohydrates by difference.

carbohydrate is measured as *available carbohydrate* defined as the sum of free sugars, dextrins, starch and glycogen, the sum being expressed as monosaccharide. The factors applied to the amounts of protein, fat and available carbohydrates are 4, 9, and 3.75 kcal/g. These were rounded by Paul & Southgate (1978) to give 17 kJ/g protein, 37 kJ/g fat and 16 kJ/g available carbohydrate expressed as monosaccharide.

In Table 3.4 a comparison is made of the metabolisable energy of the same food as estimated by the methods used by Merrill & Watt in the USA and Paul & Southgate in the UK. There are very large differences particularly with respect to fruit and vegetables. Values for the estimated intake of metabolisable energy must differ according to which tables are employed to convert weights of items of diet into terms of metabolisable energy. When diets contain high proportions of fruit and vegetables the discrepancy can be considerable. Experimental studies in which intake and excretion were measured accurately and in which three diets varying in their content of unavailable carbohydrate were employed, showed that reasonable accuracy could be achieved by using available carbohydrate rather than carbohydrate by difference in the calculation (Southgate & Durnin 1970). Obviously in interpreting data on human energy budgets in which metabolisable energy intake has been estimated from records of consumption and tabulated values for the energy of foods or nutrients, it should be considered how far the choice of factors as well as the intrinsic accuracy of

Table 3.4. *Comparison of the metabolisable energy (in kJ/100 g) of human foods on a fresh weight basis as estimated by methods used in the UK (Paul & Southgate 1978) and the USA (Merrill & Watt 1955).*

Food	Metabolisable energy		Ratio of USA/UK values
	UK	USA	
Cereals			
Bread, brown	948	1050	1.11
Bread, white	991	1012	1.02
Flour, wholemeal	1351	1368	1.01
Flour, white	1433	1477	1.03
Oatmeal	1698	1669	0.98
Rice, polished	1536	1540	1.00
Dairy products			
Butter	3041	3130	1.03
Cheese, cheddar	1682	1732	1.03
Eggs	612	707	1.16
Milk, fresh whole	272	285	1.05
Meat			
Beef, corned	905	966	1.07
Beef, steak, raw	736	740	1.01
Liver, raw	642	602	0.94
Fruit			
Apples, eating	196	230	1.17
Apricots, dried	776	1243	1.60
Bananas	337	431	1.28
Currants, black, raw	121	330	2.73
Gooseberries, green, raw	73	146	2.01
Grapefruit	95	134	1.41
Oranges	150	205	1.37
Vegetables			
Beans, butter, raw	1162	1464	1.26
Beans, runner, raw	83	129	1.56
Cabbage, Savoy, raw	109	126	1.15
Carrots, old, raw	98	134	1.36
Peas, fresh, raw	283	339	1.20
Nuts			
Peanuts	2364	2410	1.02
Walnuts	2166	2177	1.02

the primary estimate of amounts consumed might influence any conclusions drawn.

3.5.2 *The uniqueness of the metabolisable energy factors*

The methods employed by Paul & Southgate (1978) to estimate the metabolisable energy of diets for man predicate that unavailable carbohydrate has no nutritive value and that the factors for protein, fat and available carbohydrate are completely independent. Although the authors made this assumption they were well aware that it is not tenable; Southgate & Durnin (1970) had in fact shown that some of the pentosans and cellulose were apparently digested by man and that an increase in the amount of unavailable carbohydrate in the diet depressed the apparent digestion of fat and protein. A similar problem relating to the independence of factors was noted with ruminant diets when attempts were made to devise factors for the estimation of metabolisable energy from complete, summative chemical analyses. One prediction equation arrived at (Wainman, Dewey & Boyne 1981) was:

$$_{s}M_{E}=0.195P+0.343E+0.122C-0.064F \tag{3.4}$$

where $_{s}M_{E}$ is metabolisable energy in MJ/kg; P is crude protein content (g/kg); E is ether extractives (g/kg); C is nitrogen-free extractives by difference (g/kg); and F is crude fibre (g/kg). This equation suggests that crude fibre has a negative effect, a conclusion which is incompatible with knowledge that in ruminants fibre is fermented by ruminal organisms, that the volatile fatty acids which are the end-products of fermentation are absorbed and that they provide a considerable proportion of the energy which the ruminant obtains from its diet. The conclusion must be that the equation as a whole does not imply that the factors are unique to the chemically-defined entities of the diet. Rather the equation reflects the complex inter-relationships between the chemical constituents and can only be regarded as empirical.

It is for such reasons that workers concerned with the estimation of the metabolisable energy of feeds for domesticated livestock have largely abandoned searches for unique factors which can be applied to particular chemically-defined entities in all feeds. Rather they have adopted empirical approaches in which the metabolisable energies of feeds in specified classes are predicted from a few chemical determinations. This approach is exemplified by the work of Wainman, Dewey & Boyne (1984) for ruminant animals, of Fisher (1982) for poultry and of Morgan, Whittemore & Cockburn (1984) for pigs. All of these approaches were based on careful metabolic study, detailed analytical work and comprehensive statistical analysis.

With human foods it can be argued that the approach originally made by Atwater, in which classes of food were distinguished and different factors used for each, was a step in the same direction. Obviously there is urgent need to re-investigate the precision of the factors used for man and to explore alternative approaches.

For animals of agricultural importance tables have been published in several countries listing the metabolisable energy of feeds. These tables are similar to those for man. The provenance of the entries in them is usually the direct laboratory determination of metabolisable energy rather than its estimation from chemical composition. Examples of such tables are: for ruminants, the UK tables (Ministry of Agriculture, Fisheries and Food 1976) and those of East Germany (Nehring, Beyer & Hoffman 1970); for pigs (in terms of digested energy) the US–Canadian tables of feed composition (US National Academy of Sciences 1969); and, for poultry the Dutch tables of Janssen *et al.* (1979).

3.6 Metabolisability

The proportion of the heat of combustion of the food consumed which is represented by metabolisable energy is termed its *metabolisability*, a not particularly euphonious term! An alternative term adopted in ecological studies is *assimilation efficiency* as defined by Petrusewicz & MacFadyen (1970). There is, however, some confusion in the literature on the energetics of wild animals since the term assimilation efficiency is also employed to denote apparent digestibility. Petrusewicz & MacFadyen (1970) defined assimilated energy explicitly as the sum of heat production and energy retention, a definition in accord with equation (3.2). To avoid confusion it seems sensible to use the two terms, apparent digestibility and metabolisability rather than the equivocal one.

The contribution to metabolisability of energy losses in faeces, urine and as combustible gas varies from species-to-species and with the nature of the food. Some values are given in Table 3.5. In almost all instances faecal loss is the major determinant of metabolisability, and this component is considerable in the instance of herbivores given highly fibrous diets. The proportional loss of energy in urine varies less between species but is greater with higher protein diets. The proportional loss of methane is greatest in ruminants in which fermentation takes place in the anterior parts of the digestive tract but is also considerable in species in which microbial fermentation takes place in the hind gut.

For a particular diet the loss of energy in faeces cannot be regarded as absolutely constant; as the amount ingested increases, the proportion excreted in the faeces also tends to increase. The effect is most apparent in herbivores and particularly ruminants in which, simultaneously, methane production declines. It is for this reason that most tabulations of the metabolisable energy of feeds for ruminants refer to measurements made at the maintenance level of nutrition.

It is common practice to estimate metabolisability from the apparent digestibility of the diet by multiplying it by a factor which takes into account

Table 3.5. *Energy losses in faeces, urine and methane by different species given a variety of diets*

	kJ/100 kJ of food			
Species and diet	Faeces	Urine	Methane	Metabolisable
Man				
Low fibre	3.5	4.1	?	<92.4
High fibre	7.5	3.3	?	<89.2
42% fat, 18% protein	5.8	4.5	0.4	89.3
Rat				
Laboratory chow	13.2	3.5	0.0	83.3
Horse				
Hay and oats	35.1	4.4	1.9	58.6
Wheat straw	61.0	4.0	3.2	31.8
Sheep				
Barley straw	60.6	2.2	6.7	30.6
Barley grain	16.7	3.3	10.6	69.6
Ox				
40% hay, 60% grain	26.0	2.7	6.2	64.4
Pig				
Kale	34.3	3.5	0.4	61.8
Maize grain	13.9	4.2	?	<81.9
Rabbit				
15% fibre diet	31.3	3.6	?	<65.1
Hen				
Layer ration	16.5	3.0	0.0	80.5

urinary and gaseous energy losses. In particular, since determination of the heat of combustion of urine is technically difficult, proportional losses of energy in the urine are often estimated from nitrogen excretion per unit of food ingested.

3.6.1 *The ratio of energy to nitrogen in urine*

The ratio of energy to nitrogen in urine, although often taken to be that in urea (22.6 kJ/g), varies widely and in herbivores given forages can be as high as 150 kJ/g (Blaxter & Wainman 1964a). Such high values are largely due to the excretion of hippuric acid (benzoylglycine) for which the ratio of energy to nitrogen is 302 kJ/g, and other aromatic acids (Blaxter, Clapperton & Martin 1966; Martin 1978). A more precise estimate of urinary energy can be made from a determination of its carbon content. The energy to carbon ratio varies from 32.2 kJ/g C for uric acid to 51.9 kJ/g C for urea, with hippuric acid having a value of 39.3 kJ/g C. Where average values of the ratio of energy to nitrogen are needed, acceptable values are

40 kJ/g N for pigs, 60 kJ/g N for ruminants and 35 kJ/g N for man. The values for pigs and ruminants are based on extensive East German studies (Hoffmann, Schiemann & Jentsch 1971). For most practical purposes the proportional losses of energy in the urine of ruminants can be estimated from the crude protein content of the diet by the equation:

$$U/I = 0.025P + 1.6 \tag{3.5}$$

where U/I is the loss of energy in urine per unit enthalpy of combustion of the food; and, P is the protein content (N × 6.25) of the food dry matter in g/kg.

3.6.2 *Methane and hydrogen production*

Methane and hydrogen are produced within the lumen of the digestive tract, the former by the reduction of carbon dioxide by hydrogen arising during the course of microbial fermentations. The gases are mostly eructed or lost in flatus but small amounts are absorbed into the blood and excreted by the lung. The proportional absorption of hydrogen is greater than that of methane because of its higher solubility. Understandably methane production has been studied intensively in ruminants, and as a proportion of the energy of the food, is broadly proportional to the apparent digestibility of the diet (Blaxter & Clapperton 1965). At the maintenance level of nutrition it can be predicted from the expression:

$$100\dot{G}/\dot{I} = 3.67 + 6.22\,(\dot{I} - \dot{F})/\dot{I} \tag{3.6}$$

where $\dot{G}$ is the energy lost as methane; $\dot{I}$ the intake of energy as the enthalpy of combustion of the diet; and, $\dot{F}$ the faecal energy. A decrease in the proportional loss occurs when higher amounts of ingested feed are given. This relationship, while it has general applicability, has been shown by Wainman *et al.* (1984) not to predict value for certain feeds, notably those which result in very low methane production, such as the products of the distilling industry ($\dot{G}/\dot{I} = 0.03$) or those causing high methane production, notably cassava ($\dot{G}/\dot{I} = 0.14$). This probably reflects differences in the amounts of readily-fermentable carbohydrate in the feeds concerned. Moe & Tyrrell (1980) however, have shown that methane production from soluble carbohydrate in the diet is considerably less that it is from cellulose.

Ruminants are fore-gut fermenters; other herbivores are hind-gut fermenters and in these methane production is a smaller proportion of the food energy apparently digested. There is indirect evidence that carnivores produce no methane or hydrogen at all (McKay & Eastwood 1984).

Methane excretion can be detected in only about 40% of the adult human population but all appear to produce hydrogen. Whether methane is produced appears to be related to the number of methanogenic bacteria present in the gut (Miller & Wolin 1986). There is little quantitative information about the excretion of either gas. Measurements of hydrogen in expired air have mostly been qualitative and undertaken to assess the extent of malabsorption of carbohydrate (Calloway, Colasito & Mathew 1966; Tadesse & Eastwood 1978). Only 10–15% of any hydrogen produced in the large bowel is absorbed and excreted in the expired air. Studies by van Es, de Groot & Voght (1986), in which subjects were confined to a respiration chamber and hence losses of methane and hydrogen from both the lung and the digestive tract could be measured, suggest that in normal subjects the loss is about 0.5 l H_2/d and that when a carbohydrate is given which results in fermentation in the large intestine, these losses increase to over 2 l/d.

3.6.3 *Fermentation heat*

It was mentioned in Section 2.4 that metabolisable energy is not a measure of the energy of organic molecules which are absorbed from the digestive tract since the method of measurement ignores the enthalpies of hydrolysis of fats, complex carbohydrates and proteins in the gut and heat arising from anaerobic fermentations. The enthalpies of hydrolysis of peptide bonds, determined by experimental study of dipeptide hydrolysis, range from 5 to 10 kJ/mol or 0.1–0.2% of the heat of combustion of the dipeptide. Calculations by Morowitz (1978) suggest that the enthalpies of hydrolysis of esters of fatty acids and glycerol, and of glucosidic bonds are of the same order as those for peptides. Heat arising from enzymatic hydrolyses in the gut is thus not a large component of the total heat arising there.

The heat arising from microbial fermentation in the gut has mostly been studied in ruminants. Stoichiometric calculations have been made for ruminal fermentation of hexose, notably by Hungate (1966). The principle has generality and can be applied to other species. From the known biochemical pathways of fermentation, equations can be written which describe the oxidation of hexose to give the steam-volatile fatty acids, acetic, propionic and n-butyric acids:

$$1\ \text{hexose} + 2H_2O = 2CH_3COOH + 2CO_2 + 4H_2 \quad (3.7)$$

$$1\ \text{hexose} + 2H_2 = 2CH_3CH_2COOH + 2H_2O \quad (3.8)$$

$$1\ \text{hexose} = CH_3(CH_2)_2COOH + 2CO_2 + 2H_2 \quad (3.9)$$

The three equations above can be weighted in proportion to the molar proportions of steam-volatile acids produced. Their sum will reveal an excess of hydrogen on the right-hand side. In species which only produce methane, this hydrogen is used to reduce carbon dioxide according to the equation:

$$CO_2 + 4H_2 = CH_4 + 2H_2O \quad (3.10)$$

Balancing the hydrogen of the weighted sum of equations (3.7), (3.8) and (3.9) by equation (3.10) gives a stoichiometric equation for fermentation. Using enthalpies of formation of products and reactants in this equation, the heat of fermentation can be estimated. For a fermentation in which the molar proportions of acetic, propionic and n-butyric acids were 62:22:16, Hungate (1966) calculated that methane production would account for 18% of the enthalpy of combustion of the hexose fermented and the heat of fermentation would account for 6.5%. The general relationships in equations (3.7)–(3.10) show that a high molar proportion of propionic acid in the end-products should reduce methane production. The calculated heat of fermentation, however, varies relatively little with the proportion of the end-products and amounts to about 180 kJ/mol hexose fermented. Direct measurement of fermentation heat in calorimetric experiments *in vitro* gave a value of 177 kJ/mol (Arieli 1986).

The approach can also be applied to the fermentation of protein and, as summarised by Webster (1978), the respective fermentation heats for carbohydrate and protein can be taken to be 7% and 2.8% of the substrate fermented. Czerkawski (1986) took matters further, and on the assumption that fermentation heats for carbohydrate and protein are fixed at these values, divided the Brouwer (1958) expression for estimating heat production into two equations, one representing heat arising from oxidation in the body proper and the other from fermentation in the gut.

$$\left.\begin{aligned} H_{Fer} &= 5.10CO_2 - 3.63O_2 - 2.15CH_4 \\ H_{Body} &= 19.82O_2 - 0.11CO_2 - 5.93N \end{aligned}\right\} \quad (3.11)$$

with the sum:

$$H_p = 16.19O_2 + 4.99CO_2 - 5.93N - 2.15CH_4 \quad (3.12)$$

where H_{Fer} is fermentation heat in kJ; and H_{Body} is heat (kJ) produced within the body proper. The gases are in litres and N is the urinary nitrogen excretion in grams. These equations partition the heat, subject only to the assumptions made about the stoichiometry. They are general ones and not specific to ruminants. They could be extended to include the production of hydrogen but do not appear to have been applied to animals other than ruminants.

4

Components of the energy budget: heat production

4.1 Direct calorimetry

Direct calorimetry is the measurement of heat loss by radiation, convection, conduction and as latent heat arising from the vaporisation of water. The physical processes involved are described in Chapter 7, in the context of the effect of the environment on heat loss. The calorimeters capable of measuring heat loss in animals and man which are presently in use are of two main types; heat sink calorimeters and gradient layer calorimeters. In the former the chamber accommodating the subject is heavily insulated to prevent heat loss and the heat produced is collected by liquid-cooled heat exchangers. Non-evaporative heat loss is calculated as the rate of flow of coolant multiplied by its temperature rise and the specific heat of the coolant. Evaporative heat loss is measured as the product of the mass flow rate of air, the increase in its humidity and the latent heat of vaporisation. In gradient layer calorimeters the chamber confining the animal is lined with a thin layer of material of constant thickness and thermal conductivity and the whole chamber is surrounded by a jacket maintained at constant temperature. Measurement of the temperature difference over the insulating layer, which, following the work of Benzinger & Kitzinger (1949) is usually made by employing a network of thermocouples in series, enables the determination of non-evaporative heat loss. Evaporative heat loss is measured by additional gradient calorimetry of the air stream, in which the air is conditioned so that the enthalpies of incoming and outgoing air are precisely balanced.

Whole animal calorimeters are complex instruments; they have been built to accommodate laboratory animals, animals the size of an ox and man. Many have been described in detail. Human calorimeters have been illustrated by Webb (1984) and an excellent account of the construction of both human and animal calorimeters, their operation and calibration is given by McLean & Tobin (1987). These aspects are not discussed here.

4.1.1 *Heat storage in the body*

Calorimeters measure heat loss not heat production, and, as briefly discussed in Section 2.3, to obtain an estimate of heat production from a calorimetric determination of heat loss, account must be taken of any heat storage within the body. Even in homeothermic animals, body temperature, as judged by rectal or deep body temperature, is not stable. The use of in-dwelling, remote-sensing thermometers has enabled the range of deep body temperatures of animals to be assessed without disturbing the subject (and incidentally affecting their temperature). Such studies, notably by Bligh & Robinson (1965) with unrestrained East African mammals, showed daily variations of 1–3 °C with curious species differences. The rhinoceros, for example, had a particularly wide range. In nocturnal animals such as rats and other rodents, deep body temperature is higher during the night. The reverse is true of animals active by day. Muscular activity increases deep body temperature. In man, for example, heavy sustained exercise is associated with rectal temperatures approaching 40 °C, and in marathon runners, those finishing in the fastest times have been shown to have rectal temperatures of up to 41.1 °C (Koivisto 1986). In birds and mammals which undergo periodic torpor, the range of body temperature is much greater than in obligatory homeotherms.

Heat storage in the body can be measured as the difference between heat loss as determined by direct calorimetry and heat production determined from gaseous exchange and urinary nitrogen excretion. Given absolute instrumental accuracy the error of the measurement is determined by the accuracy of the factors used to estimate heat production. This accuracy can be regarded as high – as judged by the fact that over a 24-h period the measured difference between heat loss and heat production tends toward zero.

Heat storage in the body during an interval of time could also be estimated precisely as:

$$\dot{s} = \sum C_{P(i)} \Delta T_{(i)} W_{(i)} \tag{4.1}$$

where the summation is over all the 'i' tissues of the body, each of mass $W_{(i)}$, the total being the total body mass; $C_{P(i)}$ is the specific heat of the 'ith' tissue; $\Delta T_{(i)}$ is the change in its temperature. Equation (4.1) allows for different rates of change of temperature in different tissues. This is necessary since in muscular exercise the temperature rise of the muscle mass is greater than that in the visceral organs, while when animals make adjustments to cold the distal parts of limbs cool more rapidly than the central core. This equation while precise has, however, never been used since it is difficult to measure the temperature of different tissues or regions of the body. Instead an alternative expression has been employed:

$$\dot{s} = W C_P \Delta T_B \qquad (4.2)$$

where W is total body mass; C_P mean specific heat of the whole body; and ΔT_B the change in mean temperature of the body during an interval of time.

4.1.2 *The specific heat of the body*

In all modern determinations of heat storage the specific heat of the body has been taken to be 3.47 kJ/(kg K). This constant appears to derive from determinations of specific heat of tissues made in the nineteenth century (Pembrey 1898) and has been accepted, apparently without much question, ever since. The specific heat of body components in kJ/(kg K) are, from more modern analyses: lipid 2.12, protein 1.25, mineral matter 0.84, carbohydrate 1.14 and water 4.18. On this basis the specific heat of the fat-free body of man (which in adults is reasonably constant in chemical composition, containing 72% water, 22% protein and 6% minerals) is 3.34 kJ/(kg K) – a value distinctly lower than the commonly accepted value for the body as a whole. The specific heat of the body of a thin man containing 12% body fat would thus be 3.20 kJ/(kg K), and that of an obese one containing 50% fat, would be 2.73 kJ/kg K).

4.1.3 *The average temperature of the body*

The method used to estimate the mean temperature of the body entails weighting rectal and skin surface temperatures. This approach assumes that rectal temperature measures the 'core' temperature and skin temperature that of the more superficial tissues that constitute the 'shell' (Burton 1935). The weighting coefficients are arrived at by solving for the weighting coefficient α in the following equation:

$$\dot{s}/W = C_P \Delta (\alpha T_R + (1-\alpha) T_S) \qquad (4.3)$$

Where $\dot{s}$ is the rate of heat storage determined as the difference between heat production and heat loss; C_P the specific heat of the whole body; T_R rectal temperature; and T_S skin temperature.

Several workers have solved equation (4.3) for α on the assumption that C_P is constant at 3.47 kJ/(kg K). In man, using rectal temperature as a measure of core temperature, values ranging from 0.65 to 0.80 have been obtained (Burton 1935; Hardy & DuBois 1938; Bonjour, Welti & Jequier 1976). Livingstone (1968) points out that the value of α in man cannot be regarded as constant since it varies with environmental temperature. In cattle using carotid artery temperature as a measure of core temperature, McLean found α to be 0.89 ± 0.027 in one set of observations and 0.85 ± 0.016 in another (McLean *et al.* 1983a; McLean, Stombaugh & Downie 1983b). Equation (4.3) can, however, be solved for both α and C_P simultaneously. From the published data of Burton (1935) it has been calculated that C_P of the whole body is 3.22 kJ/kg K, a value which agrees with the theoretical calculations given above for a thin man, and that α has a value of 0.746. Published data by other authors do not permit such a computation.

Provided that environmental and nutritional circumstances are kept constant, surface temperature and deep body temperatures show little variation when measured at 24-h intervals and the heat storage term is negligible. Heat loss measured calorimetrically is then equal to heat production. It is only during short periods of observation or when environmental conditions change that the term is of quantitative significance. This is well illustrated by the experiments of Webster *et al.* (1986) in which calorimetric periods of 7.5 h showed considerably poorer repeatability than did runs of 24 h.

4.1.4 *Heat of warming food and water and cooling of excreta*

The measured heat loss in a calorimeter excludes heat given up by the body to bring ingested food and drinking water to body temperature. It includes the heat lost by the excreta as they cool to calorimeter temperature. In long-term experiments these two terms tend to cancel but in the short term each component can represent considerable proportions of the current rate of heat production. Corrections have then to be applied which take into account rates of warming and cooling. These have been discussed fully by McLean *et al.* (1982).

4.2 **Indirect calorimetry**

The term indirect calorimetry is employed to describe those methods of estimating heat production which are based on determinations of gaseous exchange and which depend on the factors discussed in Section 2.2. The equipment used is of two main types. In the first type the whole animal is confined in what is termed a *respiration chamber* (and should never be called a calorimeter) for the quantitative measurement of the gaseous exchange. In the second, masks, hoods or tracheal cannulae are employed to measure the gaseous exchange from the lungs alone. Experiments in respiration chambers are usually, but not invariably, of long duration; the various 'mask' methods are ideally suited to short-term studies lasting minutes. Considerable ingenuity has been displayed in constructing equipment of both types and many instruments have been reviewed over the years (Abderhalden 1924; Paechtner 1931; Swift & French 1954; European Association for Animal Production 1958; van Es 1984; McLean & Tobin 1987).

These instruments, whether of the respiration chamber or mask type, generally work on one of three principles; the closed circuit or Regnault & Reiset principle, the open circuit or Pettenkofer principle and the confinement or Laulanie, principle – the names being those of the workers who first devised them. In closed circuit instruments air is circulated within a closed system, whether a chamber and its pipework or a mask and its attached spirometer. The carbon dioxide produced by the animal is absorbed by suitable absorbents and weighed and the oxygen consumed by the animal is measured as it is replaced. In open circuit systems outdoor air of known composition is passed through the system. Its rate of flow is determined and the increase in its carbon dioxide content and decrease in oxygen content are measured. In confinement instruments, the animal is kept in a hermetically-sealed chamber (or breathes through an air-tight mask into a hermetically-sealed receptacle) and the change in composition of the known volume of air in the system is ascertained. In recent years modern methods of air analysis have replaced laborious chemical methods, new methods of measuring air flow have been devised, data collection has been automated and the numerical work computerised. At the same time methods of calibration have been improved. These developments and a description of the methods employed are given by McLean & Tobin (1987) and are not further discussed here.

4.2.1 *The whole body and pulmonary gas exchange*

In snakes up to 35% of the carbon dioxide produced in metabolism is lost through the skin and in lizards up to 10% is lost. In endotherms there is a much smaller loss (Feder & Burggren 1985). This is of consequence when metabolism is measured using mask or hood methods because the overall respiratory exchange is then underestimated. In man the oxygen taken up by the skin surface may amount to as much as 1.9% of the total consumption and the carbon dioxide lost through skin can be up to 2.7% of the total (Fitzgerald 1957). The loss is augmented at high temperature and is entirely by diffusion. Bats, with their large membraneous wings, lose about 11% of the carbon dioxide they produce through their skin. In these animals the diffusional loss per unit surface area is the same as in other endotherms but the area for diffusion is much greater. Mask methods, which are commonly used for man and for larger animals, thus underestimate the total gaseous exchange.

For the largest of the land mammals, the elephant, the shape of the head and the presence of a trunk makes the construction of a mask somewhat difficult! Observing that elephants appeared to breathe solely through their trunks, led Brody, Proctor & Ashworth (1934) to devise a mask that fitted over the trunk alone. The soundness of this approach was later tested by Benedict (1936) who showed that the respiratory exchange measured in trunk-breathing trials was 30% lower than in those conducted in a respiration chamber.

A tracheostomy obviates the need for a mask or hood but in all species this approach interferes with the water vapour and heat exchange in the nasal passages (see Section 7.4.2). In ruminants gaseous exchange measured in this way underestimates carbon dioxide and methane production (Blaxter & Joyce 1963a). Considerable carbon dioxide and virtually all the methane produced by anaerobic fermentations in the rumen are belched. Belching takes place with the glottis open and these gases are inhaled on inspiration to be mixed with carbon dioxide from the lung. A tracheostomy interrupts this normal flow. In animals in which fermentation takes place in the hind gut gas is lost in flatus. In Benedict's classic studies with the elephant three-quarters of the methane production – 500 to 746 l/d – were lost in this way (Benedict 1936).

In the human subject, when mouth-pieces and nose clips are employed to facilitate quantitative collection of expired air, care has to be taken to avoid over-ventilation. If this occurs carbon dioxide production is overestimated due to the depletion of blood bicarbonate. It seems probable that much of the early controversy about the fuel of muscular exercise arose from this technical problem since carbon dioxide production was overestimated.

4.2.2 *Isotopic methods of estimating carbon dioxide production*

Alternative methods of estimating carbon dioxide production involve the use of isotopes. One which has been developed for use with large grazing animals is termed the *carbon dioxide entry rate method.* If the total body pool of carbon dioxide is labelled by continuous administration of $^{14}CO_2$ at a constant rate then the specific activity of carbon dioxide will be an index of carbon dioxide production (Whitelaw 1974). Continuous collection of a sample for radioassay provides a measure of the overall rate of carbon dioxide production. Usually, radioactive bicarbonate is infused subcutaneously and the collected urine provides an integrated sample for the assay of radioactivity. The precision of the method has been tested over periods of 24 h by comparing estimates of carbon dioxide production with those determined in a respiration chamber. An overestimate of 2–4% was found. This was accounted for by the slow equilibration of carbon dioxide with carbonate in bone. The problem of assuming a respiratory quotient in order to estimate heat production, discussed in Section 2.3.3, is a limitation.

A second method of estimating carbon dioxide production is the doubly-labelled water method. This method is based on the finding by Lifson, Gordon & McClintock (1955) that there is an isotopic equilibrium, due to activity of carbonic anhydrase, between the oxygen of carbon dioxide and that of water. Labelling the hydrogen of water with deuterium (^2H) or tritium (^3H) provides a measure of the rate of elimination of water. Labelling the oxygen of water with ^{18}O results in half the label being present in body water and half in carbon dioxide. The basic equations for estimating rates of production of carbon dioxide and water (mol/d) from measurement of rate constants for the disappearance from body water of deuterium (k_2) and of ^{18}O (k_{18}) are (Lifson & McClintock 1966):

$$\left.\begin{aligned} r_{HO} &= k_2 P \\ r_{CO} &= (k_{18} - k_2)P/2 \end{aligned}\right\} \qquad (4.4)$$

where P is the body-water pool size in moles. In practice this equation has to be modified to allow for fractionation of the isotopes on changing state from liquid to gas. This arises from differences in the mass of the labelled carbon dioxide and water relative to the normal compounds. The usual procedure employed in using the doubly-labelled water technique is to give an oral dose of labelled water and measure its concentration during the ensuing days. The method is essentially one for relatively long-term measurement.

The basic assumptions involved in the use of the method have been explored and show that with care and due allowance for changes with time in the size of the pool of body water, the method gives a good estimate of carbon dioxide production (Lifson & McClintock 1966; Nagy & Costa 1980). Validation studies were made with small laboratory animals in the 1960s (see Gessaman 1973). It was then thought that costs for animals larger than one kilogram in weight would be prohibitive. However, using minimal enrichment of body water and highly sensitive mass spectrometers, costs have been reduced so that the method can be applied to man. Several laboratories have provided evidence that the method is valid for the human subject and it is now being applied to people living their normal lives.

The doubly-labelled water method of estimating carbon dioxide production is ideally suited to estimating metabolism in free-ranging animals and birds, that is what is called *existence metabolism* or *field metabolic rate* (see Section 9.3.2). An early use of the method was to estimate carbon dioxide production during flight (LeFebre 1964; Utter & LeFebre 1970). To convert the estimate of carbon dioxide production to one of heat production entails making assumptions about the RQ. The error arising from this source has been dealt with in Section 2.3.3.

4.3 Pulse rate and oxygen consumption

It was mentioned in Chapter 1 that as early as the 1780s Lavoisier noted the correlation between pulse rate and metabolism. This relationship has been studied in many species, particularly wild ones, since by integration of records of pulse rate it seemed likely that metabolism could be predicted over long periods. Apart from the small cutaneous oxygen exchange, all oxygen is transported to the tissues by the blood and the oxygen consumption of the body is the product of the frequency of contraction of the heart, the stroke volume of the ventricle and the arterio–venous (A–V) difference in oxygen concentration over the tissues. All three determinants of oxygen consumption change with increase in metabolism, the greatest changes being in the A–V difference, and the frequency of contraction. Furthermore the relation between the three varies with posture, environmental temperature, the nature of the stimulus to increase oxygen consumption and from individual to individual. The relation between pulse frequency and oxygen consumption can thus not be expected to be highly precise.

When species are compared it is evident that larger ones have the slower pulse rates. Such observations have been generalised in the form of allometric equations (Calder 1981; Lindstedt & Calder 1981; Schmidt-Nielsen 1984):

For eutherian mammals:

$$f_P = 241\ W^{-0.25} \tag{4.5}$$

For marsupial mammals:

$$f_P = 106\ W^{-0.27} \tag{4.6}$$

For birds:

$$f_P = 156\ W^{-0.23} \tag{4.7}$$

where f_P is the pulse rate per minute and W is body weight in kilograms. Between species varying widely in size, heart mass and stroke volume both vary directly with body weight, although, in small shrews and humming birds weighing less than 5 g, heart weight is proportionally greater than would be expected by comparison with larger species. Under standard conditions the arterio–venous difference in oxygen content is constant with body size at about 60 ml O_2 /l blood. Taking f_P to vary with $W^{-0.25}$ see equation (4.5) it follows that oxygen consumption of the whole body, measured under standard conditions, should vary with $W^{0.75}$. As discussed in Chapter 8, it does.

These considerations suggest that the oxygen consumption of all placental animals should be constant at 0.05 ml per unit of the product, pulse rate/minute times body weight in kg. While this generalisation applies between species kept under standard conditions it cannot be used within a species to estimate oxygen consumption under different circumstances. Body posture, environmental temperature, the nature of the stimulus to increased metabolism and individual factors all alter the relationship between the three components concerned in oxygen provision to the tissues – pulse frequency, stroke volume and arterio–venous difference in oxygen concentration. In some individuals there is little relationship between pulse rate and oxygen consumption at all, while in others close relationships are obtained with slopes that do not accord with the theoretical one (Brockway & McEwan 1969). Prediction of oxygen consumption from pulse rate is thus not an accurate method of estimating heat production. In some circumstances it can give an estimate with a precision of 5–10%. To obtain such precision it is essential that the relationship between oxygen consumption and pulse frequency is established for each individual studied and under conditions which are similar to those in which the relationship is to be used.

4.4 Ventilation rate and heat production

An estimate of oxygen consumption and hence heat production in man can be obtained simply by measuring the volume of air expired. This relationship was examined by Durnin & Edwards (1955) who found a direct proportionality; heat production in kJ/min was '*a*' times the pulmonary ventilation in l/min. The direct proportionality factor, *a*, varied from 0.75 to 1.25 kJ/l, with a mean value of 0.97 kJ/l. In man the oxygen content of expired air is usually about 16.5% by volume and varies little with the rate of oxygen consumption. Man is not a panting animal and relies little on respiratory tract evaporation for cooling the body under warm conditions. In panting animals, such as sheep, the oxygen content of expired air varies from 15.6% to 20.2% (Joyce & Blaxter 1964a). At the same rate of oxygen consumption pulmonary ventilation in sheep can vary by over four-fold, being very high in hot humid conditions and low in cold ones. In man, estimation of heat production from the pulmonary ventilation rate alone may in some circumstances be of value; indeed it has been found to be a more accurate method than one based on pulse frequency (Datta & Ramanthan 1968). The method cannot be applied to animals which rely on panting for thermal homeostasis.

An extension of the method of estimating oxygen consumption or heat production from pulmonary ventilation is to estimate it from respiration frequency. Pulmonary ventilation is the product of respiratory frequency and tidal volume, and oxygen consumption is the product of the difference in oxygen concentration between inspired and expired air and pulmonary ventilation rate. It would be remarkable if there was a precise proportionality between respiration frequency and oxygen consumption, if only because of the variation in the rate of oxygen abstraction from inspired air and in tidal volume with factors unrelated to oxygen consumption. It is generally thought unlikely that recording of respiratory frequency in free-living animals can be used to estimate the metabolic rate (Hargrove & Gessaman 1973).

4.5 Insensible loss of weight and heat production

Early studies with man showed that the heat lost to the body by evaporation was a fairly constant proportion (0.24) of the heat produced, provided that the subject was kept at a thermoneutral temperature, was resting and received no food and that heat production was measured with the subject in the basal state. This finding led to the suggestion by Benedict & Root (1926) that basal metabolism could be estimated very easily from determinations of the insensible loss of body weight. The loss of weight of a subject over a short period of time, in which no food or water is ingested and no urine or faeces voided, is largely due to the evaporation of water. However, oxygen is consumed and carbon dioxide is lost to the body; it is only when the weight of oxygen gained equals the weight of carbon dioxide lost – when the ratio of their volumes or the respiratory quotient is 0.727 – that insensible weight loss equals the weight of water lost. In fasting subjects the respiratory quotient, uncorrected for protein metabolism is usually about 0.85. Under these conditions the amount of water evaporated is 89% of the observed insensible loss. The basal heat production of man might thus be expected to be:

$$\dot{H}_B = a \times 0.89(\dot{W}_{IL})/0.24 \qquad (4.8)$$

where $\dot{H}_B$ is the basal metabolism (kJ/min); a is the heat equivalent of vaporised moisture (kJ/kg); and $\dot{W}_{IL}$ is the insensible loss of body weight (kg/min). The data that Benedict & Root (1926) presented, however, did not show such a direct proportionality. Firstly the relationship between heat production and corrected insensible loss had a large intercept term and secondly the errors were of the order of $\pm 10\%$.

The method has been applied successfully to babies by Hey & Katz (1969) who measured metabolism during short periods when the babies were resting and without food in a constant environment and also predicted it from the insensible weight loss measured simultaneously. In these babies the proportion of heat production accounted for by water vaporisation ranged from 20 to 25% of the total. The method has been applied to rabbits and cattle with less success, largely because the proportion of heat lost by vaporisation under basal conditions varies with the thickness of the coat and is not so invariant under constant temperature conditions as in man.

The relation between metabolism and insensible weight loss applies only to basal metabolism. It does not apply to conditions in which metabolism is increased above the basal level by effects of food or exercise. This is understandable; when metabolism increases in a defined thermoneutral environment virtually the whole of the additional heat produced is lost by the vaporisation of moisture and major departures from the standard 24% factors thus occur. The limitations of the approach have been well documented by Gump (1980).

5

Components of the energy budget: energy retention

5.1 Comparative slaughter methods

The most direct estimate of the energy retained in the body of an individual animal would be obtained as the difference between a determination of the enthalpy of combustion of the body at the beginning and again at the end of a period of time. This is clearly impossible; an animal can only be killed and analysed once. An alternative is to determine the enthalpy of combustion of the body of a precisely similar animal at the beginning of the period and of the experimental animal at the end. This might be done using monozygous twins but a more usual approach is to kill a group of similar animals at the beginning to provide a mean value of the initial body energy. This method of determining energy retention is termed the comparative slaughter method and it was probably first used by Lawes & Gilbert (1861) in their classic studies on the dietary source of energy for fat deposition. The method is particularly well suited to small laboratory species. It has been used with large animals including cattle, and obviously cannot be employed with man.

The techniques used to estimate total body energy all involve removal of the contents of the digestive tract followed by comminution of the carcase to obtain samples for determinations of enthalpy of combustion. With small animals, obtaining a representative sample is fairly simple – a common procedure is to autoclave the carcase in a sealed vessel to provide an homogenate (Lofti & McDonald 1976). With large animals, dissection is involved and various components are analysed separately. Losses of weight occur during these procedures, largely reflecting evaporation of water, and care has to be taken in allowing for them in the final calculations.

Allowing for sampling error and with the knowledge that adiabatic bomb calorimetry is highly precise, the errors attached to a determination of the enthalpy of combustion of the whole body of an individual animal can be expected to be ± 1–2%. The error attached to estimates of the initial enthalpy of the body of an animal is, however, much higher because of animal-to-animal variation. In studies with sheep in which lambs from the same flock constituted the initial slaughter group Thomson & Cammell (1979) found coefficients of variation of body energy content to be ± 15.2% (Fig. 5.1). With rats, the data of Rothwell & Stock (1982) indicate a coefficient of variation of body energy content of ± 10.8% for rats weighing 309 ± 12 g and an even higher variation for smaller rats. Coefficients of variation reflect errors attached to a single observation. Error attached to the determination of initial energy content can be reduced by averaging results for several similar animals – a sample size of six animals probably reduces 'initial error' to about ± 5% irrespective of species. Final error in an individual animal can be taken to be about ± 2% of the enthalpy of combustion of the body. Thus the difference between initial and final energy contents is likely to be more than ± 3% *of the final enthalpy of combustion of the body* independent of the magnitude of retention or duration of the experiment.

5.1.1 *Indirect methods based on changes in body composition*

If the enthalpy of combustion of the body could be estimated indirectly, energy retention could be estimated without killing the animal. Enthalpy of combustion of the body of the same animal could be determined, both at the beginning and end of a period of time and the difference

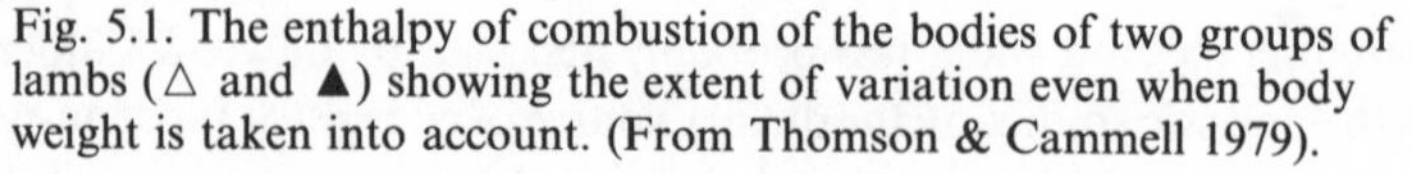

Fig. 5.1. The enthalpy of combustion of the bodies of two groups of lambs (△ and ▲) showing the extent of variation even when body weight is taken into account. (From Thomson & Cammell 1979).

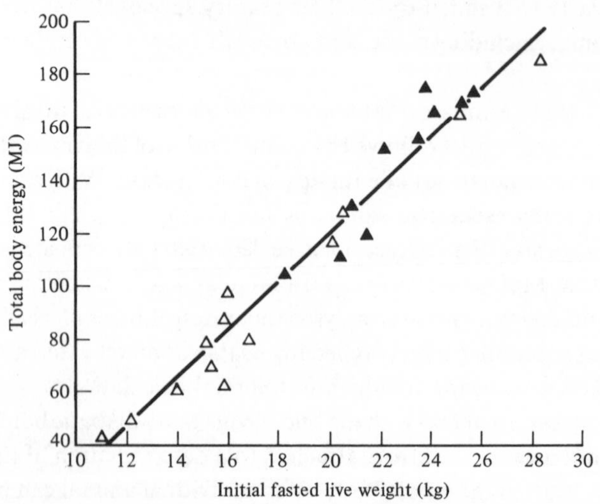

would be a valid estimate of energy retention. Considerable work has been done, particularly with human subjects, to enable this to be done and some of the methods are described below. However none of these attempts to measure the enthalpy of combustion of the body directly; rather they are concerned with estimating its chemical composition, since with a knowledge of the specific enthalpies of the chemical components, the enthalpy of combustion of the total body can be calculated.

The energy retained by the body – and retention can be positive or negative – represents the enthalpy of combustion of organic compounds stored or lost from it. These can be classified in the conventional way into lipids, proteins and carbohydrates. Each of the classes contains numerous compounds; thus the lipid extracted by the chloroform–methanol procedure includes many triacylglycerols differing in fatty acid composition, phospholipids, glycolipids and sterols; the proteins include the different classes of collagen – which account for over 25% of body protein – the keratins, the contractile proteins of muscle and a very large number of enzymes. Additionally, when protein is estimated from its nitrogen content by multiplying it by the factor 6.25 (assuming all proteins contain 16% nitrogen), the resultant *crude protein* includes nucleic acids, glycoproteins, and simple peptides together with many other nitrogen-containing, water soluble compounds such as free amino acids, creatine and urea. Glycogen is the most abundant body carbohydrate, but free glucose also occurs. It cannot be assumed that the retention of lipid, or protein or carbohydrate represents retention of a material of constant composition.

Despite this difficulty, an even simpler division of the animal body has proved to be useful. Schematically the body can be regarded as consisting of two components – fat and non-fat. The former consists of all material soluble in lipid solvents and should not be confused with the fatty tissue which can be dissected. No distinction is made between the structural lipids associated with cell membranes and the lipid of depots; all lipid is included. The non-fat material consists of water, the minerals of bone and soft tissue, carbohydrate, nitrogen-containing compounds, and, in the living animal, the contents of the digestive tract. The non-fat component of the body is usually referred to as the *lean body mass* or *fat-free mass*. Many studies, commencing with those of Murray (1922) and embracing a wide range of adult species, have shown that the chemical composition of the fat-free body is approximately constant. The wide range of composition of animals is largely, but not entirely, due to variation in the proportion of fat. This implies that the enthalpy of combustion of the body can be estimated if the proportion of fat (and hence of non-fat) can be determined. Alternatively, composition in terms of fat and crude protein ($N \times 6.25$) could be used to

Table 5.1. *The composition of the fat-free body of different species*

Species	Body weight (kg)	Water (%)	Protein (N × 6.25) (%)	Ash (%)
Rat	0.35	73.7	22.1	4.2
Hen	2.5	71.9	22.0	3.9
Rabbit	2.6	72.8	23.2	4.0
Cat	4	74.4	21.0	4.6
Man	65	72.8	19.4	7.8
Sheep	80	71.1	21.9	4.2
Pig	125	75.6	19.6	4.7
Ox	500	71.4	22.1	6.0
Horse	650	73.0	20.5	5.8

predict the enthalpy of combustion if crude protein is an index of fat-free mass. This approach has long been used as an alternative to the direct determination of enthalpy of combustion of the whole body by bomb calorimetry by applying specific enthalpy factors to the determined crude composition. For example, in sheep the enthalpy of combustion of the ether extracted (crude) fat is 39.1 kJ/g and of fat-free organic matter 23.2 kJ/g. For cattle the corresponding values are 39.5 kJ/g and 23.0 kJ/g, respectively (Agricultural Research Council 1980). The enthalpy of combustion of the fat-free organic matter in these two species is very close to that of crude protein (N × 6.25) which is 23.6 kJ/g. In these two species application of the respective factors to chemical analytical data predicts enthalpy of combustion extremely well. Factors differ little from species-to-species and there are many examples of their validation in laboratory species.

5.1.2 *The composition of the fat-free body*

Table 5.1 gives data on the composition of the fat-free body of different species, which reflects the broad constancy referred to above. It will be noted that there is no entry for carbohydrate. Glycogen is found in the liver where its concentration ranges from considerably less than 1% to over 5%, and in muscle where its concentration ranges from less than 1% to about 4%. For the body as a whole, the mean concentration of glycogen is usually less than 1%. In a normally-fed man, not undergoing severe exercise, the total amount of glycogen in the body is 300–500 g, or 0.4–0.7% of body weight. Liver and muscle glycogen are both depleted in starvation and by severe sustained exercise. Glycogen is stored in association with 3–4 times its weight of water (Olsson & Saltin 1970). By measuring energy retention and change in body weight twice daily for 10 d, Garrow (1978) showed that body weight increased by 1 g for every 4.2 kJ of energy retained. His results, which are shown in Fig. 5.2, suggest that the short-term fluctuations in body weight could well have been due to short-term gains and losses of glycogen and its accompanying water, associated with the pattern of food consumption and exercise.

This variation in the glycogen content of the body necessarily implies that the composition of the non-fat component cannot be regarded as constant. In addition there are changes in its composition with age and with pregnancy. Moulton (1923) developed the concept of chemical maturation of the body to describe the fact that the composition of the fat-free bodies of animals at birth are characterised by higher water and sodium contents and lower protein, potassium, calcium and phosphorus contents than are found in the same species at maturity. Figure 5.3 depicts results obtained by Spray & Widdowson (1950) for rats, rabbits, cats and pigs, demonstrating the gradual approach to mature values. The same phenomenon is seen in other species. In pregnancy in man, sheep, cattle, pigs, and rats there is an increase in the extracellular fluid compartment of the body which again alters the water content of the fat-free mass (Robinson 1986). Robinson indeed recorded that the water:protein ratio in prolific sheep could increase by 10% during pregnancy. In some species the hydration of the body during pregnancy persists into lactation (Robinson 1986).

The assumption of constancy of the composition of the fat-free mass of the body is central to most indirect methods of determining bodily composi-

Fig 5.2. The results of Garrow (1978) showing the short-term changes in the rate of energy retention of man in relation to concomitant changes in body weight. Body weight increased by 1 g for every 4.2 kJ energy retained. This is consistent with deposition of glycogen and its associated water.

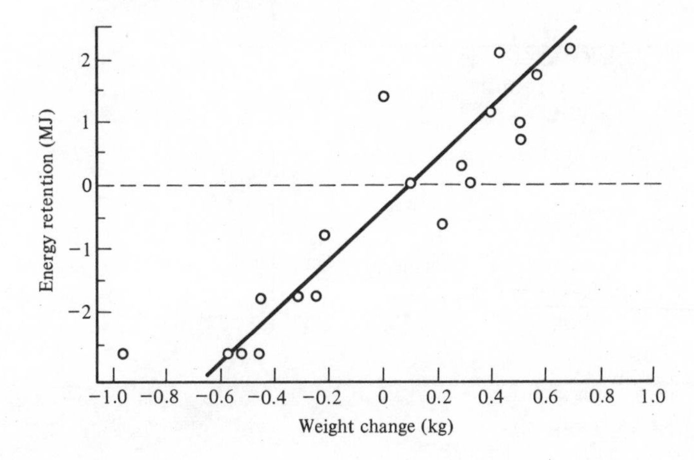

tion and hence energy value. Methods based on estimation of total body water imply that the ratio of water:fat-free mass is constant. Methods based on total body density assume that the fat-free mass has constant composition and hence constant density, while methods based on total body potassium assume that the potassium:fat-free mass ratio is also constant. None of these assumptions is entirely justified.

5.1.3 *Accepted methods for estimating body composition*

Considerable ingenuity has been displayed in devising methods for estimating body composition in living animals. The three most accepted methods are: measurement of body water by the dilution principle; measurement of the ^{40}K radiation of the body; and, estimation of body density. These methods have all been used extensively with man. The first two have also been applied in studies with large animals, and a number of studies aimed at validating them have been made with small animals.

If an animal is given a precise amount of substance, which distributes throughout the total water of the body and which is then eliminated at

Fig. 5.3. The effect of age on the chemical composition of the fat-free bodies of rats, rabbits, cats and pigs as determined by Spray & Widdowson (1950). For rats and rabbits males are represented by X and females ●. The sexes of cats and pigs are not distinguished.

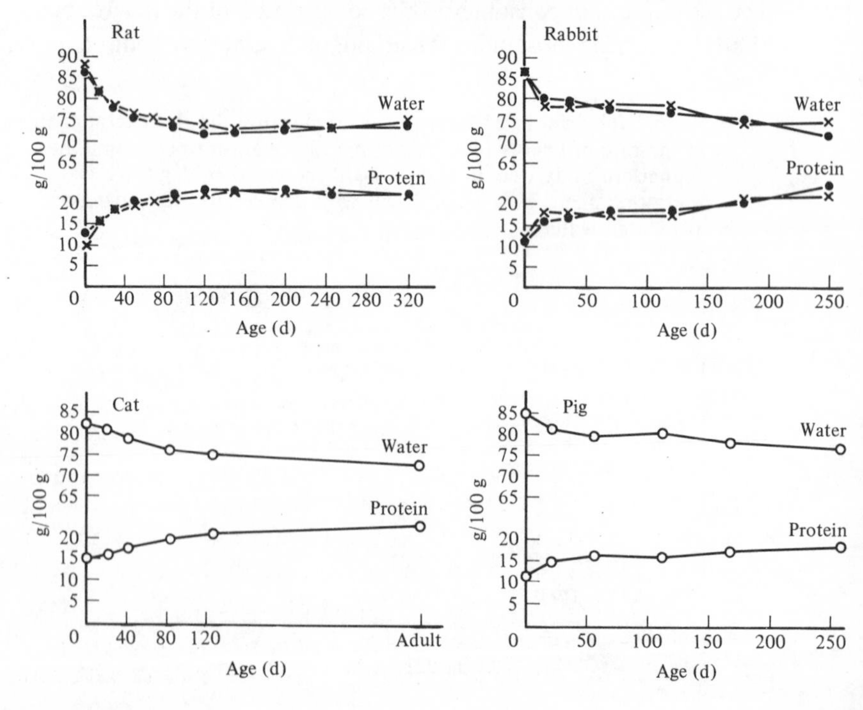

constant rate, its concentration in a sample of the body water can be used to estimate total body water. In practice the substance is given at time zero and its concentration in the water of the blood is measured thereafter. After a short period the rate of fall of the logarithm of the concentration becomes constant and extrapolation to time zero gives the initial concentration in the total body water. Knowing the amount of 'tracer' given, the volume of body water can be estimated. The tracers used for this purpose have included urea, antipyrine, n-acetyl-4-amino antipyrine, and, more recently, tritiated water (^3H_2O) and deuterated water (^2H_2O). Many studies have been made with man using ^2H_2O and the technique and precautions of Halliday & Miller (1977) are thought highly satisfactory.

Obviously the method cannot be tested directly in man; estimates of its accuracy must come by killing animals and comparing the estimate of body water with that directly determined. An example of such a test is that made by Houseman, McDonald & Pennie (1973) with pigs. They found that the chemically-determined total body water (including that in the lumen of the digestive tract) could be estimated from the ^2H_2O space by the equation:

$$\text{body water} = 0.989(^2H_2O \text{ space}) - 0.5 \quad (5.1)$$

where both the body water and deuterated water spaces are measured in kilograms. The residual error was $\pm 1.9\%$ of the mean. The ^2H_2O space was 2.2% greater than the measured total body water, which reflects the fact that deuterium exchanges with the hydrogens of organic molecules in the body. The regression of the fat-free mass on ^2H_2O space had a residual error of $\pm 3.2\%$ of the mean. The lipid content of the body was estimated as the difference between body weight and the weight of the fat-free mass, given by the ^2H_2O space. When this was related to the chemically-determined lipid content of the body the residual error was $\pm 6.7\%$ of the mean. From similar studies with other species, notably those of Searle (1970) with sheep, it appears that the error of estimating body energy content from the volume of distribution of a tracer for body water is about $\pm 5\%$.

The natural radioactivity of the body arises mainly from γ-radiation of the naturally occurring ^{40}K. This can be measured using a gamma spectrometer suitably calibrated using a 'phantom' – that is, a physical model of the body containing potassium – to count the 1.46 MeV γ-rays. A phantom model of the body can be complex for it is designed to simulate the spatial distribution of an element (Bewley 1986). Total body potassium can be calculated from radioactive potassium if the abundance of the isotope is known, and the fat-free mass can then be estimated from the ratio K:fat-free mass, which is 2.7/(g kg) or 65 mmol/kg – a value which is not constant but varies with sex and body fat. The method has been validated with human infant cadavers (Garrow 1965) and with other species by comparing the estimated body composition with chemically-determined values. Errors are again of the order of ±5%.

The density of body fat is 0.900×10^3 kg/m^3 while that of the fat-free body is about 1.100×10^3 kg/m^3 (Keys & Brozek 1953). As Behnke realised in 1942 (see Behnke 1965), a determination of body density provides an estimate of the proportion of fat. The density of the fat-free body obviously depends on its composition; at body temperature the densities of water, average protein and bone mineral are 0.993, about 1.34, and about 3.00×10^3 kg/m^3, respectively. For the estimation of the total energy of the body, determination of density is as dependent as are the previous two methods, on the basic assumption that the compositions and enthalpies of combustions of the two conventional divisions of the animal body are invariant.

Archimedes principle is commonly employed to estimate the density of the body in man. The person is weighed in air and in water. A difficulty arises in that the underwater weighing includes air in the lungs and respiratory passages. This can be overcome by measuring the change in volume of a sealed air space enclosing the head, when pressure is changed (Garrow *et al.* 1979). Alternatively, body volume can be measured by air or gas displacement methods. The method is not suitable for animals since the cooperation of the subject is essential. The final computation of body fat content from density is usually made using Siri's (1961) formula:

$$\text{fat (\%)} = 450 \, \frac{1.1 - 1}{d} \qquad (5.2)$$

where d is density of the body, weight/volume (kg/m^3).

5.1.4 *Other techniques for determining body composition*

A number of other techniques have been or are being investigated to provide estimates of the composition of the body. These have not necessarily been developed to provide ways of estimating the enthalpy of combustion of the body and thus estimates of energy retention. The composition of the body in health and disease is a subject of interest in its own right. The more modern methods involve apparatus and techniques of considerable sophistication and have mostly been developed within departments of physics in medical schools with applications to the human patient in mind.

An early method recognised the difference in the solubility of nitrogen, the noble gases and of anaesthetic gases in body lipid and in body water. If an animal breathes air enriched in these gases, eventually a tissue equilibrium is established which reflects their solubilities. Measurement of the amounts retained by 'washing out' the body gas stores through breathing normal air provides an estimate of the amounts of fat and water in the body. Unfortunately the time taken to reach equilibrium in both the uptake and washout stages are long, even in small laboratory species, and the method has not been employed to any appreciable extent.

A series of methods for the measurement of the elemental composition of the body has been devised based on neutron activation analysis. The body is irradiated with neutrons from a source. For the estimation of body nitrogen, the nitrogen-prompt γ-ray emitted after neutron capture, or alternatively the induced activity of ^{13}N are measured (Burkinshaw 1982, 1985; Burkinshaw & Oxby 1986). For estimation of body carbon, a pulsed neutron source is used and the 4.44 MeV γ-ray arising from neutron-inelastic scattering is measured. Analogous methods can be employed for potassium, calcium, sodium and chlorine. These methods have considerable potential; however, it appears that at present the errors attached to them are of the order of 7–10%, values which will no doubt reduce with further development. Work has indeed shown that protein can be estimated with an error of $\pm 4.2\%$ and fat with an error of $\pm 6\%$ by a combination of tritiated water dilution and neutron activation analysis.

Other methods at present being developed are based on the fact that the electrical conductivities of extracellular water and lean tissue far exceed those of fat. Measurements of total body electrical conductivity, or electrical impedance, provide an index of fat content. Nuclear magnetic resonance (NMR) measurement techniques, photon activation by direct irradiation, near-infrared spectroscopy of body surfaces and several other approaches are currently being investigated. Nuclear magnetic resonance using imaging techniques is a useful adjunct to metabolic studies because it can show the localisation of lipid within the body.

5.2 Metabolic balance methods of estimating energy retention

A method of estimating the retention of energy which has been in use for more than a century is the carbon and nitrogen balance. It is based on the assumption that the only energy-yielding compounds stored by the body are fat and protein and that these have fixed chemical compositions and enthalpies of combustion. Measurement of the intake of carbon, its losses in faeces, urine, methane and as carbon dioxide provide a measure of the carbon lost or gained by the body, that is the carbon retention or balance. Measurement of nitrogen intake less the loss in faeces and urine provides a measure of the nitrogen retention or balance. Protein retention can be calculated from nitrogen retention and the carbon contained in this protein deducted from the observed carbon retention to give an estimate of the carbon retained as fat. Knowing the carbon content of fat, the fat content can be calculated. Early estimates of the carbon contents of fat and protein were based on extracted and purified depot fat and muscle protein and they varied considerably. Atwater & Benedict (1903), in their classic studies with man, used carbon contents of 76.08% and 53.0% for fat and protein, respectively. The nitrogen content of body protein was taken to be 16.0%. Brouwer's (1965) more recent factors are 76.70% C in body fat and 52.0% C in body protein. The latter values together with enthalpies of combustion of fat and protein give the following expression for estimating energy retention from the simultaneous retention of carbon and nitrogen:

$$R = 51.83\ C - 19.40\ N \qquad (5.3)$$

where R is energy retention in kJ; C is carbon retention in g; and N is nitrogen retention in g.

Realising that extracted fat and muscle protein were not necessarily representative of the fat and protein of the body as a whole, Blaxter & Rook (1953) used a statistical definition of body fat and protein. The nitrogen and carbon contents and heats of combustion of a series of samples of body tissues were determined. Regression analysis of these gave the heats of combustion and carbon contents of a theoretical protein containing 16% N and a theoretical fat containing none. This approach takes into account the fact that the body stores both nitrogen-free and nitrogen-containing compounds other than depot triglycerides and purified muscle proteins. The analogous equation for estimating energy retention, derived from these statistically defined factors, was:

$$R = 52.51\ C - 28.87\ N \qquad (5.4)$$

where the symbols are the same as in equation (5.3) above.

The carbon and nitrogen balance approach can be extended to include calculations based on the retention of hydrogen. This can be computed from the difference between intakes of hydrogen in food and water and output in organic solids and in water. The observed retention of hydrogen includes hydrogen retained as fat (which contains 11.80% H) and protein (which contains 7% H). Any retained hydrogen which is not accounted for by retained fat and protein, called the *corrected hydrogen retention*, must largely represent water which accompanies protein deposition in the body. In addition it represents a gain or loss of body glycogen and its associated water. Determination of oxygen retention then allows separation of the water component from the carbohydrate component of the corrected hydrogen retention. This has rarely if ever been done; nor has the simultaneous balance of carbon, nitrogen and hydrogen often been determined.

5.3 Change in body weight as an index of energy retention

It has already been noted in Section 5.1.2 that changes in weight during a 24-h period can be referred to changes in the body content of glycogen and its associated water, implying that the minimal value for the energy content of a change in body weight is 4.2 MJ/kg. Body lipid is stored in adipocytes without concomitant storage of water, implying that the maximal value for the energy content of a change in weight is 39 MJ/kg. This almost 10-fold range in the enthalpy of combustion of a gain or loss of body weight is however a considerable underestimate.

One of the factors contributing to variation in the energy content of a change in body weight is variation in the weight of the gut contents. In ruminant animals and indeed in most herbivores, changes in the amount of food consumed lead to marked changes in the weight of the gut contents. Thus when sheep are given food *ad libitum* their body weights immediately increase to about 122% of those noted during fasting. In non-herbivores the changes are much smaller. Even when herbivores given constant food are weighed at the same time of day, thus obviating variation in weight due to the pattern of meals, a variation of $\pm 2\%$ in body weight is still noted from day-to-day. This can be attributed to aperiodicity in the voiding of faeces and urine. Again, in non-herbivores this variation in body weight is less noticeable. Gump (1980) for example, has shown that over 24 h the variation in body weight of a normal man is $\pm 0.7\%$ of mean body weight. Gains or losses of body weight measured over short periods of time thus do not necessarily reflect changes in the body tissues.

When change in body weight is measured over periods of time which are sufficiently long to minimise effects of gut-fill on body weight and to ensure stability of the glycogen content, it might be expected that the enthalpy of combustion of body gain would be reasonably constant, reflecting that of the fat and protein deposited. As discussed in Section 11.7.2, in young, normal animals, gains have an enthalpy of combustion of about 8 MJ/kg. As normal growth proceeds, these rise to maximal values of about 30 MJ/kg. These values are lower than that of pure triglycerides of depot fat and reflect concomitant deposition of protein. Nevertheless, enthalpies of combustion of unit gain can be considerably outside this range. Thus, as reviewed by McCracken (1986) and commented on in Section 12.3.2, rats and pigs given low-protein diets gain fat and lose protein plus water. This results in apparent heats of combustion of gain being three times that of pure fat. Again, when rapidly growing animals are given restricted amounts of an adequate diet, gain in weight can occur due to deposition of protein and water but with sufficient loss of body fat to result in a loss of energy from the body. Enthalpies of combustion of gains are then negative. In lactation in a number of species the energy retention is usually negative due to fat mobilisation. This may not be associated with any loss of body protein. Clearly change in body weight is not a good index of energy retention.

5.4 The conservation of mass and energy in animals

The idea of the conservation of mass is implicit in Newton's third law; Newton was also responsible for the idea of the conservation of kinetic energy or '*vis viva*', but the concept of conservation of energy in all its aspects was undoubtedly due to von Helmholtz (see Feather 1963). The physical sciences had accepted these concepts by the mid-nineteenth century. Even so, experiments to show that mass was conserved in chemical reactions continued until the early years of the present century when it was shown that the difference between the masses of the reactants and the products of a chemical reaction was less than one part in ten million. In biology, and particularly human biology, things were rather different. There was a view that the phenomenon of life was essentially different from those of the physical world. A vital principle was evoked to lead to the view that physics and chemistry were not sufficient to explain living things. The initial results obtained by Lavoisier and Laplace did not help to refute such ideas, for they had shown that they could not account for the whole of the heat produced by an animal from the amount of carbon oxidised – having neglected the combustion of the hydrogen of organic compounds. Vitalism was not a single belief; it postulated some force additional to those known to physics; whether this force could be measured or not by the reductionist processes of physics and chemistry was debated.

Conservation of mass is a concept that is now completely accepted in biology. It is the basis of all balance techniques for the measurement of animal requirements, in that the difference between the intake and excretion of a mineral element or of nitrogen is axiomatically the retention of that element in the body. Table 5.2 shows explicitly that total mass is conserved; the small discrepancy is accounted for by the errors of weighing the living, uncooperative animal.

Conservation of energy could be demonstrated by a table analogous to Table 5.2. There should be entries for initial and final enthalpies of combustion of the body, any inputs of energy as food, and the energy losses in excreta and as heat. The heat term should be that determined directly together with heat storage terms, not the heat production measured indirectly from the gaseous exchange, for the latter brings into consideration a series of assumptions as discussed in Section 2.2. No such experiment has

Table 5.2. *The conservation of mass in an experiment with a calf confined in a respiration chamber for four days*

Input parameters	Input weight (g)	Output parameters	Output weight (g)
Weight of animal	36450	Weight of animal	35846
Food (milk)	12040	Faeces	272
Water	1600	Urine	10164
Oxygen	2480	Water vapour	3596
		Carbon dioxide	2640
		Methane	14
		Hair and skin debris	28
Total	52570	Total	52560

Note:
Discrepancy = 10 g = 0.019% of input.

been traced; the limitation of published work in this respect is the absence of accurate data on the initial and final enthalpies of combustion of the body. There have, however, been a number of experiments which purport to be absolute demonstrations that energy is conserved in living organisms.

5.4.1 *Rubner's and other demonstrations of energy conservation*

In 1894 Rubner measured the amounts of fat and protein catabolised by a fasting dog through measurements of its respiratory exchange, calculated their enthalpies and showed that the sum agreed with the heat measured directly (see Rubner 1902). Laulanie in 1896 carried out similar experiments with rabbits, pigs, dogs and ducks with similar results (Berraud 1975). These demonstrations of the validity of the conservation principle may be criticised in that the amounts of fat and protein oxidised were computed from the respiratory exchange and their enthalpies of combustion inferred. The extensive experiments of Atwater & Benedict (1903) involved measurement of the enthalpies of combustion of food and excreta and the direct determination of heat. The change in the enthalpy of combustion of the body was estimated from the carbon and nitrogen balances as described above. The subjects undertook work on a bicycle in the calorimeter and the external work done was recorded as heat by the calorimeter. The results of these very careful experiments are given in Table 5.3. They suffer from the defect evident in those of Rubner and Laulanie; change in the enthalpy of combustion of the body was estimated indirectly and is subject to errors of assumption as discussed above. The mean discrepancy found was 50 kJ/d or +0.6% of the heat production deter-

Table 5.3. *A summary of the calorimetric experiments of Atwater & Benedict (1903) classified by individual subject both with and without exercise (all values in kJ/d)*

Number of trials	Intake energy	Faeces energy	Urine energy	Change in body		Heat produced	Discrepancy	
				Protein	Fat		Actual	S.D.
Trials without exercise								
13	11 288	456	586	−55	766	9 552	16	234
4	11 042	423	485	−88	582	9 778	138	322
4	10 837	448	552	−71	1046	8 924	62	84
1	10 539	460	565	−151	−196	10 029	168	—
All (22)	11 125	448	561	−67	736	9 498	51	226
Trials with exercise								
3	15 928	690	540	25	−1611	16 154	−130	301
6	14 573	452	544	−205	−1108	14 828	−62	196
14	20 171	870	598	−305	−2510	21 485	−33	359
All (23)	18 158	736	577	−238	−2025	19 054	−54	310
Fasting trials								
4	0	0	439	−1937	−7916	9 150	−264	288

Note:
The mean discrepancy in the 22 trials without exercise was 0.6% of the observed heat production and in those with exercise 0.3% of the heat production.
The values were calculated directly from the tabulated results of the individual experiments, converting calories to joules using a factor of 4.184.

mined for resting subjects and −54 kJ/d or −0.3% for exercising subjects. Adoption of the Brouwer factors for computing energy retention from the carbon and nitrogen balances would reduce the mean discrepancies slightly to 40 and −49 kJ/d for the rest and work experiments, respectively. Atwater and Benedict realised the limitations of the estimates of change in body retention of energy particularly with respect to the fasting experiments in which glycogen was likely to have been depleted. Webb's (1980) contention that the discrepancies observed represent some 'unidentified quantity of unmeasured energy' evident in under-nutrition may be due to this cause. Webb however included in his analysis of Atwater and Benedict's data two dubious experiments (numbers 50 and 55); in one the electrical output from the calorimeter broke down and in the other the alcohol checks of its functioning were grossly in error due to failure of an aliquoting pump. These experiments were rejected by Atwater and Benedict and their inclusion in Webb's analysis is the main factor responsible for his contention. If explanation of the discrepancies is required, then, as shown in Table 5.3, the standard deviations of the observations about their respective means are those to be expected in calorimetric work as a result of analytical and instrumental error alone. The conclusions which Atwater and Benedict drew are consonant with their results for they realised the difficulties. They wrote 'For practical purposes we are therefore warranted in assuming that the law [the first law of thermodynamics] obtains in general to the living organism as indeed there is every reason *a priori* to believe that it must' (Atwater & Benedict 1903).

5.4.2 *Validation of indirect methods*

Many different methods have been employed to estimate one or several components of the overall energy budget and the extent of agreement between methods has been assessed. The first comparison to be considered is the agreement between determinations of heat production by direct calorimetry and those by indirect calorimetry based on the gaseous exchange. Given absolute instrumental and analytical accuracy this comparison is in effect a test of the validity of the factors adopted to estimate heat from oxygen consumption, carbon dioxide production and urinary nitrogen excretion as discussed in Section 2.2 and of the validity of the calculations of the heat storage terms as discussed in Section 4.1.1. With sheep, comparison of direct gradient layer calorimetry with closed circuit indirect calorimetry resulted in a mean difference of 29 kJ/d or 0.36% of the mean value (Blaxter 1980). In man, estimation of heat production by heat sink calorimetry and has been compared with estimates made from the gaseous exchange over periods of 12 d (Murgatroyd *et al.* 1985). The difference between the two was 0.39% of the mean.

A second comparison which can be made is between energy retention determined calorimetrically or by the carbon and nitrogen balance technique. In Atwater & Benedict's (1903) experiments the calorimetric estimate of retention was based on heat production determined directly. More usually the calorimetric determination of retention is obtained as the difference between metabolisable energy and heat production determined indirectly. In 37 experiments with sheep Blaxter & Graham (1956) found a mean difference of 18 kJ/d, which represents 0.1% of the energy of the food. In cattle the results obtained with the Copenhagen respiration chamber (Mollgaard & Lund 1929) gave similar results and a mean discrepancy for 19 experiments of 0.29% of the energy of the food.

A third comparison is that between energy retention measured in a respiration chamber and that estimated by comparative slaughter methods. Studies by H.L. Fuller, Dale & Smith (1983) in chicken showed a mean discrepancy with no bias amounting to 1.2% of the energy intake as food. M.F. Fuller & Boyne (1972) showed by comparative slaughter methods that

Table 5.4. *Agreement of estimates of fat loss from the bodies of 19 obese women made using different methods, together with the between-subject standard deviation which includes the error associated with the measurement technique*

Method	Mean fat loss (± S.D.) (kg)
Direct and indirect calorimetry with correction for nitrogen loss	2.77 (0.71)
From body weight and an assumed weight equivalent of nitrogen loss	2.69 (1.23)
From body density determination made using body plethysmography	2.83 (2.32)
From determination of total body water by deuterium oxide dilution	2.37 (2.38)
From total body potassium, counting the 1.46 MeV γ-radiation	2.90 (3.54)

Note:
The standard errors of the means can be obtained by dividing the standard deviations by 4.36.
Source: Garrow *et al.* (1979).

pigs growing from 20 to 90 kg retained 1079 MJ. The same pigs were used in respiration chamber studies and their energy retention during the period was 1038 MJ, giving a mean discrepancy which, while it represents 3.9% of the energy retained, is considerably less than 1% of the metabolisable energy intake.

A further comparison that can be made is between retention of fat as determined by balance techniques employing direct calorimetry and deposition of fat as estimated from body composition determined by various indirect methods. This was done by Garrow *et al.* (1978, 1979). Nineteen obese young women were given low-energy diets under metabolic ward conditions such that they lost weight. The losses of body fat were determined in different ways and, as Table 5.4 shows, the five methods gave statistically indistinguishable results. The errors given in the table have a mixed significance since they include between-subject variation in fat loss. They do however suggest that the calorimetric method was the most accurate.

Generally, these comparative tests show that, when properly conducted, the methods employed in energy metabolism studies are consistent with one another. There is no reason to suppose that direct calorimetry gives significantly different results from those obtained using indirect approaches or that an energy retention determined calorimetrically is not precisely related to a change in the energy content of the body. If any discrepancies of magnitude are found, attention should be paid to the precision of the analytical and instrumental work involved rather than to questions relating to the validity of the conservation laws.

6

Intermediary metabolism and energy exchanges

6.1 The second law of thermodynamics

Animal cells contain an enormous array of interlocking chemical processes catalysed by enzymes. In the context of bioenergetics, the second law of thermodynamics is concerned with the factors which limit the passage of energy from one reaction to another or which govern the use of energy derived from a reaction to accomplish mechanical work (as in muscular contraction) or chemiosmotic work (as in the maintenance of the so-called ion pumps associated with cell membranes). The second law simply states that part of the energy change associated with a particular reaction is not available to do work or to be passed to some other reaction. This attribute of a system is called *entropy*. To provide continuity with Chapter 2 the concept of entropy, S, which is the essence of the second law, can be combined with the first law to give the expression:

ΔH	$= \Delta G$	$+ T\Delta S$	(6.1)
change in enthalpy	change in Gibbs free energy	absolute temperature × change in entropy	

The change in the Gibbs free energy, ΔG, is the measure of the *maximal* amount of energy that can be passed on by a reaction; an analogous measure is the Helmholtz free energy which is defined in terms of change in internal energy, ΔE, rather than change in enthalpy, ΔH, (see equation (2.4)). Entropy is expressed as Joules/degree mole (J/K mol) and is the measure of the inability of the energy of a reaction to be passed on, that is to do work in a thermodynamic sense.

The derivation of the concept of entropy is given in most textbooks of equilibrium thermodynamics and thermochemistry (i.e. Nash 1970;

Morowitz 1978). The approaches generally involve a consideration of heat engines and the Carnot cycle or derive from statistical mechanics. Such derivations are not given here. Rather some of the attributes of the free energy function are described so as to enable an appreciation of the problems of relating biochemical events to the overall energy exchange of the whole body.

6.1.1 *Free energy of formation and entropy of formation*

The free energy change in a reaction depends on the conditions under which it occurs. As with enthalpy, to avoid questions about what these conditions might be, standard conditions are imposed so to arrive at *standard free energies*, ΔG^0, as distinct from non-standard ones, ΔG. These conditions are a temperature of 25 °C (298.16 K), one atmosphere pressure and the reactants and products in their standard states. Since most biochemical reactions take place in solution further conditions are that products and reactants are present at 1.0 molal activity. For most practical purposes, molal activity can be regarded as the observed molality. This creates difficulties if the hydrogen ion is involved for standard conditions for one mole of hydrogen ion means the pH is 0. The convention is to compute free energy changes for a pH of 7.0 and the *standard free energy change* is then usually designated $\Delta G^{0\prime}$ to indicate that the only departure from standard conditions is that the pH is 7.0. As with enthalpy, the free energies of the elements under standard conditions are taken to be zero and, in a way analogous to the computation of the enthalpies of formation (see Chapter 2) the free energies of formation, ΔG_F, can be computed. Tables of free energy of compounds of biochemical interest are given by Kaye & Laby (1973) and by Weast (1985).

A similar notation is used for standard entropy change, ΔS^0. A standard entropy implies that reactants and products are in their standard state, the pressure is atmospheric and the temperature is 25° C (298.16 K). The third law of thermodynamics, or the Nernst heat theorem, states that the entropy of all single crystals of pure substances is zero at 0 K. This allows the *absolute entropy*, S^0, to be estimated:

$$S^0 = \int_0^{298.16} \frac{C_P \, dT}{T} \qquad (6.2)$$

with additional terms for the entropy of fusion and vaporisation included where there are changes in state from solid to liquid and from liquid to gas over the range of temperature, 0–298.16 K. Absolute entropies can be used to calculate entropies of formation if the absolute entropies of the elements are known. The method can be illustrated for L-alanine. L-Alanine contains 3 C, 7 H, 2 O and 1 N. Its absolute entropy determined from its specific heat at constant pressure (C_P) is 129.2 J/(K mol). The absolute entropies of C(s), H_2(g), O_2(g) and N_2(g) are 7.7, 130.6, 205.0, and 191.48 J/(K mol), respectively. The entropy of formation of L-alanine is the sum of the absolute entropies of the elements it contains less the absolute entropy of the L-alanine itself, giving $\Delta S_F = 651.8$ J/(K mol).

Given free energies of formation and entropies of formation, the standard free energy and entropy changes for particular reactions can readily be computed. Tables of absolute entropy and entropy of formation are given by Kaye & Laby (1973) and by Weast (1985).

6.1.2 *Free energy and chemical equilibria*

While tabulations of free energies of formation enable the standard free energy changes of reactions to be calculated, a more usual approach is to compute them from chemical equilibria, indeed this is how many of the tabulated values were derived. One of the generalisations arising from the second law is that the free energy change accompanying a chemical reaction is:

$$\Delta G = \Delta G^0 + RT \log_e \Phi \qquad (6.3)$$

where ΔG is the free energy change; ΔG^0 the standard free energy change; R the gas constant (8.314 J/(K mol)); T the absolute temperature; and Φ a reaction quotient representing the products' concentration divided by the reactants' concentration. Concentrations of products and reactants can be expressed in terms of partial pressures, mole fractions or activities.

For a chemical reaction at equilibrium, $\Phi = K_{EQ}$, the equilibrium constant. At equilibrium the change in free energy is axiomatically zero since no work can be done. Equation (6.3) can thus be written:

$$0 = \Delta G^0 + RT \log_e K_{EQ}$$

and (6.4)

$$\Delta G^0 = -RT \log_e K_{EQ}$$

Equation (6.4) enables the standard free energy change of a reaction to be determined from the observable equilibrium constant. Table 6.1 tabulates this relationship, and shows that when the equilibrium constant is greater than 1.0 the free energy change is negative; when it is less than 1.0 the value is positive. An example clarifies the relationships. Consider a biochemical reaction in which molar quantities of chemical species A and B react to form C and D:

$$A + B \rightleftharpoons C + D$$

Table 6.1. *The relationship between the equilibrium constant, K_{EQ} and the standard Gibbs free energy change, ΔG^o;* $\Delta G^o = -RT \log_e K_{EQ}$

K_{EQ}	ΔG^o (kJ/mol)
0.001	17.1
0.01	11.4
0.1	5.7
1.0	0
10	−5.7
100	−11.4
1000	−17.1

If the value of ΔG^0 is negative then the reaction will proceed from left to right, and the more negative the value of ΔG^0, the further towards the right will the reaction proceed. Such a reaction is termed *exergonic*. If the value of ΔG^0 is positive the reaction proceeds from right to left and the reaction is termed *endergonic*.

The above relationships relate to standard conditions; in cells the conditions are very different. Equation (6.3) can be used to estimate ΔG when ΔG^0 and the concentrations of products (C,D) and reactants (A,B) are known. Then:

$$\Delta G = \Delta G^0 + RT \log_e \frac{[C][D]}{[A][B]} \tag{6.6}$$

In practice it is not necessarily easy to determine the concentrations of reactants and products for particular reactions in cells.

6.1.3 *Free energy changes in the dissimilation of carbohydrate*

The standard free energy change on the oxidation of glucose by gaseous oxygen to gaseous carbon dioxide and liquid water may be computed from the standard free energies of formation of the products and reactants (kJ/mol):

$$\left.\begin{array}{ll} & \text{Glucose} + 6O_2 \rightarrow 6CO_2 \quad + 6H_2O \\ \Delta G^0{}_F = -910.3 & + (6 \times 0) \rightarrow (6 \times -394.4) + (6 \times -237.2) \\ \multicolumn{2}{l}{\text{Products less reactants} = \Delta G^0 = -2879.3 \text{ kJ/mol}} \end{array}\right\} \tag{6.7}$$

The enthalpy change for this oxidation (the heat of combustion of glucose)

Table 6.2. *The individual steps of glycolysis and the tricarboxylic acid cycle showing the ATP formed from ADP and inorganic phosphate*

Reaction	ATP (mol)
1 Glucose → glucose 6-phosphate	−1
2 Glucose 6-phosphate → fructose 6-phosphate	
3 Fructose 6-phosphate → fructose 1,6-bisphosphate	−1
4 Fructose 1,6-bisphosphate → 2 (glyceraldehyde 3-phosphate)	
The subsequent reactions are duplicated, corresponding to the two moles of glyceraldehyde 3-phosphate formed in the preceding reaction	
5 Glyceraldehyde 3-phosphate → 1,3-diphosphoglycerate, NADH	2×2=4
6 1,3-Diphosphoglycerate → 3-phosphoglycerate	2×1=2
3-Phosphoglycerate → 2-phosphoglycerate	
8 2-Phosphoglycerate → phosphoenolpyruvate	
9 Phosphoenolpyruvate → pyruvate	2×1=2
10 $Pyruvate_{cyt}$ → $pyruvate_{mit}$	−2×0.25=0.5
11 Pyruvate → acetyl-CoA, NADH	2×3=6
12 Acetyl-CoA + oxaloacetate → citrate	
13 Citrate → isocitrate	
14 Isocitrate → α-oxoglutarate, NADH	2×3=6
15 α-Oxoglutarate → succinyl-CoA, NADH	2×3=6
16 Succinyl-CoA → succinate	2×1=2
17 Succinate → fumarate, FADH	2×2=4
18 Fumarate → malate	
19 Malate → oxaloacetate, NADH	2×3=6
Total ATP	35.5

is −2803.1 kJ/mol and hence from equation (6.1), given in its more usual form below:

$$\Delta G = \Delta H - T\Delta S \tag{6.8}$$

the temperature times entropy term, $T\Delta S^0$, is $\Delta H^0 - \Delta G^0 = 76.1$ kJ/mol and the entropy change, ΔS^0, is 263.2 J/(K mol). The sign of the free energy term and its numerical magnitude indicates that the reaction proceeds from left to right and that the equilibrium which will be reached is far to the right.

The oxidation of glucose depicted in equation (6.7) is simply a summary of what occurs in a cell, for in the cell the oxidation occurs in a series of steps set out in simplified form in Table 6.2. In the cell cytoplasm glucose is oxidised by the *Embden–Meyerhof pathway* to yield two moles of pyruvate. The pyruvate enters the mitochondrion where it is converted to acetate as acetyl coenzyme A, (acetyl-CoA) which is completely dissimilated within the mitochondrion by the cyclical enzymic sequence called the *Krebs cycle* (synonyms of which are the *tricarboxylic acid cycle* (TCA cycle) and the *citric acid cycle*). The initial step is the condensation of acetyl-CoA with oxaloacetate to form citrate. During passage through the cycle two carbon atoms are released as carbon dioxide and the four hydrogen atoms of the acetate can be regarded as reducing specific coenzymes which in the *respiratory chain* of enzymes (electron transfer chain) eventually combine with oxygen to produce water. Specific labelling of the carbon and hydrogen atoms of the acetate show that these are not released on a single turn of the cycle; the dissimilation is a net one. No error is incurred by assuming that one turn of the cycle results in the complete oxidation of the acetyl-CoA which enters. Oxidation of glucose to carbon dioxide and water involves 20 steps excluding those related to electron transport in the respiratory chain. That only 19 are shown in Table 6.2 reflects the fact that the 3-C compounds formed from fructose-1,6-bisphosphate are dihydroxyacetone phosphate and glyceraldehyde 3-phosphate, in equal yields. The enzyme triose phosphate isomerase leads to conversion of the former to the latter.

The extent of the simplification of Table 6.2 is evident when reaction 1 is considered. In full this reaction, catalysed by the enzyme hexokinase, is:

$$\text{ATP}^{4-} + \text{glucose} \rightarrow \text{glucose 6-phosphate} + \text{ADP}^{3-} + \text{H}^{+}: \quad \Delta G^{0\prime} = -16.7 \text{ kJ/mol} \qquad (6.9)$$

where ATP^{4-} is adenosine 5′-triphosphate; and ADP is adenosine 5′-diphosphate. This reaction can be considered as consisting of two separate reactions:

$$\text{ATP}^{4-} + \text{H}_2\text{O} \rightarrow \text{ADP}^{3-} + \text{HPO}_4^{2-}: \quad \Delta G^{0\prime} = -31.0 \text{ kJ/mol} \qquad (6.10a)$$

and:

$$\text{glucose} + \text{HPO}_4^{2-} \ \text{glucose 6-phosphate} + \text{H}_2\text{O}: \quad \Delta G^{0\prime} = +14.3 \text{ kJ/mol} \qquad (6.10b)$$

The first reaction (6.10a), in which the standard free energy change is negative; ‘pulls’ the second reaction (6.10b), in which the free energy change is positive, such that the whole reaction in equation (6.9) has a negative free energy change and proceeds from left to right. The two reactions are coupled.

Similar relationships are seen in other reactions in which ATP participates. The hydrolysis of phosphoenolpyruvate by water to yield pyruvate, inorganic phosphate and a proton has a negative value of ΔG^0. It is coupled with the reverse of the reaction shown in equation (6.10a) in which ATP and water are formed from ADP, inorganic phosphate and a proton. The net reaction has a ΔG^0 of -25.5 kJ/mol.

Another example can be given. The production of lactate from glucose by glycolysis follows the sequence of reactions shown in Table 6.2 up to the formation of pyruvate. Then pyruvate is reduced to lactate by the reduced nicotinamide adenine dinucleotide (NADH) arising from reaction 5. The overall chemical change is:

$$C_6H_{12}O_6 \rightarrow 2CH_3CHOHCOO + 2H^+: \quad \Delta G^{0\prime} = -198.3 \text{ kJ/mol} \qquad (6.11a)$$

This can be written more fully:

$$\text{Glucose} + 2\text{ADP} + 2\text{HPO}_4^{2-} \rightarrow 2\text{H}^+ + 2 \text{ lactate} + 2\text{H}_2\text{O} + 2\text{ATP}: \quad \Delta G^{0\prime} = -135.6 \text{ kJ/mol} \qquad (6.11b)$$

That the free energy change is more negative in equation (6.11a) than in (6.11b) is understandable; ATP is being synthesised from ADP and inorganic phosphate. The difference between the two (62.7 kJ/mol glucose) is the standard free energy when two moles of ATP are formed from ADP as given by equation (6.10a).

This relationship, in which glucose is broken down anaerobically and ATP is simultaneously formed, can be regarded as an energy conservation. Of the free energy of the chemical change (−198.3 kJ), −62.7 kJ can be accounted for by the concomitant synthesis of ATP from ADP. The efficiency of free energy capture in this instance is $100 \times 62.7/198.3 = 32\%$.

6.1.4 *Oxidative phosphorylation*

The above two examples represent *substrate level phosphorylations*, that is, the formation of ATP from phosphorylated intermediates. These and many similar ones occur in the cytoplasm of the cell. While substrate level phosphorylation is important as a source of ATP, most of the ATP formation in the body takes place in the mitochondria of cells. The process involved is termed *oxidative phosphorylation.*

Although the components of the electron transfer chain are known, it is now usual to regard the chain as consisting of three complexes; first, the reduced nicotinamide adenine dinucleotide dehydrogenase complex (NADH dehydrogenase); second, the ubiquinol dehydrogenase complex; and lastly the cytochrome C oxidase complex. These complexes are located on the inner membrane of the mitochondrion which encloses its matrix. In the matrix are the enzymes of the TCA cycle and those concerned with the oxidation of fatty acids.

The elucidation of the mechanism of oxidative phosphorylation is still not complete. It was shown in the 1950s that there were three separate sites for the formation of ATP during the passage of two electrons through the chain and these agree with the present concept of three enzyme complexes. A search was made for putative covalent intermediates which could be phosphorylated in a way analogous to substrate level phosphorylation and thus lead to the formation of ATP, but none were found (see Jones 1981 for an account of these early developments). At present it is generally accepted that generation of ATP is linked to movement of protons across the inner mitochondrial membrane establishing an electrochemical gradient. The pumping of protons out of the matrix establishes a difference in potential between the inner and outer membrane; movement back leads to generation of ATP and reduction of the potential difference. For each pair of electrons that passes through each enzyme complex it appears that four protons move across the inner membrane and that on moving back one ATP is formed by the enzymic action of ATP synthase (McGilvery & Goldstein 1983). One ATP is formed at each of the enzyme complexes, giving a yield of three ATP for each pair of electrons on its passage through the electron transport chain. The last stage is the transfer of electrons from reduced cytochrome C to oxygen with the formation of water. The number of ATP formed in the oxidative sequence in the mitochondrion, terminating in the formation of water, is usually expressed as the P:O ratio – and in oxidative phosphorylation the ratio is 3.0.

There is at present, however, some uncertainty about the precise proton stoichiometry at the three complexes, that is the number of protons transported per electron pair and the number of protons transported to generate one ATP. This necessarily implies some uncertainty about whether the P:O ratio is always 3.0 (Ferguson & Sorgato 1982). The implications of a wide range of stoichiometries have been explored theoretically by Livesey (1984,1985). Studies which support the 'traditional view' that in normal cells the P:O ratio is very close to 3.0 include Scholes & Hinkle (1984) and Flatt (1985).

There are, however, cells in which respiration is uncoupled from phosphorylation, that is the electrochemical gradient is dissipated without the formation of ATP. The most interesting of these are the cells of brown adipose tissue which are responsible for the non-shivering thermogenesis seen in some species of animal exposed to cold (see Section 10.3.9). In these cells a proton conductance pathway is opened which bypasses the pathway in the mitochondrial membrane where phosphorylation of ADP to ATP occurs (Nicholls & Locke 1983; Himms-Hagen 1985).

6.1.5 *The 'high energy phosphate bond' concept*

The concept of phosphate bond energy was introduced by Lipmann in 1941 when he proposed that ATP is the intermediary linking exergonic and endergonic processes. Endergonic processes include mechanical movement, creation of concentration gradients, and the synthesis of cellular constituents. Exergonic processes include the oxidation of food constituents and, in starvation, of body tissues. The term *bond energy* is used as a shorthand to designate the fact that when ATP is hydrolysed to form ADP a large negative standard free energy change occurs and that this hydrolysis is readily coupled to reactions in which the standard free energy change is positive. ATP is not the 'donor' in all coupled systems. Other nucleoside phosphates are employed for particular endergonic purposes – guanosine triphosphate, cytidine triphosphate and uridine triphosphate. These can be regarded as equivalent to ATP. Some endergonic processes require specific reducing agents, notably the reduced forms of nicotinamide adenine dinucleotide (NAD), nicotinamide adenine dinucleotide phosphate (NADP) and flavin adenine dinucleotide (FAD). In an energy-accounting sense the free energy of these reduced coenzymes can also be expressed as ATP since they can all act as electron donors in oxidative phosphorylation, NADPH being converted to NADH for this purpose. It is thus possible to speak of the net yield of ATP on the complete or partial oxidation of a nutrient and of the net ATP requirement to effect a synthesis, maintain an ionic gradient or furnish the energy for movement.

Within a cell the respiration rate is controlled primarily by the rates of the reactions which feed electrons into the electron transport chain. This is exerted by their effects on the ratios of reduced and oxidised components of the chain, notably those of the NADH dehydrogenase complex and the ubiquinol dehydrogenase complex. The ratios concerned are NADH/NAD and QH_2/Q. It is also controlled by the rates of reactions consuming or producing ATP, which determines the cytosolic phosphorylation potential or ATP:ADP ratio. Control of the ratio of reduced to oxidised forms of the complexes is largely determined by the availability of oxidisable substrates, the regulation of pathways such as glycolysis or fatty acid oxidation, and through calcium ion activation of intramitochondrial dehydrogenases (Brand & Murphy 1987).

6.2 The yield of ATP from oxidations

As shown in Table 6.2 and discussed in Section 6.1.2, when one mole of glucose is oxidised in the cell, the net yield of ATP in the cytoplasm (ATP_{cyt}) is 35.5 moles of ATP. The standard free energy change for the complete oxidation of glucose is -2879.3 kJ/mol (see equation (6.7)). Using the standard free energy of the formation of ATP from ADP and inorganic pyrophosphate shown in equation (6.10*a*), the efficiency with which the free energy of glucose is conserved is $(35.5 \times 31)/2879 = 38\%$. Similar calculations to those for glucose can be made for the dissimilation of other nutrients and body constituents. These are given below.

6.2.1 *The oxidation of long-chain fatty acids*

Details of the steps of fatty acid oxidation are given in standard textbooks of biochemistry; here a brief account is presented with emphasis on the number of ATP formed. The first step, which takes place in the cytosol, is the combination of the fatty acid with coenzyme A to form an acyl-CoA. In this reaction ATP is hydrolysed to AMP (adenosine monophosphate) and pyrophosphate. This is equivalent to the 'expenditure' of two high energy phosphate bonds. The acyl-CoA is too large to pass through the inner membrane of the mitochondrion but does so as O-acyl carnitine to reform acyl-CoA in the matrix of the mitochondrion. Oxidation then takes place from the carboxyl end of the molecule, removing two carbons at a time as acetyl-CoA. Each of these steps involves four reactions and leads to the production of one mole of $FADH_2$ and one of NADH. $FADH_2$ enters the respiratory chain at the ubiquinol-complex stage and thus leads to the formation of two ATP. This, with the three ATP arising from NADH, gives a yield of 5 ATP formed from ADP for the removal of each two-carbon unit as acetyl coenzyme A. Each acetyl-CoA is then oxidised in the Krebs cycle yielding 12 ATP. The overall reaction for palmitic acid (16 carbons) thus involves 7×5 ATP for chain shortening plus 8×12 ATP for oxidation of the two carbon fragments less 2 ATP for the initial formation of the acyl-CoA, giving a total of 129. The complete equations for palmitic acid and for oleic acid (18 carbons – the removal of the double bonds of a fatty acid involves two additional enzymic steps) are:

Palmitic acid:

$$C_{16}H_{32}O_2 + 23O_2 + 129(ADP + P_i) \rightarrow 16CO_2 + 129ATP + 16H_2O \quad (6.12)$$

Oleic acid:

$$C_{18}H_{34}O_2 + 25O_2 + 144(ADP + P_i) \rightarrow 18CO_2 + 144ATP + 17H_2O \quad (6.13)$$

With triacylglycerols, account must be taken of the oxidation of the glycerol. Schulz (1978) has systematised the calculation from the stoichiometry of the oxidations in the form of the following equation for a triacylglycerol in which the esterified fatty acids are all the same and contain an even number of carbon atoms

$$ATP_{cyt} = 25.5n - 1.0 - 6U \quad (6.14)$$

where n is the number of carbon atoms in the fatty acid; and U is the number of double bonds. Table 6.3 gives the yields of ATP from the oxidation of various fatty acids and of glycerol.

Table 6.3. *The yield of ATP on the complete oxidation of fatty acids and of glycerol to carbon dioxide and water, and the enthalpies of combustion per mole of ATP formed*

Fatty acid	$-\Delta H_c$ (J/mol)	ATP (mol)	$-\Delta H$/ATP (kJ/mol)
Acetic	874.5	10	87.5
Propionic	1 527.3	18	84.9
n-Butyric	2 183.5	27	80.9
Myristic	8 722	112	77.9
Palmitic	10 031	129	77.8
Stearic	11 339	146	77.7
Palmitoleic	9 876	127	77.8
Oleic	11 194	144	77.7
Glycerol	1 655	21	78.8

6.2.2 *The oxidation of amino acids*

The stoichiometry of the oxidation of amino acids to carbon dioxide, water and urea is given in all textbooks of biochemistry. These should be consulted for details. Table 6.4 summarises the pathways involved and also the production of ATP from each amino acid (see Krebs 1964; Schulz 1978; McGilvery & Goldstein 1983). The oxidations all follow a similar pattern. The amino acids are deaminated and ammonia enters the Krebs–Hensleit urea cycle in which four moles of ATP are used to provide the free energy for the synthesis of one mole of urea. The carbon-containing moiety is then metabolised to yield compounds which readily enter the tricarboxylic acid cycle, either directly as α-oxoglutarate, succinate or pyruvate. It is the fact that carbohydrate, fat and amino acids all enter the Embden–Meyerhof–tricarboxylic acid sequence at one point or another that has led to its being called *the final common pathway*.

The yield of ATP on oxidation of intact protein is the sum of the ATPs formed on oxidation of its constituent amino acids. This is obtained by multiplying the molar proportions of amino acids in the protein by the yields of ATP listed in Table 6.4. The hydrolysis of protein to its constituent amino acids in the gut does not result in the formation of ATP. In the cell, however, there is a minor requirement for ATP in that short-lived intracellular proteins covalently conjugate with ubiquitin before degradation.

The conservation of free energy on the oxidation of a number of amino acids to carbon dioxide, water and urea can be computed where free energy data are available. The range of efficiences is from 30–40%. In Table 6.4 the enthalpy changes rather than the free energy changes for these oxidations

Table 6.4. *The yield of ATP on oxidation of amino acids to carbon dioxide, water and urea and the enthalpies of these oxidations, including the enthalpy of solution of urea*

Amino acid	Molecular weight	Intermediary	$-\Delta H$ (kJ/mol)	ATP (mol)	$-\Delta H$/ATP (kJ/mol)	Mol ATP/mol amino acid	Mol ATP/kg amino acid
Alanine	89.1	Pyruvate	1297	15.5	83.7	15.5	174
Arginine	174.2	α-Oxoglutarate	2446	28	87.4	28	161
Aspartate	133.1	α-Oxoglutarate	1288	15.5	83.1	15.5	116
Asparagine	132.1	Oxaloacetate	1282	13	98.6	13	98
Cysteine	121.2	Pyruvate	1938	12.5	155.0	12.5	103
Glutamate	147.1	α-Oxoglutarate	1920	24.5	78.4	24.5	167
Glutamine	146.2	α-Oxoglutarate	1938	22	88.1	22	150
Glycine	75.1	(Serine) pyruvate	650	6	108.3	6	80
Histidine	155.2	α-Oxoglutarate	2056	22.5	91.4	22.5	145
Isoleucine	131.2	Succinate & acetate	3261	40.5	80.5	40.5	309
Leucine	131.2	Succinate & acetate	3260	39.5	82.5	39.5	302
Lysine	146.2	(Crotonate) acetate	3037	36	84.4	36	246
Methionine	149.2	Succinate	3064	21.5	142.5	21.5	144
Phenylalanine	165.2	Fumarate	4322	37.5	115.3	37.5	227
Proline	115.1	α-Oxoglutarate	2728	29.5	92.5	29.5	256
Hydroxyproline	131.1	α-Oxoglutarate		26.5		26.5	202
Serine	105.1	Pyruvate	1132	12.5	90.6	12.5	119
Threonine	119.1	Succinate	1778	20.5	86.7	20.5	172
Tryptophan	204.2	(Crotonate) pyruvate	4996	42	118.9	42	206
Tyrosine	181.2	Fumarate	4106	41.5	99.0	41.5	229
Valine	117.1	Succinate	2597	31.5	82.4	31.5	269

have been listed. The values include the enthalpy of solution of urea. Additionally the enthalpy changes per mole of ATP formed from ADP are given. These show a number of systematic relationships. For the monocarboxylic monoamino acids, enthalpy/mole decreases with increasing length of the carbon side-chain. Thus, glycine, with no side chain, has a value of $-\Delta H$/mol ATP of 108 kJ, alanine with two carbons in the side-chain has a value of 84 kJ and valine with four side-chains a value of 82 kJ. Similar decreases in ΔH/mole ATP occur with the monoamino hydroxy acids, the monoamino dicarboxylic acids and the carboxamide amino acids. The aromatic amino acids have high values and so too do the sulphur-containing amino-acids. The use of these 'enthalpy equivalents of ATP' is dealt with in Sections 12.2.1e and f.

6.3 Synthesis of body constituents

Synthesis of the organic constituents of the body is an endergonic process. Some of the syntheses involve the formation of polymers from simple monomers, examples being the synthesis of glycogen from glucose and the formation of proteins from amino acids. Others involve the synthesis of complex molecules from simple precursors; examples are the synthesis of the pyrrole ring of the porphyrins from succinyl-CoA and glycine and the synthesis of the pyrimidine ring of the pyrimidine nucleotides from aspartate, glutamine and carbon dioxide. Still others involve what might appear to be, but are not, reversals of the metabolic pathways involved in oxidation, an example of which is the formation of glucose from pyruvate. Only those of major importance in terms of the overall energy exchange are dealt with here – the formation of glycogen, the synthesis of fatty acids and fats, the synthesis of proteins and gluconeogenesis.

6.3.1 *Glycogen synthesis*

Glycogen is made by adding one glycosyl residue at a time to an existing glycogen molecule. The first step involved is the formation of glucose 6-phosphate which is in equilibrium with glucose 1-phosphate. Glucose 1-phosphate reacts with uridine triphosphate (UTP) to form UDP-glucose and pyrophosphate. The C-1 position of the glucose, thus activated, donates a glucose residue to the hydroxyl group of C-4 of the glycogen to form an α-1,4 linkage. One molecule of UDP is released and one molecule of water formed. The UDP is then phosphorylated by ATP to form UTP to recommence the cycle. It follows that the addition one molecule of glucose glucose to glycogen is equivalent to the 'consumption' of two moles of ATP. This scheme would result in a linear polymer of glycogen. In fact, branches are introduced through formation of 1,6 linkages, each of which also requires two moles of ATP. From the frequency of branching it appears that approximately 2.1 ATP are required to provide the energy for the addition of a glucose residue to glycogen.

In passing it is instructive to note that, in the utilisation of glycogen as a fuel source, glucosyl residues are removed from glycogen as glucose 1-phosphate by the action of the enzyme phosphorylase. About 5% of the residues, however, will be released as free glucose from hydrolysis of 1,6 linkages. The regulation of phosphorylase is complex and involves the action of protein kinases which add phosphate groups (by transfer from ATP) and phosphatases which remove these groups. Such kinases and phosphatases are common to a number of regulatory enzymes. It depends on the enzyme whether it is active or inactive when phosphorylated: generally catabolic enzymes are activated when phosphorylated; for anabolic ones the reverse is true. These phosphorylations are endergonic and the energy donor is ATP. The amounts of ATP are very small when compared with the amounts of metabolites which are subject to catalysis by the enzyme. The same is true of the role of cyclic AMP acting as a messenger conveying signals from the cell surface to the interior. While there is an energy requirement in terms of ATP, the number of moles required is minute relative to the magnitude of the metabolic flux through the activated system.

6.3.2 *Synthesis of fat from carbohydrate*

Fatty acids are synthesised by the addition of two-carbon units to a primer unit which for even carbon-numbered fatty acids is acetyl-CoA and for uneven carbon-numbered fatty acids is propionyl-CoA. The two-carbon unit is malonyl-CoA obtained by carboxylation of acetyl-CoA in the cytosol. The lengthening of the fatty acid chain, which is catalysed by the fatty acid synthase complex, consists of a condensation followed by a reduction in which the electron donor is NADPH. The overall reaction for the formation of palmitate from acetyl-CoA is thus:

$$\left.\begin{array}{l} \text{7 acetyl-CoA} + 7CO_2 + 7ATP \rightarrow \text{7 malonyl-CoA} + 7ADP + 7P_i \\ \text{acetyl-CoA} + \text{7 malonyl-CoA} + 14NADPH \rightarrow \\ \qquad\qquad 8CoASH + 14NADP^+ + \text{palmitate} \end{array}\right\} \quad (6.15a)$$

The electron carrier, NADPH, is largely generated by oxidative carboxylation of malate. In this process formation of each mole of acetyl-CoA in the cytoplasm from pyruvate is linked to the formation of one mole of NADPH. Additional sources of NADPH are from the oxidative carboxylation of isocitrate or from the pentose phosphate pathway. The final relationship with glucose as the initial energy source for the acetyl-CoA and NADPH is then

$$4.5 \text{ glucose} + 4O_2 + 5(ADP + P_i) \rightarrow \text{palmitate} + 11CO_2 + 5ATP \quad (6.15b)$$

To form tripalmitin, glycerol as glycerol 3-phosphate and three palmitates must be provided:

$$13.5 \text{ glucose} + 12O_2 + 15(ADP + P_i) \rightarrow 3 \text{ palmitate} + 33CO_2 + 15ATP \quad (6.15c)$$

$$0.5 \text{ glucose} + 4ATP \rightarrow \text{glycerophosphate} + 4(ADP + P_i)$$

The net reaction is therefore:

$$14 \text{ glucose} + 12O_2 + 11(ADP + P_i) \rightarrow \text{tripalmitin} + 33CO_2 + 11ATP$$

The ATP required to synthesise tripalmitin from glucose can be estimated from the above stoichiometry to be $(14 \times 35.5) - 11 = 486$ mol/mol tripalmitin. The ATP yield on complete oxidation of tripalmitin is 407 mol/mol. Thus, considerably more free energy as ATP has to be expended to effect the synthesis of tripalmitin than is realised on its complete oxidation.

6.3.3 *Synthesis of protein*

The minimal energy cost of protein synthesis is that of forming the peptide bonds which constitute the 'backbone' of protein molecules. There are other energy costs which are dealt with later; first the energy cost of forming the peptide bond, that is the amide link (CO-NH) is considered.

Each of the amino acids to be linked is first enzymically activated to form a specific amino acyl-transfer ribonucleic acid (tRNA). Two moles of ATP are used to create the covalent bond between the carboxyl of the amino acid and a hydroxyl of the terminal adenylic acid residue of the tRNA. Synthesis of the amide bonds takes place on the surface of the ribosome to which is bound messenger ribonucleic acid (mRNA). The mRNA provides information about the sequence in which the amino acids are to be incorporated in the polypeptide chain. The repetitive sequence involves the formation of two moles of guanosine diphosphate (GDP) from guanosine triphosphate (GTP) for each amide bond formed. The overall cost in terms of ATP equivalents is thus 4/mol peptide bond. The standard free energy of a peptide bond, estimated from free energy changes on the hydrolysis of simple peptides is about 13 kJ/mol. Taking the free energy of hydrolysis of ATP at 31 kJ/mol, the efficiency with which the peptide bond is formed is low – about 10%.

By knowing the number of amino acid residues in a protein the overall requirement of ATP for its synthesis can be computed. This is not, however, the sole cost. Firstly, many of the polypeptides synthesised are much larger than the final ones. They are synthesised as pre-pro-proteins. This is particularly true of proteins which are secreted by the cell. Signal peptides are removed during passage through the cell membrane, while for proteins such as digestive enzymes, or hormones other sequences of amino acids are removed from the pro-protein before it is in its active form. For example, insulin is synthesised as one long chain, pre-pro-insulin. A peptide of 24 residues is removed before entry into the Golgi apparatus of the beta cell and subsequently, on secretion into the blood, a C-terminal peptide of 31 residues, two arginines and two lysines are removed. In the final active molecule there are 49 peptide bonds; in the original pre-pro-insulin molecule there were 119. The cost of synthesising the peptide bonds of these additional sequences is an additional charge which can be attributed to the final protein. Secondly, a number of post-translational modifications of proteins take place. These involve formation of covalent bonds and many are endergonic. Examples include the phosphorylation of serine and threonine residues, methylation of carboxyl side chains, glycosylation to form glycoproteins and the formation of disulphide bridges. Thirdly, it cannot be assumed that the amino acids available in the cell for polypeptide synthesis are necessarily present in the proportions in which they are needed. Nutritionally non-essential amino acids may have to be synthesised from smaller carbon-containing molecules and ammonia. The usual starting point is an oxoacid and the source of the ammonia is glutamate. However, taking glucose as the starting point for all newly-synthesised amino acids, it is clear that the overall syntheses are usually endergonic. A further factor related to the energy cost of protein synthesis concerns the formation of mRNA. Each amino acid is coded by a sequence of three nucleotides, termed a *codon*, which is transcribed from the complementary base sequence of DNA. mRNA is formed from the four ribonucleoside phosphates (adenosine, guanosine, cytidine and thymidine 5′-triphosphates; ATP, GTP, CTP and TMP) and is directed by the base sequence of the DNA. As each ribonucleoside phosphate is added, one mole of pyrophosphate is released, implying that for each amino acid in a protein the equivalent of 6 moles ATP are expended. The energy cost of synthesis of a protein must include a portion of this cost of formation of the codon of the messenger RNA. The proportion depends on how many polypeptides are formed from one mRNA. Calculations suggest that the energy cost of synthesising mRNA, rRNA and tRNA contributes about 10% of the energy cost of the synthesis of the peptide bond on the ribosome.

Thus, to state that the energy cost of protein synthesis is simply that calculated from peptide bond formation at 4 mol ATP/mol peptide bond, is certainly incorrect. Additional energy costs are involved but these are at present difficult to compute without making many assumptions.

6.3.4 *Gluconeogenesis*

Glucose is the preferred fuel of many tissues in the body. Brain tissue is an example of such a tissue, although it can use ketones – acetoacetate and β-hydroxybutyrate – to provide energy. The formation of glucose from non-carbohydrate sources is known as gluconeogenesis. However, not all non-carbohydrate sources can be so used. The commonly used are lactate, the glucogenic amino acids, or rather their corresponding oxo-acids, and three-carbon compounds, notably propionic acid. The Cori cycle shown in Table 6.5, is a good example of gluconeogenesis. Here lactate formed from glycogen in muscle during exercise is transported by the blood to the liver where gluconeogenesis takes place to release glucose into the blood. Glycogen is formed in muscle from this glucose to complete the cycle.

Glycolysis is not reversible; additional steps have to be inserted to circumvent those enzymic steps which are irreversible. Table 6.5 shows that 6.5 moles of ATP equivalent are used to form one mole of glucose. The yield of ATP on the oxidation of glucose to give two moles of lactate is only two. Gluconeogenesis is thus expensive in energy terms. Schulz (1978) gives similar calculations for the formation of glucose from amino acids.

6.4 **The turnover of body constituents**

Following upon the discovery of the mass isotope of nitrogen in the 1930s, workers at Columbia University, New York showed that body proteins are being continuously synthesised and degraded (Schoenheimer & Rittenberg 1940; Schoenheimer 1942). It had been thought – and this view persisted in part until the mid-1950s – that the turnover of body proteins was solely accounted for by protein extrusion from cells and by their life and death. The work of the Columbia group, continued by others, has shown without doubt that the many proteins within cells are in a dynamic state.

Proteins differ considerably in the rate at which they turn over. Some proteins, notably the globin of haemoglobin, and also the porphyrin moiety of its haem, exhibit lifetime kinetics – the haemoglobin molecule has a fixed lifespan. Indeed the lifespan of the erythrocyte as a whole behaves in the same way, existing in the circulation for a time which varies with the species, being about 12 d in ducks, 55 d in rats and about 11 months in turtles. Most other proteins once labelled are degraded at a rate which is a negative exponential function of time. If there is no net change in the content of

Table 6.5. *Gluconeogenesis from lactate showing intermediaries formed and the ATP requirement. These reactions occur in the liver during the Cori cycle. The lactate is produced in the muscle from glycogen breakdown. The cycle is completed by resynthesis of glycogen in muscle from the glucose exported by the liver*

Reaction	NADH (mol)	ATP–ADP (mol)
2 Lactate → 2 pyruvate (in cytosol)	2	
2 Pyruvate (in cytosol) → 2 pyruvate (in mitochondrion)		0.5
2 Pyruvate → 2 oxaloacetate		2
2 Oxaloacetate → 2 phosphoenolpyruvate		2
2 Phosphoenolpyruvate → 1 fructose bisphosphate	−2	2
1 Fructose bisphosphate → 1 glucose		
Net production:		
2 lactate 1 glucose	0	6.5

protein of the tissue, or organ or animal considered, then synthesis must equally be exponential. It is conventional to refer to the rate in terms of a *half-life* or *turnover rate*, that is the time taken for half the protein to have been degraded or synthesised. The range of turnover rates of individual proteins is considerable. Thus, many inducible enzymes have very short turnover times, an example being tryptophan pyrolase with a turnover time of 2–3 min, while the proteins of supporting structures such as the mature collagen and elastin of tendons have turnover times of weeks. Procollagens have a much shorter half-life and so too do the types III and V collagens of soft tissues. The myosin and actin of rabbit muscle turn over at the same rate (Lobley & Lovie 1979) and the myofibrillar proteins of the hen turn over at the same rate as do the sarcoplasmic proteins (Schiemann *et al.* 1983). This suggests, but does not prove, that the contractile mechanism of the muscle may well be replaced as a single unit.

It is clear that in the cell relatively long-lived proteins, with half-lives close to the lifetime of the cell as a whole, coexist with proteins which have short and variable half-lives – as short as 1% of the generation time of the cell. The question of how these proteins, which are mostly regulatory, are selectively degraded is of interest. Bachmair, Finley & Varshavsky (1986) have demonstrated that in yeast the amino-terminal sequence of the protein is the determinant of the half-life. β-Galactosidase with different single amino acid residues at their amino termini had strikingly different half-lives. N-Terminal methionine, serine, alanine, threonine, valine and glycine had half-lives of more than 20 h; phenylalanine, leucine, aspartic acid, lysine and arginine had half-lives of 3 min. Other N-terminal amino acids had intermediary half-lives. No doubt this 'N-end rule' concept will be further developed.

While the major developments in this field have been concerned with the study of the biochemistry of synthesis and degradation of individual proteins, there has been a parallel interest in the estimation of protein turnover in the whole body of animals and man. Methods using labelled amino acids have been devised to measure this (Waterlow, Garlick & Millward 1978). These have been applied to situations in which protein and energy metabolism have been experimentally perturbed. It can be argued that because of differences between turnover rates of different proteins, because some proteins are synthesised within the cell and exported from it and because of differences between the amino acid compositions of different proteins, estimates of protein synthesis by the whole body would be difficult to interpret. This is not so; considerable advances in knowledge of growth and its disturbance have resulted from application of such methods.

Emphasis has been given above to turnover of body proteins. There is no turnover of body triglycerides of similar magnitude. The rate of fatty acid synthesis can be determined from the rate at which tritium from ^3H_2O is incorporated into the fatty acid chain (Windmueller & Spaeth 1966). Such experiments show that this rate is virtually the same as the net rate of fat accretion (Reeds, Wahle & Haggarty 1982).

6.5 Substrate cycles

Many of the pathways for the oxidation of metabolites contain steps in which the reaction is far from equilibrium and thus not readily reversible through changes in the concentrations of reactants and products. These constitute considerable free energy barriers. The device used to circumvent them is the employment of an alternative enzyme and an input of free energy, usually through the ATP–ADP reaction. The best example is gluconeogenesis from lactate. The process can be depicted in general terms as follows:

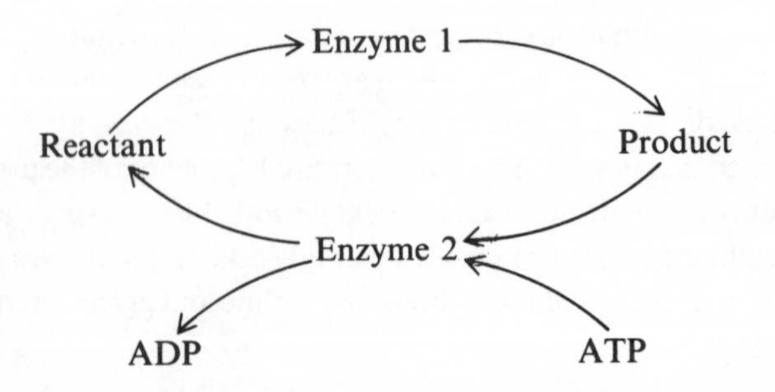

In the dissimilation of the reactant the pathway is via Enzyme 1 (pathway 1); reversal of this to give formation of reactant from product involves a pathway via Enzyme 2 (pathway 2). There is however the possibility that cycling could occur – that both pathways could be operating simultaneously without any net flow from reactant to product or vice versa. This is termed a *substrate cycle* and its magnitude is assessed as the difference in flux between pathways 1 and 2. There is a number of such cycles. One is between glucose and glucose 6-phosphate in the liver. The forward reaction requires ATP which is hydrolysed to ADP and catalysed by glucokinase. The backward reaction is catalysed by the enzyme glucose 6-phosphatase and releases inorganic phosphate from the glucose 6-phosphate. Another similar reaction is the formation from fructose 6-phosphate of fructose 1,6-bisphosphate catalysed by phosphofructokinase. The reverse reaction is catalysed by fructose diphosphatase. Other examples are the cycle between phosphoenolpyruvate and pyruvate, between a triacylglycerol and its constituent fatty acids and glycerol, and between a fatty acid and its CoA derivative. All these cycles apparently accomplish nothing other than the dissipation of energy provided by ATP. It is for this reason that they are often called 'futile cycles'. They are not futile, however, if considered from the point of view of the control which they exert on the extent of the net reaction across the cycle. Newsholme & Crabtree (1976) showed that such cycles can increase the sensitivity of enzymes to changes in substrate concentrations.

What is not clear is the extent to which futile cycles operate in the whole animal and whether they in fact constitute a major 'sink' for ATP. Newsholme (1982) has calculated that if the fructose 6-phosphate – fructose 1,6-bisphosphate cycle was fully active it alone could account for half of the animal's daily energy expenditure, and on this basis suggests that 'stimulation of a large number of such cycles could result in a considerable rate of conversion of chemical energy into heat'. Reeds, Fuller & Nicholson (1985) have calculated from *in vitro* results that at most the fructose 6-phosphate – fructose 1,6-bisphosphate cycle could account for only 0.6% of the ATP generated in the metabolism of a whole animal.

It could be argued that protein turnover equally represents a futile cycle and so too does the Cori cycle in that their activities all result in the maintenance of a final state identical to an initial one but with considerable energy expenditure. These processes are, however, measurable; their existence is unequivocal rather than putative.

7

Energy exchanges by radiation, convection, conduction and evaporation

This chapter examines the factors affecting the loss of heat from the body. The relation between heat loss and heat production was discussed in Section 2.3 and is exemplified by the equation:

$$\dot{H}_{P} - \dot{w} = \dot{H}_{L} + \dot{s} \tag{7.1}$$

where $\dot{H}_{P}$ is the heat produced in metabolism; $\dot{w}$ is the work done on the environment; $\dot{H}_{L}$ is the heat emitted; and $\dot{s}$ is the storage of heat in the body. All are expressed as rates per unit time. The work and storage terms were discussed in Section 4.1.1. In this chapter, consideration is given to heat exchange with the environment by energy fluxes due to *net radiation* ($\dot{R}_{NET}$) *convection* ($\dot{C}$), *conduction* ($\dot{K}$) and the evaporation of water ($\dot{E}$).

$$\dot{H}_{L} = \dot{R}_{NET} + \dot{C} + \dot{K} + \dot{E} \tag{7.2}$$

The term net radiation recognises that the loss of energy from the body by radiation represents a balance between the gain of radiant energy from incoming infra-red and solar radiation and its loss by infra-red radiation from the body surface. The magnitude of each of the terms obviously varies with attributes of the environment and is governed by physical laws. Animals, however, can make physiological and behavioural adjustments to conserve heat in some environments and augment its loss in others. Some consideration is given here to these, but body temperature regulation is not discussed (see Monteith 1973 and Mount 1979).

Much of the physics in this chapter stems from developments from classical physics made by chemical engineers, aerodynamicists and other applied workers. Some of their studies have led to empirical solutions which have great generality. The conceptual framework provided by physics and engineering is certainly the correct one in which to analyse effects of the physical environment on the heat loss of animals and man. However, animals are far more complex in shape and surface characteristics than are the heat exchangers of chemical engineers or the structures considered by hydrodynamicists. Furthermore animals can make physiological adjustments to alter the characteristic dimensions, temperatures and properties of the surfaces which exchange heat with the environment. Nevertheless, despite the animal's ability to alter the characteristics of its skin surface, the actual exchanges of heat from the skin surface to the environment are governed entirely by physical laws in which biological processes play no part.

7.1 Radiation

All materials, whether solids, liquids or gases, at temperatures above absolute zero emit photons as a result of changes in the vibrational, rotational and electronic states of their constituent atoms and molecules above the ground state. Although the electromagnetic spectrum is continuous it is convenient to consider separately the short wave solar radiation (R_S) and the long wave infra-red radiation (R_I).

7.1.1 *Infrared radiation*

The physics of radiation exchange apply to all wavelengths and to both infra-red and solar radiation. For example, although equation (7.3) below is expressed in terms of infra-red radiation, it applies to all radiation. As far as animals are concerned and, as will be evident later, simplifications can be made when dealing with solar radiation.

The amount and spectral distribution of radiation from a perfect radiator depends on its temperature. In 1879 Stefan showed experimentally that the energy emitted by a perfect radiator, called a *black body*, is proportional to the fourth power of its absolute temperature, and five years later Boltzmann derived the same law from thermodynamic considerations. Morowitz (1978) gives an elegant proof of the Stefan–Boltzmann law in terms of quantum theory. The radiant energy emitted by a black body per unit time, can thus be stated to be

$$\dot{R}_I(T) = \sigma A T^4 \quad (7.3)$$

where A is the radiating area of the body; T its absolute temperature in Kelvin; and σ the Stefan–Boltzmann constant which has the value 5.670×10^{-8} W/(m^2 K^4) (Quinn & Martin 1985).

Table 7.1 gives values for black body radiation at temperatures which include those that are encountered in the environment or occur at the surfaces of animals. If the surfaces of animals and of their environments were perfect radiators, this expression could be used to calculate the net radiation exchange by difference. Thus if the surroundings had a tempera-

Table 7.1. *Black body radiation from surfaces at different temperatures*

Temperature (°C)	W/m^2	$kJ/(h\ m^2)$	$MJ/(d\ m^2)$
0	315.7	1137	27.3
5	339.5	1222	29.3
10	364.5	1312	31.5
15	391.0	1407	33.8
20	418.8	1508	36.2
25	448.1	1613	38.7
30	479.0	1724	41.4
35	511.3	1841	44.2
40	545.3	1963	47.1
45	581.0	2092	50.2
50	618.4	2226	53.4
55	657.6	2367	56.8
60	698.6	2514	60.4

ture of 25 °C and the surface of the animal had a temperature of 35 °C, values from Table 7.1 show the net loss radiant energy per square metre of body surface would be $44.2-38.7=5.5$ (MJ/d). But animals and their surroundings are not necessarily perfect radiators. When radiant energy at any wavelength impinges on a body, part can be transmitted through it (Υ), a part is absorbed by it (α) and the remaining part is reflected (ζ);

$$\Upsilon+\alpha+\zeta=1.0 \tag{7.4}$$

Kirchhoff's law states that the *emissivity* of a body ϵ – its power to radiate at a given temperature – is equal to its *absorptivity*. A good absorber of radiation is an identically good emitter. The emissivities of animal surfaces and their surroundings are not, however, equal to 1.0. Table 7.2 gives some typical values for the infra-red emissivities for surface temperatures likely to be encountered in the environment. In estimating the net infra-red radiation exchange ($R_{I(NET)}$) the emissivities of both the surroundings and the animal's surface have to be taken into account, together with the geometrical relationships of these surfaces. Emissivities and geometric relationships can be combined into an emittance factor, F_ϵ, to give the general relationship:

$$R_{I(NET)}=F_\epsilon\sigma(T_1^4-T_2^4) \tag{7.5}$$

Values of F_ϵ have been formulated by engineers for flat surfaces exchanging radiation, for spheres or cylinders in spherical or cylindrical enclosures and for other geometrical configurations. For animals in enclosures of different shapes, Christiansen's equation is the best approximation of the emittance factor;

Table 7.2. *Total emissivities in the infrared (ϵ) of terrains, surroundings and animal surfaces*

	ϵ
Terrains	
Fresh snow	0.89
Ice	0.96
Dry sand	0.89
Wet sand	0.89
Concrete	0.95
Moist soil	0.97
Grass surface	0.96
Man-made structures	
Red brick	0.92
Planes wood	0.90
White plaster	0.91
White paint	0.93
Aluminium paint	0.55
Galvanised iron	0.28
Aluminium foil	0.08
Animal surfaces	
Human skin	0.95
Mole skin	0.97
Sheep fleece	0.97

Sources: From Monteith (1973); Mount (1979) and other sources.

$$F_\epsilon = \frac{1}{\dfrac{1}{\epsilon_1} + \dfrac{A_1}{A_2}\left(\dfrac{1}{\epsilon_2} - 1\right)} \tag{7.6}$$

where A_1 and A_2 are the areas of the radiating surfaces.

Table 7.3 lists the emittance factors for objects with an infra-red emissivity of 0.95 – close to that of the coats or skins of most animals – in enclosures of different sizes and constructed of different materials. The interesting point arises that as the enclosure becomes larger relative to the object, so the emittance factor approaches the value of the emissivity of the object alone. This is understandable. Only a small part of the radiation from a large enclosure actually strikes the object; most of it strikes the opposing walls and is reflected back. Detailed discussion of the emittance factor and the radiative exchange of different-shaped objects, are given in works on heat transfer in relation to engineering (i.e. Jakob & Hawkins 1957).

Table 7.3. *The effect of the emissivity of the walls of an enclosure and its size relative to the size of an animal contained within it on the radiation exchange factor,* $F\epsilon$. *For simplicity both the animal and enclosure were assumed to be spherical and the emissivity of the animal surface was assumed to be 0.95*

Material	Emissivity of the walls of the enclosure	Radius of enclosure relative to radius of the enclosed body				
	ϵ	2	4	8	16	32
White paint	0.93	0.933	0.946	0.949	0.95	0.95
Aluminium paint	0.55	0.795	0.906	0.939	0.95	0.95
Galvanised iron	0.28	0.590	0.824	0.915	0.941	0.95
Aluminium foil	0.08	0.255	0.565	0.811	0.911	0.940
Volume of body relative to volume of enclosure		8	64	512	4096	32 768
Approximate analogy		Animal in burrow	500 kg ox in a room 4m × 4m × 2m	60 kg sheep in a room 4m × 4m × 2m	Dog in a room 4m × 4m × 2m	Rabbit in a room 4m × 4m × 2m

In the derivation of the emittance factor it is implicit that the walls of the enclosure all have the same temperature and emissivity. In houses this is not so; glass windows are usually colder than insulated walls and radiators are hotter. In such circumstances the estimates of the radiant heat exchange of an occupant involves consideration of the geometrical relationships between the particular wall surfaces and the occupant. These can be complex, and while they have been computed for geometrically simple situations they have only rarely been computed for people in rooms (see for example, Fanger 1970).

7.1.2 *Radiation out-of-doors*

Considerations of animals in enclosures obviously apply to animals in burrows, to housed domesticated livestock and to people in offices and houses. Solar radiation is absent and the only radiation to be considered is that in the infra-red. Most animals, however, spend their lives out-of-doors where for part of the day they receive short wave solar radiation and long wave infra-red radiation from the sky and surrounding terrain.

The sun, the source of short wave radiation, is a very small object in the surroundings of an animal – about 5×10^{-6} of the 4π radians around it and questions of back radiation can safely be ignored. The solar heat absorbed by the coat of an animal or the outer surface of human clothing is thus simply the radiant energy which impinges upon it multiplied by the proportion absorbed, that is 1.0 minus the *reflection coefficient*, or *albedo*, of the animal's surface. Reflection coefficients for animal surfaces and for the terrains in which they live are presented in Table 7.4. Light-coloured coats reflect more than dark ones and people with light skins more than those with dark skins. The differences due to skin reflectance are not small; a Negro would absorb about 25% more solar radiation than a Eurasian in the same environment and a brunette Eurasian about 15% more than a blonde.

The overall radiation balance of an animal out-of-doors receiving solar radiation and radiating in the infra-red from the sky and terrain can be expressed:

$$\dot{R}_{NET} = (1-\rho)(\dot{R}_{S(D)} + \dot{R}_{S(I)} + \dot{R}_{S(R)}) + (\dot{L}_U + \dot{L}_D)/2 - \sigma T_F^4 \qquad (7.7)$$

where $(1-\rho)$ is the proportion of the solar energy absorbed; $\dot{L}_D$ is the incident long wave radiation from the sky and $\dot{L}_U$ is long wave radiation from the terrain – the subscripts implying 'up' and 'down'. In the equation the solar radiation impinging on an animal consists of three components: the first is that contributed from the solar beam, that is, the 'direct' component, $\dot{R}_{S(D)}$; the second is the 'indirect' component, $\dot{R}_{S(I)}$, which represents the scattered or diffuse radiation from the vault of the heavens; the third, $\dot{R}_{S(R)}$, is the solar radiation 'reflected' on the animal from the terrain.

Table 7.4. *Reflectance of solar radiation (ρ) from the surfaces of animals and their surroundings*

Terrains		Animals	
Surface	ρ	Surface	ρ
Clipped grass	0.24	Black human skin	0.18
Rye grass sward	0.25	White human skin	0.35
Heather	0.14	Black cattle coat	0.10
Bracken	0.24	Eland coat	0.22
Dry sand	0.40	Hartebeest coat	0.42
Dry tar macadam	0.12	Clean sheep fleece	0.79
Dry concrete	0.24	Dirty sheep fleece	0.26
Fresh snow	0.85	Newly-shorn fleece	0.42
		Red deer coat	0.69
		Black pig skin	0.07
		White pig skin	0.51
		Red squirrel coat	0.51
		Mole coat	0.19

a. The amount of solar radiation

The amount of solar radiation received at the earth's surface depends on the elevation of the sun and the extent to which the radiation received at the outer margin of the atmosphere is depleted by passage through the atmosphere. The radiation, which strikes the top of the atmosphere (the *solar constant*, or $\dot{Q}_O$), varies slightly with variation in the mean distance of the sun from the earth and averages 1360 W/m^2. The extent of the depletion by absorption and by scattering by atmospheric gases, dust and water vapour depends on the 'depth' of the atmosphere and its composition, particularly its water vapour, dust and aerosol contents. The depth, called the *optical air mass* (m) measured to sea level, is given a value of 1.0 when the sun is directly overhead. This is a minimum and at lower solar elevations the column of air through which the radiation passes increases. The optical air mass can be approximated by the expression:

$$m = \operatorname{cosec} \theta \tag{7.8}$$

where θ is the elevation of the sun above the horizon. The approximation is of no account until solar elevation is less than 10° when the curvature of the earth's surface introduces an error. Solar elevation of course depends on latitude, time of year and time of day.

Analysis of the depletion of extraterrestrial radiation by the atmosphere is somewhat complicated by variations in water vapour and dust content and because each small interval of wavelength of the solar spectrum has to

Table 7.5. *Calculated amount of solar radiation incident upon a horizontal surface for different solar elevations. The atmospheric transmission coefficient was taken to be 0.7*

Solar elevation (θ)	Direct component (Q_D)	Indirect component (Q_I)	Total radiation (W/m^2)
5	2	53	55
10	30	92	122
15	89	116	205
20	164	130	294
25	247	138	385
30	333	142	476
35	419	146	565
40	502	147	649
45	581	147	728
50	654	147	801
55	721	146	867
60	780	146	926
65	832	145	977
70	874	144	1018
75	908	144	1052
80	932	143	1075
85	947	142	1089
90	952	143	1095

Source: From Clapperton *et al.* (1965).

be considered separately. An approximation of sufficient accuracy for biological purposes is:

$$\dot{Q}/\dot{Q}_O = a^m \tag{7.9}$$

where $\dot{Q}_O$ is the extraterrestrial solar radiation (1360 W/m^2); $\dot{Q}$ is the incident radiation; a is the *atmospheric transmission constant* which varies with water vapour content and cleanliness of the atmosphere.

The direct solar radiation which falls on a horizontal surface of the earth ($\dot{Q}_D$) is thus:

$$\dot{Q}_D = \dot{Q}_O\, a^m \sin\theta \tag{7.10}$$

and since about 9% of the extraterrestrial radiation is absorbed by the atmosphere and about half that scattered is scattered towards the earth, the indirect solar radiation flux on a horizontal surface is

$$\dot{Q}_I = (0.91\, \dot{Q}_O - \dot{Q}_D)\,(\sin\theta)/2 \tag{7.11}$$

Table 7.5 gives computed values for the radiation flux on a horizontal surface using an atmospheric transmission coefficient of 0.7 which applies to the high atmospheric moisture content of air in Britain. Values for other transmission coefficients are easily computed. Other formulations can be used to estimate the incidence of solar radiation on a horizontal surface when the skies are clear. These have been considered by Tracy *et al.* (1983) to show that they give slightly different results. Obviously in any particular application solar radiation incidence should be measured. Suitable instruments have been described by Monteith (1972).

Cloud reduces the amount of incoming solar radiation. The extent of the reduction depends on the type and thickness of the cloud and can be estimated for different types. Monteith (1973) gives a graph, derived from work by Lumb, allowing such a computation. Thick nimbostratus or stratocumulus cloud reduces total solar radiation transmitted through the atmosphere to about $\frac{1}{8}$ of that observed with clear skies.

b. The solar radiation incident on animals

The amount of body surface that intercepts the solar beam is termed the *solar radiation profile*. The shadow which it casts on a horizontal surface is an equivalent measure, the two being related

$$\text{profile area} = \sin\theta \times \text{shadow area} \tag{7.12}$$

Solar radiation profiles have been determined by photographing animals and people from different angles, and Figure 7.1 shows the results of

Fig. 7.1. The solar radiation profile of an adult male subject as determined by Underwood & Ward (1966). The profiles were determined by photographing the subject from different angles simulating the elevation of the sun.

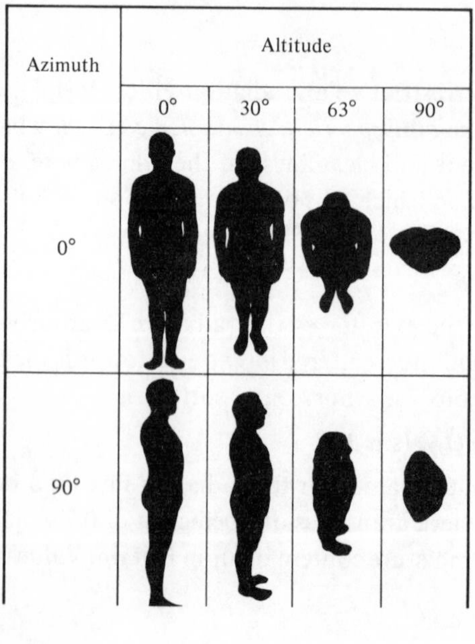

Underwood & Ward (1966) for adult males. The solar radiation profile diminishes markedly as the sun moves overhead. Man can be approximated by a vertical cylinder 1.65 m high and with a diameter of 0.234 m. Similar photographic studies have been made with other species, an example being those for sheep with and without their fleeces (Clapperton, Joyce & Blaxter 1965).

Determinations of the radiation profile allows estimation of the total radiation impinging on the animal, and, with knowledge of the reflectivity of the surface of the coat or clothing, the total heat gained. The components of solar radiation are the direct, indirect and reflected ones, the latter involving an estimate of the reflectivity of the terrain. Such a calculation is given for a sheep in Table 7.6. The results agree with direct measurements made on a model sheep exposed to the sun.

Calculations of this type reveal some interesting aspects of the magnitude of the heat load on animals in different latitudes. Table 7.7 shows the results of calculations of the total heat load on sheep with fleeces and shorn sheep summated over 24 h in different latitudes at different times of the year, assuming the same atmospheric transmission coefficient for each site. What is remarkable is that at the equator the total heat load in 24 h is less than it is at midsummer in high latitudes. This arises because the radiation profile is maximal at 55° N and because the length of daylight at the equator is less than that at high latitudes. This applies generally to animals which can be regarded as horizontal cylinders; with man and other upright animals, regarded as vertical cylinders, the difference is even greater because of the considerable dimunition in the radiation profile at high solar elevations. There is usually a much higher heat load on animals in the dry tropics than in northern temperate regions; this reflects the low levels of water vapour in the atmosphere and consequent smaller depletion of the solar beam.

7.1.3 *Incoming long wave radiation*

As was shown earlier, when a man or an animal is within a large enclosure, the incoming long wave radiation is determined solely by the temperature of the walls of the enclosure and their emissivity can be ignored. In some instances the temperature of the walls of an enclosure can be taken to be the same as that of the air it contains, but this is not universal. One can feel very cold in a poorly insulated house if the air temperature outside is below freezing and the walls are cold, despite the fact that the air temperature within the room is indicative of comfort.

Out-of-doors the upward and downward fluxes of infra-red radiation have to be considered. The upward flux is simply determined; it is the radiation calculated from measurement of the temperature of the

Table 7.6. *The solar radiation from clear skies incident on a fleeced sheep with its side at right angles to the solar beam. Radiation is expressed per square metre of skin surface area (W/m^2)*

Solar elevation (θ)	Direct component ($\dot{R}_{S(D)}$)	Indirect component ($\dot{R}_{S(I)}$)	Reflected component ($\dot{R}_{S(R)}$)	Total ($\dot{R}_S$)	Heat load ($(1-\rho)\dot{R}_S$)
5	14	29	5	48	36
10	79	34	12	125	94
15	150	43	19	212	159
20	208	48	27	283	212
25	252	51	35	338	254
30	285	52	44	381	286
35	312	53	52	417	313
40	324	55	60	439	329
45	331	55	67	453	340
50	336	55	74	465	349
55	340	54	80	474	356
60	344	53	86	483	362
65	346	53	91	490	367
70	347	53	95	495	371
75	348	52	97	497	373
80	349	52	99	500	375
85	351	52	100	503	377
90	352	52	101	505	379

Notes:
The calculations were based on the computed amount of radiation incident on a horizontal surface as given in Table 7.5, using the observed radiation profile of the sheep and its skin surface area. The reflectivity of the terrain was taken to be 0.25 and of the sheep's fleece, 0.45. The calculations are based on the following:

$\dot{R}_{S(D)} = \dot{Q}_D P$

$\dot{R}_{S(I)} = \dot{Q}_I X$

$\dot{R}_{S(R)} = (\dot{Q}_D + \dot{Q}_I)\rho_T X$

where P is the profile area; X the cross-sectional area of the animal; and Q is the incoming solar radiation, with subscripts representing the direct, indirect and reflected components.

Source: From Clapperton *et al.* (1965).

groundcover with allowance made for its emissivity. The downward flux of radiation depends on the distribution within the lower atmosphere of temperature, water vapour and other gases. From meteorological records Swinbank (1963) showed that the incoming radiation on cloudless nights could be estimated from air temperature, T_A, according to the relationship:

$$\dot{L}_D = 1.2\sigma T_A{}^4 - 171 \qquad (7.13)$$

where $\dot{L}_D$ is the incoming infrared radiation in W/m^2; and temperatures are in Kelvin. Values calculated from this empirical relationship are given in

Table 7.7. *The total amount of radiant energy incident on fleeced and shorn sheep during 24 h expressed as MJ/(d m²) skin surface area. The assumptions are that skies are cloud-free and that the atmospheric transmission coefficient is 0.7.*

	Latitude and time of year		
	Winter solstice 55° N	Summer solstice 55° N	Spring or summer solstice equator
Sheep with winter fleece (ρ=0.25)			
Incident radiation (R_S)	1.7	18.6	14.8
Heat absorbed $(1-\rho)R_S$	1.3	13.9	11.1
Sheep with newly-shorn fleece (ρ=0.45)			
Incident radiation (R_S)	1.2	12.5	8.9
Heat absorbed $(1-\rho)R_S$	0.7	6.9	4.9

Source: From Clapperton *et al.* (1965).

Table 7.8. This table also includes equivalent radiant temperatures of the sky. These represent the temperatures of a large surface which would emit the same amount of energy per unit area as that radiated from a clear sky. The table shows that the radiant temperatures of the sky are all below those of the air. It is usually stated that for clear skies the radiant temperature is on average 20 °C below that of the air. This is broadly true for air temperatures ranging from about 5 to 25 °C, but at very low temperatures, the difference is much greater.

When the sky is covered with cloud there is little error in assuming that the radiant temperature of the sky is the same as that of the air. The incoming infra-red radiation is then that given in Table 7.1 for the equivalent temperature.

The remaining factor necessary to calculate the incoming long wave radiation which the animal receives is the area of the body on which the radiation impinges. Using the cylindrical model, this can be shown to be half the surface area for both downward and upward components (see Monteith 1973). The average of the two fluxes per square metre of a horizontal surface is the amount which the animal receives per unit cross section area. This is the origin of the figure '2' in the radiation balance equation (7.7).

7.1.4 *First power and radiant heat transfer coefficients*

The infra-red radiation loss from the surface of an animal (T_F) to the walls (T_W) of its cooler surroundings is proportional to the difference

Table 7.8. *The incoming long wave radiation from clear skies calculated from Swinbank's (1963) formula and the equivalent 'radiant temperatures'*

Observed air temperature (°C)	Calculated incoming radiation (W/m^2)	Calculated 'radiant temperature (°C)
−30	67	−88
−20	108	−65
−10	155	−45
0	208	−28
10	266	−12
20	331	3
30	404	17

between the fourth powers of their temperatures. The Stefan–Boltzman constant is the coefficient of proportionality. For infinitesimally small differences between the temperatures of the animal surface and the walls, the radiant heat transfer can be expressed in terms of the temperature difference alone. The *first power radiant heat transfer coefficient* is then $4\sigma T^3$. More generally, by factorising $(T_F^4 - T_W^4)$ by $(T_F - T_W)$ the first power radiant heat exchange factor (h_R) can be shown to be:

$$h_R = \sigma(T_F^3 + T_F^2 T_W + T_F T_W^2 + T_W^3) \quad (7.14)$$

where temperatures are in Kelvin and from which it can be seen that the factor $4\sigma T^3$ is the limiting case when the temperatures of the walls and of the animal's surface are close to each other.

The limiting value for the radiant heat exchange factor is used fairly extensively as a good approximation when the differences between surface and wall temperatures are not too great; an example relates to the definition of operant temperature of environments discussed in Section 7.3.2.

Table 7.9 lists values of h_R calculated from equation (7.14) for different surface and wall temperatures. Values on the diagonal are for conditions in which wall and animal surfaces are equal. The table also includes the *convective heat transfer coefficient* $h_{C(F)}$ for an air temperature which is 10 °C below the surface temperature of the animal. For a constant gradient between surface and wall temperature the first power radiant heat exchange coefficient declines as the mean temperature falls. The last column of the table shows that the corresponding convective heat exchange coefficient increases as air temperature decreases. This increase is due to the increase in the kinematic viscosity of air when its temperature falls. Generally, as temperature falls, radiative losses decline and convective losses increase per

Table 7.9. *First power radiative heat transfer coefficient calculated from equation (7.14) in the text and the corresponding heat exchange coefficient for forced convection at air temperatures equal to wall temperatures. The convective heat loss applies to an animal with a characteristic dimension,* L, *of 0.5 m and an air velocity of 0.5 m/s. All heat transfer coefficient values are in W/(m² °C)*

Temperature of surface (T_W) (analogous to 'walls') (°C)	Heat transfer coefficient Temperature of animal coat surface (°C)							Convective coefficient (h_R)
	−10	0	10	20	30	40	50	
−40	3.5	3.7	3.9	4.2	4.5	4.7	5.0	
−30	3.7	3.9	4.2	4.4	4.7	4.9	5.3	
−20	3.9	4.1	4.4	4.6	4.9	5.2	5.5	4.5
−10	*4.1*	4.4	4.6	4.9	5.2	5.5	5.8	4.4
0	4.4	*4.6*	4.9	5.2	5.4	5.7	6.0	4.3
10	4.6	4.9	*5.1*	5.4	5.7	6.0	6.3	4.3
20	4.9	5.2	5.4	*5.7*	6.0	6.3	6.7	4.2
30	5.2	5.4	5.7	6.0	*6.3*	6.6	7.0	4.1
40	5.5	5.7	6.0	6.3	6.6	*7.0*	7.3	4.0
50	5.8	6.1	6.4	6.6	7.0	7.3	*7.6*	3.9

degree gradient in temperature between the animal's surface and its environment. The compensation is not, however, complete.

7.1.5 *Radiation balances of animals*

The above account of solar and long wave radiation exchanges can be placed in perspective by consideration of the gains and losses of heat by the body in different radiation circumstances (equation (7.7)). The radiation balance considers radiant energy losses and gains only and the balance can be positive or negative. Very generally, the balance is positive during the day when the solar energy receipt – even from overcast skies – together with the high incoming long wave radiation from the cloud base is greater than the sum of the solar energy reflected by the animal and the long wave energy it radiates. At night, when skies are clear, the radiation balance is usually negative, particularly under cold conditions.

7.2 **Convection**

The convective transfer of heat depends on the movement of air, or indeed of any other fluid, around an object. It involves the transfer of molecules from close to the surface of the object to its surroundings. The extent of this transfer depends on the temperature of the object, its shape and the nature of its surface, the temperature and pressure of the surrounding fluid (or air), and the velocity of the fluid which impinges on it.

7.2.1 *The physics of convective heat transfer*

Two types of convective heat transfer can be distinguished – *forced convection*, and *natural*, or *free convection*. The first applies to animals in air streams and the latter to animals under still conditions when movement is imparted to the air close to the body by its thermal expansion. Buoyancy of the air results in its upward movement; indeed over a standing man air movement due to free convection can be detected 1.5 m above his head. Similarly the air flow can be *laminar* or *turbulent*. Transitions take place between free and forced convection and between laminar and turbulent flow as air velocity increases. These transitions depend on the shape and surface characteristics of the object concerned.

The convective loss of heat from the body is proportional to the gradient of temperature between the surface of its fur or clothing and the surrounding air, $(T_F - T_A)$. The *coefficient of proportionality*, or the *convective heat transfer coefficient per unit surface area* (h_C) is not an absolute constant since it depends on the various factors listed above. All bodies in air have a laminar film of still air at their surface called the *boundary layer*. This can be regarded as having a thickness, *d*. The heat transfer coefficient can thus be expressed as the *thermal conductivity* (*k*) of undisturbed air (rate of heat flow per unit thickness for a one degree gradient of temperature) divided by the thickness term. Thus the heat loss by convection per unit surface area of the body can be written:

$$\dot{C}/A = h_C(T_F - T_A) \tag{7.15}$$

$$\dot{C}/A = (k/d)(T_F - T_A) \tag{7.16}$$

The thermal conductivity of air is 2.41×10^{-2} W/(m °C) at 0 °C and 2.64×10^{-2} at 30 °C. Equation (7.16) can be applied to cases in which the boundary layer is modified by air movement, when *d* becomes not the real thickness of the boundary layer but a nominal one, that is, one which is the still air equivalent.

Central to a consideration of convective transfer is the contention that geometrically similar shapes behave in a similar way and that aspects of convective heat flow can be expressed in terms of dimensionless groups. The critical dimensionless group in convective heat transfer is the *Nusselt number*. In equation (7.16) the thickness of the boundary layer, *d* – which cannot be measured – is substituted by a characteristic dimension of the object (*L*) which can. This characteristic dimension for spheres and horizontal cylinders under conditions of forced convection is their diameter. Thus the equation above becomes:

$$\frac{\dot{C}}{A}=\frac{Lk}{dL}(T_F-T_A) \tag{7.17}$$

The ratio, L/d, is the *Nussėlt number* (Nu). The convective heȧt transfer coefficient can be estimated provided that the dimensionless group, the Nusselt number is known, since:

$$h_C=Nu\ k/L \tag{7.18}$$

Estimation of the Nusselt number from independent information enables the heat transfer coefficient for convection to be estimated, and in engineering practice this can be done for both forced and free convection.

a. Forced convection

The Nusselt number for forced convection can be estimated from two other dimensionless quantities, the ***Reynolds number*** (***Re***) and the ***Prandtl number*** (***Pr***). The latter represents the physical properties of the air itself and can be regarded as a constant within the range of environments likely to be encountered by an animal. The Reynolds number is a dimensionless representation of air velocity which is defined as:

$$Re=VL/\nu \tag{7.19}$$

Where V is air velocity in m/s; L is the characteristic dimension of the object in metres (the diameter in the instance of a cylinder); and ν is the kinematic viscosity of the air in m^2/s. This latter quantity varies with air temperature. At temperatures of −20, 0, 20 and 40 °C it has the values 1.16, 1.32, 1.50, and 1.69×10^{-5} m^2/s, respectively. The Nusselt number and the Reynolds number are related by the following general and empirical relationship:

$$Nu=aRe^n \tag{7.20}$$

where the coefficient, a, and the exponent, n, are not constant but depend on the magnitude of the Reynolds number as shown in Figure 7.2. The experimental methods employed to establish these relationships are well illustrated in a biological setting by the studies of McArthur & Monteith (1980a). Over a wide range of Reynolds number – which would encompass high air velocities over small animals and low air velocities over large ones – the value of a can be taken to be 0.24 and of n to be 0.60.

The convective heat transfer coefficient by forced convection per unit surface area ($h_{C(F)}$) can thus be obtained by combining the equations relating it to the Nusselt number and the Nusselt number to the Reynolds number:

$$h_{C(F)} = (akV^n)/(\nu^n L^{1-n}) \tag{7.21}$$

In Figure 7.3 equation (7.21) has been used to compute the convective heat transfer coefficient per unit surface area (m^2) of cylinders of different sizes which might be regarded as simulating animals. The figure shows that at the same air velocity small animals lose more heat per degree centigrade difference between their surfaces and the air than do large ones and that heat loss is not directly proportional to air velocity.

It is reasonable to ask whether such a theoretical approach derived from engineering experiments predicts the convective heat transfer coefficient in real animals. Convective heat transfers have been measured indirectly using calorimetric methods and have been discussed by Monteith (1973) and D. Mitchell (1974) who conclude that agreement is reasonable. Similar analyses using calorimetric methods have been employed by Mount (1977, 1978) to estimate the total loss of heat by non-evaporative pathways in baby pigs. He concluded that, using a cylinder to represent a pig, the theoretical calculations gave results with a 'useful order of accuracy'.

Agreement between an observed convective and radiative heat loss by an animal with that which can be computed on the assumption that it is a cylinder is not entirely convincing. An alternative approach has been to

Fig. 7.2. The factors *a* and *n* in the equation relating the Nusselt number to the Reynolds number, $Nu = aRe^n$, for forced convective heat transfer. For animals varying in size from about 10 kg to 300 kg little error is incurred by adoption of constant values of $n = 0.6$ and $a = 0.24$. The figure was derived by D. Mitchell (1974) from published studies.

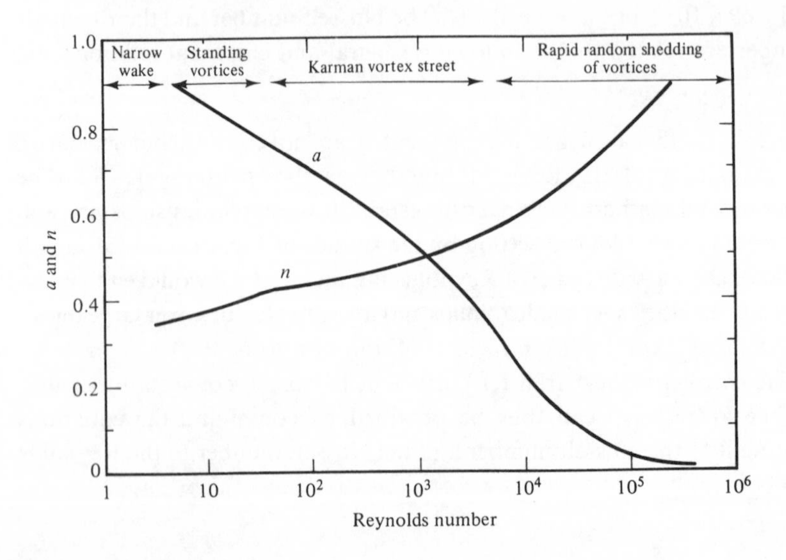

make more realistic physical models with internal heat sources and to measure the effects on heat loss of an increase in air velocity. This has been done for sheep (McArthur & Monteith 1980a) and for chicken (Wathes & Clark 1981). These studies show that the cylindrical analogy is far from perfect. McArthur & Monteith (1980a) calculated Nusselt numbers for the hind legs, fore legs, head and trunk. The weighted sum was then used to estimate the convective heat transfer coefficient for the whole animal. Furthermore the results for a smooth cylinder did not agree with those obtained when the cylindrical model was covered with a pelage, in part because it is difficult to define an interface when there is a wispy surface.

Loss of heat by convection in a real animal is difficult to determine directly. This has however been done by measuring the increase in the enthalpy of the air passing over a chicken in a specially constructed calorimeter (M.A. Mitchell 1985). Mitchell's results show that the actual convective heat transfer is considerably greater than that calculated either

Fig. 7.3. The convective heat transfer coefficient for forced convection, $h_{C(F)}$, at different air velocities computed for different sizes of animals regarded as cylinders of different sizes. The characteristic dimension chosen is given for each animal.

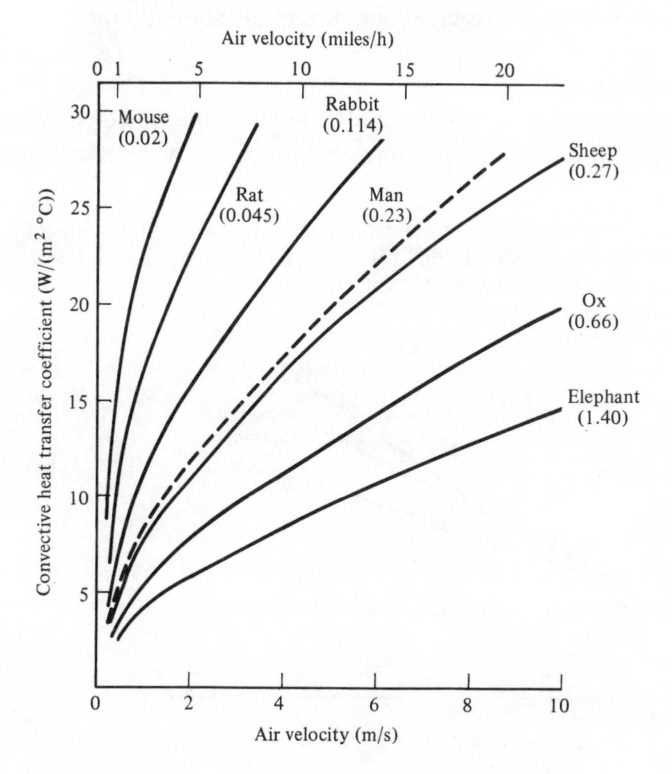

on the assumption that a chicken can be approximated in its shape by a sphere or even when more realistic models are used (Figure 7.4). This lack of agreement does not imply that the physical approach to heat transfer by convection is inappropriate. But it does indicate that there are subtleties in convective heat transfer from animals which demand more investigation.

In passing it is perhaps useful to comment on the air velocities which are likely to be encountered by animals and man under natural conditions. The wind velocities published by meteorologists as part of weather information services refer to velocity at a height of 10 m. Wind speed vanishes at some non-zero height close to the ground surface and the velocity at any height is related to the logarithm of the height above this minimum. A mouse scuttling in thick grass is unlikely to be subjected to any considerable air movement, and even a large animal such as an ox grazing on close pasture will be subjected to an air velocity which is only about 40% of that recorded at meteorological height. The feet and legs of an animal, being closer to the ground, are subject to a lower air velocity than is the trunk. In sheep, for example, air velocity at mid-shank level (0.15 m) is about 60% of that at mid-trunk level (0.46 m). Birds are obviously subjected to high air velocities of their own making when they fly, and their speeds and hence the velocity of air over their bodies can reach about 70 mph or about 30 m/s.

b. Natural or free convection

As with forced convection, so with natural convection, the Nusselt number can be estimated from other non-dimensional groups – in this instance from

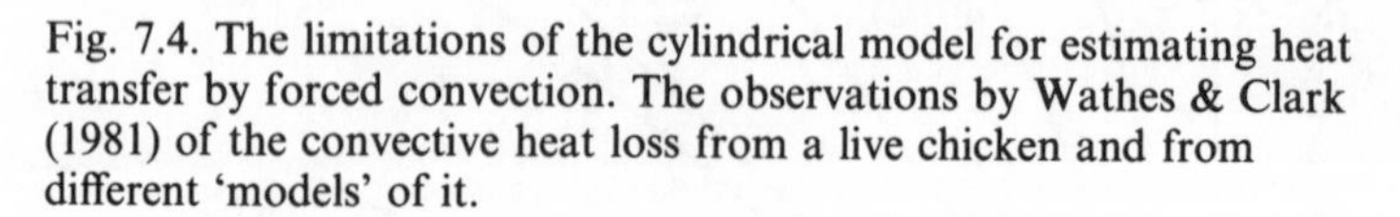

Fig. 7.4. The limitations of the cylindrical model for estimating heat transfer by forced convection. The observations by Wathes & Clark (1981) of the convective heat loss from a live chicken and from different 'models' of it.

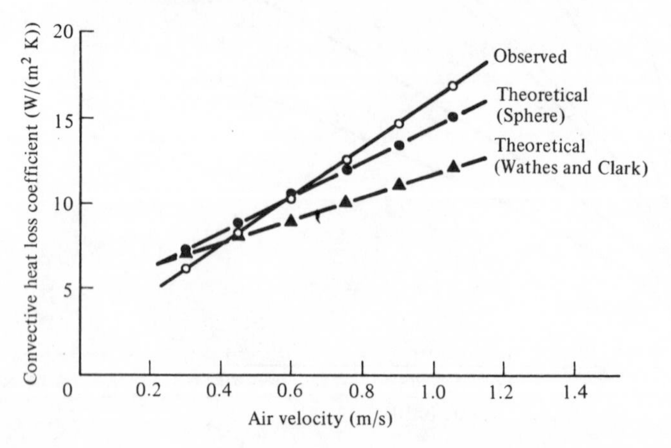

the Prandtl number and the *Grasshof number* (*Gr*). As before the Prandtl number can be taken to be constant for air. The Grasshof number is a dimensionless representation of the effects of buoyancy and inertial forces relative to viscous forces acting on the boundary layer of air, with the extent of the temperature difference between the object and the air taken into account as well. It is defined as:

$$Gr = agL^3(T_F - T_A)/\nu^2 \tag{7.22}$$

where a is the coefficient of thermal expansion of the air (1/273); g, the acceleration due to gravity; L the characteristic dimension of the object, T_F its surface temperature; T_A the temperature of the air; and ν the kinematic viscosity of the air. The larger the Grasshof number the greater the free convection. The characteristic dimension, L, is the vertical dimension, which is of most importance in free convection. For animals which can be regarded as analogues of horizontal cylinders the diameter is taken – as it is for forced convection – but with an upright animal, which can be regarded as analogous to a cylinder standing on its end, the characterisic dimension is the height. The same characteristic dimension has of course to be used in the relationship between the Nusselt number and the coefficient of convective heat transfer. The relation between the Nusselt number and the Grashof number is:

$$Nu = bGr^m \tag{7.23}$$

where b and m, which vary with shape, are experimentally determined.

The value of m for laminar flow is 0.25 and the value of b for upright cylinders is 0.58 and for horizontal ones 0.48. By combining the equation relating the convective heat transfer coefficient per unit surface to the Grasshof number (equation 7.22) with that relating the Nusselt number to the Grasshof number (equation 7.23) and inserting appropriate numerical values for the constants, the natural convective heat loss can be predicted. Values for cylinders as models for animals of different sizes are listed in Table 7.10. For example for a man 1.5 m tall, the natural convective heat loss ($\dot{C}_N$) in W/m^2 can be predicted to be:

$$\dot{C}_{(N)}A = 1.53(T_S - T_A)^{1.25} \tag{7.24}$$

For a four-legged animal with the same body mass, regarded as a horizontal cylinder with a diameter of 0.30 m, the corresponding equation has the coefficient 1.89.

At low air velocities some forced convection obviously has a component of natural convection and it is of importance to know how these combine. The convention is to regard the total convective loss as the larger of the natural and forced components.

Table 7.10. *The loss of heat by natural convection from cylinders approximating the dimensions of animals*[a]

Simulated animal	Body weight	Characteristic dimension L (m)	Heat loss (W/m²) for surface to air temperature gradient (°C) 5	10	20	30
Mouse	25 g	0.020	28	66	157	261
Rat	250 g	0.045	23	54	129	213
Rabbit	2500 g	0.114	18	43	102	169
Sheep	50 kg	0.270	15	35	82	136
Ox	500 kg	0.656	12	28	66	109
Elephant	5000 kg	1.400	10	23	55	91
Man	70 kg	1.650	11	27	63	105
Radiation loss (W/m²)			32	66	139	220

Notes:

[a] Animals: $\dot{C}/A = \frac{1.40}{L^{0.25}}(T-T)^{1.25}$

Man: $\dot{C}/A = \frac{1.69}{L^{0.25}}(T-T)^{1.25}$

In passing it may be noted that equation (7.24) shows that, as far as natural convection is concerned, heat loss is proportional to the difference in temperature raised to the power 1.25 or 5/4. This is clearly at variance with Newton's law of cooling. Newton was the first to investigate heat lost by a body in air, and as every schoolboy knows, he found that heat loss is directly proportional to the difference in temperature between the body and its surroundings. This is only true for excess temperatures of 20–30 °C and for forced convection. Dulong and Petit indeed found long ago that the rate of cooling in the absence of forced convection was proportional to the 5/4th power of the excess temperature.

7.3 Conduction

Conductive heat transfer is the process by which the energy of random motion of molecules in one substance is transferred to molecules of another substance at a lower temperature.

The usual formulation for the conductive heat transfer between two bodies represents the situation in a steady state, that is, when both temperatures and heat transmissions do not change with time:

$$\dot{K} = \frac{(kA')}{x}(T_2 - T_1) \qquad (7.25)$$

where $\dot{K}$ is the rate of conductive heat transfer; k is the thermal conductivity of a flat slab of thickness, x, and area, A', perfectly insulated around its edges; T_2 is the temperature of the hotter body; and T_1 is the temperature of the colder one. This formulation is often applied to analysis of the barrier to heat loss occasioned by a pelage or by clothing. Animals are not however flat slabs; they are complexly shaped bodies. Their shapes can be approximated by regarding them as spheres or cylinders or 'Doolittles', that is, cylinders with hemispherical ends (Swan 1972). Heat flows from them radially and this results in a different formulation to that for a flat slab. This problem is best dealt with by calculating the *apparent thickness* of the conducting layer. The apparent thickness varies with the geometry of the conducting layer. Integration of the differential equations involved leads to a number of solutions of relevance to animals. The apparent thickness of the conducting layer for a flat slab is the same as its observed thickness, x. For a sphere it is $rx/(r+x)$, where r is the radius of the sphere. For a cylinder it is $\log_e(1+x/r)$. Texts on heat conduction in solids deal with methods for arriving at other solutions for different-shaped bodies with insulating layers. For most biological applications it is usually accepted that animals can be regarded as cylinders. This is very much an approximation. Problems arise not only from the initial assumption of comparability of shape but because the thickness of the insulating layer is not constant over the whole body surface. The application of what may be called the *conductive approach* to the insulation provided by the tissues of the animal, its hair coat or other protective layer, and their interface with the environment is discussed in Section 10.2.

7.3.1 *The non-steady state problem*

When one sits on a stone bench, heat is lost by conduction to the stone, initially at a high rate. Gradually an equilibrium is reached when the loss of heat from the bench is equal to that from ones buttocks. One probably rises before this steady state is reached, and then it is found that the stone is appreciably warmer than it was before. Heat has been conducted from the body and has established a temperature gradient within the stone. This type of conductive heat loss obviously depends on the area of the body which is in contact (A'). Heat loss can be expressed as

$$K = kA'\,(\partial T/\partial x) \tag{7.26}$$

where the partial derivative is used to indicate that time is also involved in the form of a rate of change of temperature with time ($\partial T/\partial t$). It may be shown (see for a proof, Jakob & Hawkins 1957) that

$$\frac{\partial T}{\partial t} = \frac{kA'}{C_P\rho}\,\frac{d^2T}{dx^2} \tag{7.27}$$

where k is thermal conductivity; ρ is the density of the conductor; and C_P is its specific heat at constant pressure. The product density and specific heat is the heat capacity of the conductor per unit volume and the term k/C_P is termed the *thermal diffusivity*. Equation (7.27) can be solved numerically or graphically but there have been few attempts to do so. Gatenby (1977) has solved the equation for the transient heat flux in sheep when they lie down. A model system was used in which a heated pelt was placed on the ground and experiments were made in which a real sheep laid on the ground above a heat flux meter.

Most of the studies of conductive heat loss have measured heat transfer in the steady state when animals and people are lying on materials with different thermal properties. Thus Hey, Katz & O'Connell (1970) determined the areas of naked babies that were in contact with a perspex floor and measured the heat flux through heat flow disks, first when the babies were in direct contact with the floor and then when they were lying on a mattress on the floor. They found that the contact area was about 10% of the total body surface area when the babies were in contact with the perspex and about 14% when in contact with the mattress. The heat losses expressed per m^2 of contact area and per °C difference between rectal temperature and the outer surface of the perspex were 73 kJ(h °C $m^2)^{-1}$ when the babies were lying on perspex and only 13 kJ(h °C $m^2)^{-1}$ when the mattress was placed between. These losses were 25% and 5%, respectively, of the total heat loss by non-evaporative pathways.

Similarly, Mount (1967) measured the heat losses of newly-born pigs to floors of different types and some of his results are shown in Figure 7.5. Mount showed that on concrete floors about 15% of a piglet's total heat loss was to the floor, but when the floor was wood this reduced to 6%. He also noted that the way in which the piglet reclined affected heat loss. When the piglet rested with the distal parts of its four limbs folded underneath it, the abdomen and thorax were kept away from the floor and the heat loss was about two-thirds to one-half that noted when the piglet reclined with the abdomen and thorax in contact with the floor. Obviously the piglet can adjust its conductive loss of heat by varying its posture. The matter becomes even more complex when pigs are kept in groups. When they are cold they huddle together so that their body surfaces are in contact. This behaviour is also seen in other species and particularly in the young of multitocous species (Vickery & Millar 1984). Bruce & Clark (1979) concluded that the pig modifies the area in direct contact with the floor by reducing its width through changing its lying posture. They further concluded that when pigs were lying together heat loss was proportional to the square root of the number in the group! They admitted that the approach was in part empirical and in part intuitive but their analysis serves to illustrate the complexities involved.

In similar studies with white-tailed deer (*Dama virginianus*), Jacobsen (1980) distinguished different areas of the body which were in contact with the ground including the upper, middle and lower areas of both hind and fore legs as well as trunk regions. He also considered the compression of the insulating layer of hair by the weight of the animal. Winter coats, which have thick undercoats, were less susceptible to reduction in their insulation than summer ones, and both in winter or summer heat loss by conduction was about 20% of the fasting heat production.

7.3.2 *Equivalent environments*

It is clear from the discussion of heat losses by radiation, convection and conduction that the specification of an environment by its air temperature alone is insufficient to indicate the thermal demand that it elicits. Several attempts have been made to combine radiative and convective components in order to specify their joint effects in terms of a single equivalent temperature. The most useful of these is that made by

Fig. 7.5. The rate of heat loss of baby pigs by conduction to the floor in relation to the gradient between rectal temperature and ambient temperature. The floor materials were: ● concrete, X wood and ○ expanded polystyrene. The results are those of Mount (1967).

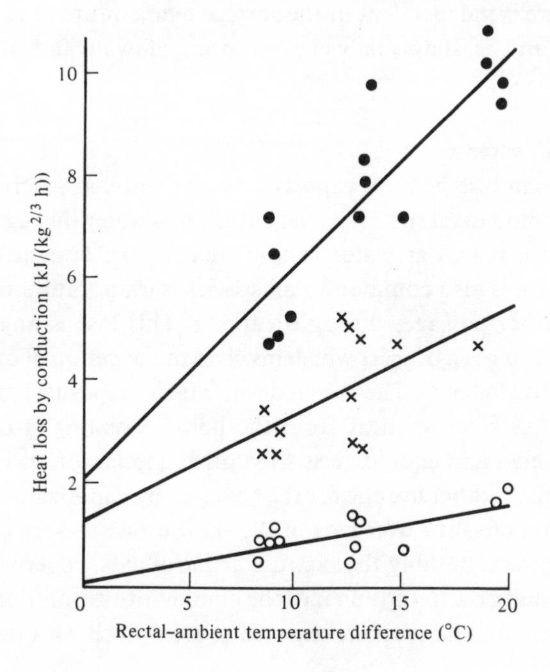

Herrington, Winslow & Gagge (1937) who devised the term 'operative temperature'. Its derivation is as follows: The total exchange of heat from the surface of an animal at temperature T_F by convection, to the air at temperature T_A, and by radiation to the walls at temperature T_W is given by the expression

$$(\dot{C}+\dot{R})/A = h_C(T_F - T_A) + h_R(T_F - T_W) \qquad (7.28)$$

where h_C is the convective heat transfer coefficient (which varies with air velocity); and h_R is the analogous radiation exchange coefficient expressed in terms of first-powers rather than fourth-powers of temperatures (see page 97).

Herrington and his associates then defined a *combined heat transfer coefficient* (h_{OP}), which when combined with an appropriate '*operative temperature*' (T_{OP}), provides the same rate of heat exchange by convection and radiation given by equation (7.28):

$$(\dot{C}+\dot{R})/A = h_{OP}(T_F - T_{OP}) \qquad (7.29)$$

By equating the right-hand sides of equations (7.28) and (7.29) one can solve for both T_{OP} and h_{OP}:

$$T_{OP} = \frac{(h_C T_A + h_R T_W)}{(h_C + h_R)} \qquad (7.30)$$

$$h_{OP} = h_C + h_R \qquad (7.31)$$

Operative temperature is independent of the surface temperature of the animal or man (T_F), and is simply a weighted-mean of wall and air temperatures.

7.4 Vaporisation of water

Animals and man lose heat by vaporising water through several avenues. First, and common to all species, is the diffusion of water through the skin and its subsequent loss as water vapour, that is, the *insensible perspiration.* A loss which is also common to all species is evaporation of water from the respiratory passages during breathing. This loss is augmented by panting. *Sweating* is a process which involves the secretion of an aqueous fluid by specialised glands. The secreted water then evaporates on the skin surface and thus removes heat from the body. Sweating is of obvious importance in man and equines, less so in other species, or, as in birds, completely absent. Another avenue for the heat loss by vaporisation relates to the distribution of saliva over part of the skin surface as seen in some rodents. Other species, notably the pig, use artificial aids, covering themselves with liquid mud or water to enhance the amount of evaporation and hence heat loss. Animals and man are exposed to rain which wets the coat or clothing. This not only reduces the insulation of the outer layers, but in addition, any water retained in the coat or on the skin surface has to be evaporated. Lastly, all animals are born completely wet and this water also has to be evaporated, occasioning an obligatory loss of heat.

7.4.1 *The physics of heat loss by vaporising water*

The loss of water vapour from a completely wet surface to the surrounding air is proportional to the area of the surface and to the gradient of water vapour concentration between the surface and the air. This gradient of concentration, usually expressed as mass per unit volume, can equally be expressed in terms of partial pressures of water vapour (by use of the gas laws):

$$\dot{m}_W/A = h_M(p_{S(sat)} - p_A)/A \quad (7.32)$$

where $\dot{m}_W$ is the mass of water vapour lost from the fully wet surface per unit time; $p_{S(sat)}$ is the water vapour pressure at the surface; p_A is the vapour pressure of the surrounding air; and h_M is a mass transfer coefficient which states the mass of water vaporised per unit surface per unit time for unit gradient in water vapour pressure.

The form of this relationship is analogous to that for convective heat transfer; boundary layer considerations are involved and the Reynolds analogy can be applied. The mass transfer coefficient is affected by air velocity in virtually the same way as is the convective heat transfer coefficient (see Section 7.2). The derivation of the coefficient and its relationship to characteristic dimensions of objects and the Reynolds number is given by Kerslake (1972). Table 7.11 gives the saturation vapour pressure of air at different temperatures from which it is evident that it increases markedly with increasing temperature. Relative humidity, ϕ, is defined as the proportional saturation of air at a particular temperature, and the basic equation above can be modified to take into account situations in which the surface is not completely wetted with water by multiplying $p_{S(sat)}$ by ϕ_S. Applied to sweating skin the concept of unsaturation at the surface can best be visualised by regarding the skin as a mosaic in which minute areas are wet and others are dry, reflecting the activity of individual glands.

The mass equation can be expressed in terms of heat by multiplying the mass of water vapour lost by its *heat equivalent*, the major component of which is the *latent heat of vaporisation* of water. The latent heat declines from 2447 J/g at 20 °C to 2402 J/g at 40 °C, and at 33 °C is 2418 J/g. The final relationship is thus:

$$\frac{\dot{E}}{A} = \frac{a\dot{m}_W}{A} h_M((p_{S(sat)}\phi_S) - p_A) \quad (7.33)$$

Table 7.11. *The saturation vapour pressure of air at different temperatures*

Temperature (°C)	Vapour pressure (kPa)	Temperature (°C)	Vapour pressure (kPa)
0	0.61	30	4.24
5	0.87	31	4.49
10	1.23	32	4.75
15	1.70	33	5.03
20	2.34	34	5.32
25	3.17	35	5.62
30	4.24	36	5.94
35	5.62	37	6.27
40	7.37	38	6.62
45	9.58	39	6.99
50	12.33	40	7.37
55	15.73	41	7.78
60	19.91	42	8.00

Note:
[a] These vapour pressures encompass those likely to be found at the surface of the skin of animals.
Source: From Handbook of Chemistry and Physics (1985).

where a is the heat equivalent of unit mass of water vapour; and h_M is the mass transfer coefficient. The relation between the convective heat transfer coefficient ($h_{C(F)}$) and the evaporative heat transfer coefficient ($h_E = ah_M$) is that when water vapour pressure is expressed in kiloPascals (kPa), h_E is equal to about $15 \times h_{C(F)}$ (Kerslake 1983).

a. The heat equivalent of vaporised water

There has been some uncertainty about the heat equivalent of the mass of water vapour lost from the surface of the body. This is of critical importance in partioning heat loss since it is usual to arrive at the sensible heat loss by deducting from the sum of the heat production and body heat storage the mass of water vaporised multiplied by the heat equivalent factor. If there is error in estimating the heat equivalent factor, the effect on estimating the sensible loss of heat could be considerable, particularly when the total loss of water vapour is high.

Hardy (1949) argued that the heat equivalent consisted of three terms. The first was the classical latent heat of vaporisation. The second was a term related to the cooling of the saturated vapour at skin temperature to saturation at air temperature, a process which requires heat since the vapour has to be heated to prevent condensation. The third component was the heat required to expand the vapour isothermally to the humidity of the surrounding air. Hardy's arguments were convincing and he stated that they were supported by studies on electrically-heated wetted cylinders. His formulation was used extensively throughout the next quarter-century. The departure of Hardy's estimate of the heat equivalent of water vapour and the latent heat term which it replaced was considerable. When body surface temperature is 33 °C – a temperature reflecting skin temperature in man – and in an environment at 27 °C with a relative humidity of 0.20, according to Hardy's calculation the latent heat term is 2418 J/g; the second term is 43 J/g and the third is 222 J/g. Their sum is 10.7% greater than the latent heat term alone and the discrepancy increases as the relative humidity of the surrounding air falls.

A thermodynamic analysis of the heat equivalent term was undertaken simultaneously by Wenger (1972) and Monteith (1973). Both concluded that the formulation was incorrect; the third term was irrelevant since in an open system no work is done and the process involved is simply one of diffusion; humidity should have no effect on the heat equivalent; the correct formulation should be:

$$\text{Heat equivalent/g} = a = \lambda - C_P(T_S - T_A) \qquad (7.34)$$

where λ is the latent heat of water vaporisation; C_P is the specific heat of water vapour at constant pressure; and T is temperature with subscripts representing surface and air temperatures. The specific heat of water vapour varies slightly with temperature but can be taken to be close to 1.9 J/(g °C), while gradients of temperature from skin to air are only great when there is a thick insulating layer of fur or clothing. The second term in the equation above is thus not likely to reduce the heat equivalent much below the latent heat term. Furthermore, it is highly probable that the heat required to meet this second term is abstracted from the air rather than from the animal. It would appear from this analysis that the latent heat term alone reflects the heat loss from the body surface of a man or animal per unit mass (g) of water vaporised.

A number of experiments with man, however, have suggested that the heat equivalent of vaporised moisture is greater than the latent heat term. Snellen, Mitchell & Wyndham (1970) showed that relative humidity has no effect on the heat equivalent of water vapour in a sweating man, thus providing an experimental basis for Wenger's contention, but they also found that the heat equivalent was 7% greater than the latent heat of vaporisation of water. Earlier work (Mitchell *et al.* 1968) gave a similar figure. These values have been criticised on statistical grounds (Wenger

1972). A more telling criticism relates to certain aspects of the measurements made and the difficulty of separating the component of heat taken from the subject from that taken from the air (McLean & Tobin 1987). It seems possible that in some circumstances what is perhaps best called 'the biological heat equivalent of vaporised water' differs from the latent heat term alone. Until more work is done the precise value is uncertain and in practical work it is least confusing to assume that the heat equivalent is equal to the latent heat term alone.

7.4.2 *Heat loss in respiration*

Inspired air passes over the wet surfaces of the upper respiratory tract and is usually expired at a higher temperature and with a higher humidity than that inspired. The heat loss in respiration can thus be expressed as the sum of two terms, one relating to evaporation and the other to the gain in enthalpy of the dry air:

$$\dot{E}_R = \lambda(\dot{m}_{W(ex)} - \dot{m}_{W(in)}) + \dot{v}C_P\rho(T_S - T_A) \quad (7.35)$$

where $\dot{E}_R$ is the rate of heat loss in the respiration; λ is the latent heat of vaporising water; $\dot{m}_{W(ex)}$ is the mass of water vapour expired per unit time; $\dot{m}_{W(in)}$ is the mass of water vapour inspired per unit time; $\dot{v}$ is the ventilation rate; T_S and T_A the surface and air temperatures, respectively; ρ the density of air and ρ its specific heat; and The product of the last two terms is 0.02 W/°C.

In non-panting animals, such as man, ventilation is directly proportional to oxygen consumption – the range in man being 23–28 l of air respired /l O_2 consumed (Kinney, Weissman & Askanazi 1985). In most animals expired air is fully saturated with water vapour. The temperature of expired air is, however, less than that of the lung since some exchange of heat takes place between inspired air and expired air. When air is inspired it is warmed by the turbinates and other upper respiratory tract structures while they themselves cool. On expiration the air is cooled and the respiratory structures rewarmed. The system is least effective in animals with short respiratory passages (Welch 1984). In the camel, during the daytime, exhaled air is fully saturated and at a temperature close to body temperature. At night, when inspired air temperature is cooler, the temperature of exhaled air is closer to that of the inhaled air and its relative humidity can fall to 0.75 saturation (Schmidt–Nielsen *et al.* 1981a,b). The surfaces of the camel's respiratory tract give off water vapour during inhalation and take up water vapour during exhalation. This mechanism obviously has considerable value in the water economy of the animal. It has not been observed in other species of desert animal.

In man, and other non-panting animals, and in panting animals at low environmental temperatures, the proportionality of ventilation rate to oxygen consumption means that heat loss from the respiratory tract under most circumstances is a constant proportion of the metabolic rate (see Brackenbury 1984).

a. Panting

Many species respond to high environmental temperatures by increasing the rate of ventilation of the upper respiratory passages through an increase in the frequency with which they respire. A familiar example is the dog. The dog at rest and cool has a low respiratory frequency. In the heat or after exercise, the respiratory pattern changes to shallow rapid respiratory movements with an open mouth and protruded tongue. The frequency of respiration accords with the resonant frequency of the chest and is proportional to the square root of the mass of the animal (Crawford 1962). The same abrupt change in respiratory frequency occurs in response to heat in the pigeon; its rate of respiration changes from 29 breaths/min in the cold to 65/min in the heat, the latter value again being commensurate with the resonant frequency of the thorax. In some birds panting is associated with gular flutter or rapid oscillation of the floor of the mouth and upper part of the throat, which is due to contraction of muscles attached to the hyoid, while in other birds gular flutter is independent of panting (Calder & King 1974).

In those species which depend on sudden changes in respiratory frequency to adjust their respiratory loss of heat, duration of panting appears to be the control mechanism adopted, although at very high temperatures the frequency and depth of respiration may change. However, most species of mammal and bird adjust their ventilation smoothly and do not suddenly change from normal breathing to panting (S.A. Richards 1970). The difference in behaviour of the 'true panters', which have either low or high respiratory frequencies, and the species which continuously adjust their respiratory frequency is illustrated in Figure 7.6.

There is considerable inter-species variation in the extent to which animals increase respiratory passage ventilation in response to heat; some species rely more on cutaneous than on respiratory evaporation. Thus at an environmental temperature of 40 °C less than 20% of the total heat production of an ox is lost by respiratory evaporative cooling (McLean 1974). In the sheep, however, at an environmental temperature of 30 °C about 60% of the total heat loss is from the respiratory passages (Brockway, McDonald & Pullar 1965).

An increase in ventilation rate does not mean that there is a proportional change in the exchange of carbon dioxide, for the shallow breathing during panting does not ventilate the alveoli of the lung. A respiratory alkalosis that is, depletion of blood bicarbonate, due to loss of carbon dioxide from the lung does, however, occur under severe thermal stress in cattle when what is termed *second phase breathing* ensues. Here respiratory frequency falls, and tidal volume increases leading to an increase in the minute volume of the respiration and a fall in blood carbon dioxide tension. Thus it appears that thermal stress can override the normal mechanisms which control blood gas tensions by altering pulmonary ventilation.

7.4.3 *Insensible perspiration*

Insensible perspiration is an obligatory loss of water vapour from the skin which is not due to the activity of sweat glands. It can be regarded as a process of diffusion of water vapour from the fluids below the epidermis, through the epidermis to the ambient air. The epidermis acts to resist the passage of the vapour. The process can be regarded (see McLean 1974) as a diffusion process and a vaporisation process in series, and can be formulated as:

$$\dot{E}_1/A = h_1(p_{S(sat)} - p_A)/A \qquad (7.36)$$

Where $\dot{E}_1$ is the rate of heat loss by insensible perspiration (W/m^2); and h_1 is a heat transfer coefficient for insensible perspiration (W/m^{2}kPa).

The coefficient for heat transfer by insensible perspiration is difficult to

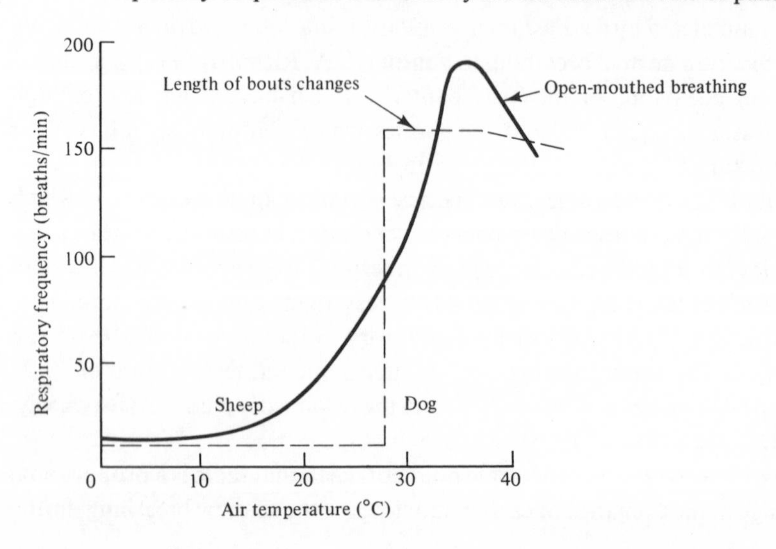

Fig. 7.6. Respiratory frequency in relation to environmental temperature in the sheep and the dog. In the dog the high respiratory frequency is at the resonant frequency of the thorax and control of respiratory evaporation is exerted by the duration of bouts of panting.

Table 7.12. *The minimal and presumptive maximal rates of loss of heat by vaporisation of water from the skin surface of different animals (W/m^2)*

	Heat loss by vaporisation of water	
Species	Below the critical temperature	Above the critical tempearture
Man (at work)	6	816
Man (at rest)	6	102
Pig	7	19
Sheep	8	35
Ox	9	98
Chicken	4	12

Source: From Ingram (1974).

determine in sweating species such as man since sweat gland activity can occur even when there is no thermal stimulus, as evidence by palmar and axillary sweating. Even so, a value of about 1.2 $W/(m^2kPa)$ has been calculated. This is considerably less than the comparable coefficient for sweating which in still air and for completely wet skin is about 50 $W/(m^2kPa)$. An estimate can be made of the heat transfer coefficient for insensible perspiration in birds (which are devoid of sweat glands) based on Richards' (1976) studies with chicken. This gives values within the range 1.0–1.8 $W/(m^2kPa)$. Measurements on sheep with a congenital absence of sweat glands (Brook & Short 1960) suggest a similar range. In both man and sheep, individuals with congenital absence of sweat glands, when kept at low environmental temperatures, lose the same amount of heat by vaporising water as do normal subjects with sweat glands. Indeed, at low environmental temperatures the cutaneous loss of heat by insensible perspiration appears to be subject to little inter-species variation (Table 7.12). However, the number of species for which information is available is small and does not include ones in which the stratum corneum is very thick.

7.4.4 *Sweating*

Sweating is the secretion onto the skin surface of fluid by specialised glands in the skin. In most mammalian species the sweat glands are associated with hair follicles but in man they are separate structures. In birds and possibly in the elephant they are absent and in many mammals they are sparse. Much of the work on sweating and particularly that with man, has often made no distinction between sweating *per se* and insensible perspiration, or indeed between the cutaneous loss of heat by vaporisation of water and the loss from the respiratory tract. In a profusely sweating species, such as man, the errors involved in failing to make the distinction are not large, but it has to be appreciated that an error is incurred (Kerslake 1983).

In non-panting animals the minimal obligatory loss of water from the body, and its associated heat, is the sum of the loss from the respiratory passages and from insensible perspiration. In panting animals the obligatory loss from the respiratory passages is that observed when the animal is cool enough not to have increased its pulmonary ventilation in response to environmental heat. Under these circumstances, evaporative heat loss is unaffected by the humidity of the ambient air. This has been demonstrated in man by Wiley and Newburgh (see Kerslake 1972) and in cattle by the results of McLean & Calvert (1972). During sweating the same is not true in either species, particularly when the skin surface is wet. The relationship which enables the heat loss by evaporation to be estimated from the vapour pressure at the skin and that of the ambient air, as given in equation (7.33) presupposes that the vapour pressure at the skin or, alternatively, the relative humidity of the skin can be measured. The latter is biologically adjusted to prevent storage or loss of heat by the body. Thus, provided the skin is not completely wet, air humidity has no effect on the rate of heat loss by sweating; all that changes is the relative humidity of the skin. When, however, the skin is completely wet, increases in air humidity affect the gradient of humidity from skin to air and sweating is reduced.

Table 7.12 shows not only the minimal losses of heat by water vaporisation under cold conditions – losses which comprise the insensible loss and the loss from respiratory passages – but also the heat losses under hot conditions. These losses include thermal sweating. The low values for chicken reflect the absence of sweat glands in birds and the low values for pigs reflect the paucity of its sweat response. The values for man are not maximums; values in excess of 2500 W/m^2 have been recorded for well acclimatised men working in environments which include strong solar radiation and air temperatures in excess of body temperature. Schmidt-Nielsen (1964) has recalled the experiments undertaken by Dr Blagden in 1775 which illustrate the ability of man and dog to evaporate moisture. Blagden, together with a dog in a basket (to protect its feet from being burned), entered a room kept at a temperature of 126 °C and remained there for 45 min. A steak he took with him was cooked but he and the dog were unaffected.

7.4.5 *Salivation and wallowing*

Rodents and some marsupials salivate in the heat and this saliva is spread over the body where its water evaporates. This method of augmenting evaporative cooling is regarded by many zoologists as a last defence against hyperthermia. Its efficacy, in terms of cooling the body proper is not known; obviously some of the heat for changing the state of the water must come from the air since the application of water is superficial.

Pigs and some other species wallow in mud and water and this has been studied by Ingram (1974). He found clean water applied to the pig's back evaporated in 15 min at a rate of 540 W/m^2, a rate greater than its rate of heat production. If mud was used instead of water the effect of a single application lasted for 2 h and the rate of loss of heat by vaporising the moisture was 700 W/m^2. Pigs have little hair and it can be presumed that much of the latent heat cooled the body.

8

The minimal metabolism

8.1 Terminology

In order to make comparisons between the heat productions of individual animals within or between species, the conditions under which the measurements are made must be standardised. This is achieved by attempting to measure a minimum rate of heat production free of the effects of any controllable factors known to increase it. Such factors include muscular exercise, the consumption of food and its subsequent metabolism, and the physical environment. The object of standardisation is to ensure comparability of estimates rather than to establish some absolute minimum value of metabolism which is compatible with life. It is understandable that varying degrees of success in controlling the variables concerned have been achieved in different species. A number of terms have thus arisen to describe these standardised measurements of 'minimal metabolism'. However, these terms are not wholly agreed upon – the glossaries suggested by Bligh & Johnson (1973), Gessaman (1973), Schutz (1984) and Simons (1988) and the definitions given by Peters (1983) and by others are not mutually consistent. The terms used here reflect current usage.

8.1.1 *Basal metabolism*

This term applies particularly to man, indeed it is now rarely used other than in human studies. Even so, the prerequisites for the measurement of basal metabolism in man are of general interest in that workers with farm, laboratory and wild species attempt to emulate them in defining the conditions they impose to estimate 'minimal metabolism'. A detailed statement of these conditions was given by Benedict (1938). They are extremely rigorous. The subject should be in complete muscular repose both before and during the measurement, which should be made with the subject lying but awake. He or she should be in a post-absorptive state and the prior food intake should have been at or about the maintenance level. The environment should be thermoneutral, eliciting no thermoregulatory effect on heat production, and the subject should have been acclimatised to such an environment. There should be freedom from emotional stress, the subject should be familiar with the apparatus, and in women, the measurement should not be made immediately before or during the menses. Most determinations of basal metabolism are undertaken in a clinical setting. The measurement is usually made after an overnight fast with the subject in bed, shortly after waking, and the duration of the measurement of the respiratory exchange from which heat production is invariably calculated is minutes rather than hours.

Basal metabolism is not a measure of the absolute minimal metabolism of man. Sleep lowers metabolic rate by about 10% (Mason & Benedict 1934) and anaesthesia and transcendental meditation also reduce it (Farrell 1980). The rate of heat production is also lowered by continued starvation, and if the condition of prior nutrition at the maintenance level is not met and the subject is underfed, metabolism determined under the remaining standard conditions is depressed. Provided all the conditions given above are strictly adhered to, the measurement is the most practical estimate of minimal metabolism.

These conditions are not always adhered to, and the term basal metabolism is increasingly being replaced by the term *standard metabolism*. This implies that metabolism is measured under a set of standard conditions which are usually but not invariably those which Benedict and earlier workers such as DuBois (1927) imposed.

8.1.2 *The problems with animals*

Animals are not as cooperative as people. They cannot be asked to remain completely at rest in a postabsorptive state at a particular time of day. Admittedly their food intake can be controlled as can their physical environment. They can also be conditioned to experimental procedures, although how far the experimental conditions imposed induce stress is difficult to discern. Most measurements of minimal metabolism made with animals do not conform with the ideal conditions necessary to determine a true value for the basal metabolism of man. With domesticated animals what is usually measured is the *fasting metabolic rate*. The fasted animal, previously accustomed to the calorimeter or respiration chamber, is allowed to move at will within its confines, and the measurement is continued for 24 h or more. In some cases the data are corrected to a standard of 12 h standing and 12 h lying. The measurement is often called a basal metabolism measurement but it is obviously not fully analogous to the basal metabolism of man because muscular movement is not controlled. This same technique can be applied to wild species kept under laboratory conditions. With small animals, however, it is usual to make similar measurements over a few hours rather than for a day or days and it is often not clear whether the animal is completely at rest.

Another approach designed to minimise the effects of muscular movement on metabolism is to make a continuous record of the heat production of the animal and from the record select those short periods in which metabolism is minimal. This measurement is termed the *least observed metabolic rate*. Obviously it is better to make the selection based on observed periods of quiescence since low values could arise from random error in the measuring devices. In all of the methods described above the environment is kept within the thermoneutral zone of the species concerned.

8.1.3 *The post-absorptive state*

A critical aspect in all studies designed to measure minimal metabolism is the length of time that animals, including man, should be fasted to ensure that the continuing effects of metabolism of previous meals are negligible and that the animal is in a post-absorptive state. The criteria used to judge this state are: firstly, whether the non-protein respiratory quotient has fallen to a level which shows that the predominant fuel is body fat (0.70–0.73); secondly, whether prolonging the fasting period results in a reduction in heat production; and thirdly, and specifically with herbivora, whether methane production, indicative of fermentation of food in the gut, has virtually ceased. It may equally be assumed that the time taken to reach the post-absorptive state is related to the mean retention time of food in the digestive tract (see Section 3.3).

In man, the conventional postabsorptive period is 10–12 h after an evening meal. This may not be a sufficient interval. Dauncey (1980) has shown that even when the diet supplied only 3.7 MJ/d – about half the normal energy expenditure – an elevation of metabolism of 6% was present 15 h after the last meal. Her observations did not extend for longer. In pigs, Breirem (1936) found that a minimal and stable metabolism was attained only after 96 h of fasting. Pig and man have similar mean retention times of food in their digestive tracts, and the divergence between the two species is curious. Furthermore, when the basal metabolism of man is determined after a fast of only 10 h the respiratory quotient is usually of the order of 0.80–0.85. Even when allowance is made for protein oxidation respiratory quotients are above that for fat. Generally, the smaller omnivorous animals reach the post-absorptive state in 10–20 h, chicken reach it in about 48 h and ruminants in 3–5 d (H.H. Mitchell 1962).

8.1.4 *Resting metabolism*

In many situations it is impractical to fast animals, yet some standardised estimate of metabolism is required. Metabolism determined without the animal being in a post-absorptive state but with the remaining conditions listed by Benedict being met, is termed the *resting metabolism*. The extent of its departure from estimates of minimal metabolism depends on the magnitude and duration of the thermogenic effect of food. As previously discussed, this can be large in the instance of ruminants.

Although the conditions for determining basal metabolism, standard metabolism, fasting metabolism, least observed metabolism and resting metabolism can all be defined, it cannot be assumed that when a particular determination is stated to have been made all the conditions have been adhered to. For example, Owen *et al.* (1986) stated that their determinations of metabolism were of resting metabolism. They clearly regarded basal metabolism as that determined in the early hours of the morning during sleep. In fact their measurements appear to have been either of basal or standard metabolism as defined above. The authors indeed compared their values with classic determinations of basal metabolism. The precise conditions imposed may not be of consequence in comparative studies within one laboratory, but are of some importance when results from different laboratories are combined. Happily, it is now an increasingly common practice to give details of the conditions under which the estimate of minimal metabolism has been made.

8.2 Effects of body size

To assess the effects of many of the variables that influence the metabolic rate, the measurement has to be referred to some base which takes into account differences in body size, both within and between species. As mentioned in Chapter 1, as early as the eighteenth century Lavoisier noted that large animals produced more heat than small ones, and work in the nineteenth century showed that small animals produced more heat per unit weight than did large ones.

Rubner is credited with making the first systematic experimental analysis of the effect of size on metabolism. He showed in 1883 that the fasting metabolism of seven dogs, varying in body weight from 3.2 to 31.2 kg, was approximately constant when expressed per unit area of body surface (m^2). Rubner's pupil, Voit, published a table in 1901 which showed that the fasting metabolisms of a number of different species were also proportional

Table 8.1. *Voit's (1901) table showing the apparent constancy of minimal (basal) metabolism per unit surface area between species*

Species	Body weight (kg)	Metabolism/m^2 surface area (kcal/24 h)	(W)
Horse	441	948	46
Pig	128	1078	52
Man	64.3	1042	50
Dog	15.2	1039	50
Goose	3.5	969	47
Rabbit	2.3	776	38
Chicken	2.0	943	46
Mouse	0.018	1188	58
Mean		988	48

to their surface areas. Voit's table is reproduced in Table 8.1, from which it can be seen that Rubner's generalisation that the fasting metabolism of all homeotherms is constant at about 1000 kcal/(m^2 24 h), apparently holds between species as well as within the range of mature sizes of dogs.

It was largely on the basis of these results and theoretical considerations (summarised by Heusner, 1985), that surface area became the attribute of body size to which basal metabolism was referred. The theoretical reasons seemed acceptable; after all, and as was evident in Chapter 7, the rate of heat loss from a body to the environment is proportional to the area of its surface. Expressing the basal metabolism of man on a surface area basis became commonplace and this convention still persists. Considerable ingenuity was shown in devising methods for the determination of surface area and relating these measurements to more easily determined attributes of size. They included the use of moulds, surface integration, photographic methods, and for animals, skinning and measuring the area of the flattened skin. The area was usually related to the two-thirds power of body weight, making the assumptions that, within a species, animals were homologous in terms of shape and that their density was constant. The surface area of all similarly shaped bodies is the square of the cube root of their volume. Meeh (1879) produced a set of coefficients for different species and these allowed calculation of surface area from body weight ($kg^{2/3}$). The Meeh coefficients of $kg^{2/3}$ grouped around a mean value of about 0.09, a value considerably greater than that for spheres with unit density (0.0484) or for cylinders with lengths twice their diameters (0.0581). This divergence largely reflects the areas of the slender limbs of animals. Different observers obtained different values for the coefficient when they adopted the Meeh approach, and statistical fitting of the relationship between surface area and body weight showed that, in some species, surface area did not vary with the two-thirds power of weight at all. It is indicative of the difficulties of measuring surface area that different formulas for predicting surface area from body weight give results which *differ* by 130% for mice, 70% for pigs and 30% for horses (Poczopko 1971)!

Kleiber, in January 1932, and Brody and Proctor, in April 1932, showed that when metabolism was related directly to body weight (neglecting the concept of surface area altogether) metabolism was proportional to a power of weight higher than two-thirds. Brody and Proctor found the power to be 0.734 and Kleiber estimated it to be 0.75. Kleiber (1961) stated that the slope as estimated from his original set of data was 0.739. In both sets of data errors were such that proportionality to the power 0.75 can be accepted as a valid estimate. In view of the virtually simultaneous nature of the two studies and the concordance of the sets of data, it seems reasonable to refer to the relationship as the Brody–Kleiber relationship. Brody's 'mouse–elephant' graph (1945) showing the relation between basal metabolism and body weight in mature animals of different species is shown in Figure 8.1. Kleiber's graph could equally well have been used.

Fig. 8.1. Brody's classic diagram showing the relation between basal (minimal) metabolic rate and body weight.

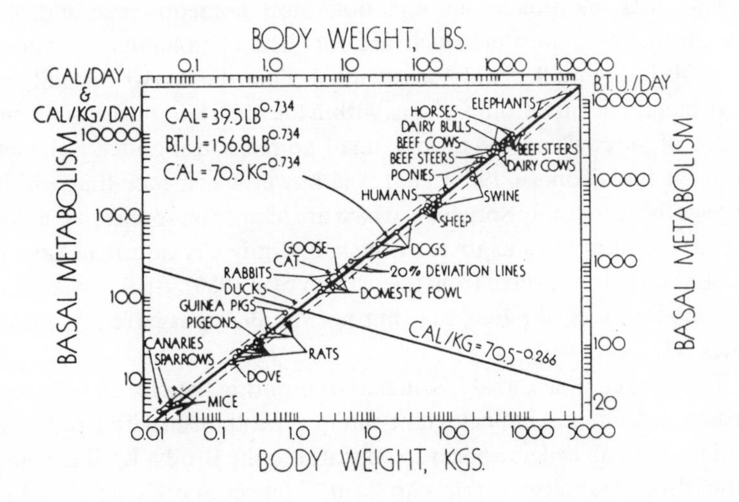

8.2.1 *Basal metabolism of mature animals*

Since the early 1930s determinations of minimal metabolism have been made in many species of animal. Notable are the investigations and reviews of Zeuthen (1947,1953) and of Hemmingsen (1950,1960). These studies showed that throughout the animal kingdom metabolism was proportional to body weight raised to the power 0.75. Hemmingsen's analysis showed that metabolism per kg $W^{0.75}$ was 4.1 W for homeotherms, 0.14 W for poikilotherms and 0.018 W for unicellular organisms. Expressed in terms of kJ/ 24 h, the coefficients are 354, 12 and 1.5 for the three groups, respectively. These results obtained over a range of body weight of 10^{18} certainly suggest that the metabolic rate of different species varies with the 0.75 power of body weight as Kleiber's and Brody's studies had shown.

The generalisation has, however limitations; the estimates of the relationships found by Brody, Kleiber and Hemmingsen were all based on relatively few species – about a dozen including both mammals and birds – and so too was the analysis made by Benedict (1938). The number of species measured has increased considerably since the first formulation of the Brody–Kleiber relationship and it has emerged that while metabolism within subgroups of both homeothermic and poikilothermic animals is proportional to the three-quarters power of body weight, the overall level of metabolism varies from subgroup-to-subgroup. Analyses of published data in 1972 gave the results summarised in Figure 8.2 (Blaxter 1972). This shows that marsupials have a lower metabolism than eutherian (placental) animals and that, within the eutherians, bats have a very low metabolism and insectivores a very high one. This type of analysis has been further extended and Peters (1983) lists 69 prediction equations for homeotherms and 46 for poikilotherms, many of which are for distinct taxonomic groups. The analysis in Figure 8.2 could be extended. Thus the shrews certainly have a high metabolic rate. Comparisons within the same laboratory of the metabolism of shrews and other very small animals, using identical methods confirm this (Tomasi 1985). There is however evidence that within the shrews the sub-family Soricinae, which are temperate species, have elevated metabolic rates but members of the sub-family Crocidurinae have metabolic rates to be expected from their size (Vogel 1980). Again, as representative monotremes, the Echidnas but not the Platypus have low metabolic rates (Hulbert 1980).

The most complete analysis, in that it comprises data for 293 species in 14 orders of mammals, indicates that minimal metabolism within orders varies and that not all orders appear to conform to the Brody–Kleiber generalisation that metabolism varies with the 0.75 power of body weight (Huyssen

& Lacy 1985). Care must be taken however in interpretation of such results. The determination of the power of weight with which metabolism varies is usually undertaken by regressing the logarithm of metabolism on the logarithm of body weight. It is likely that the errors attached to estimates of metabolism are relative – that is, percentage errors. The regression slope is thus dependent on the range of weights studied. When the range is small large errors can occur. It is thus understandable that values for the power function for different groups of animals range from less than 0.6 to over 0.9 (Calder 1987).

It seems reasonable, when making species comparisons, to retain the scaling factor of 0.75 even though within a genus or even within a class some slightly different power provides a better statistical fit. When this is done, other types of sub-classification of the mammalia can be made to explain the variation about the Brody–Kleiber proportionality. For example, McNab (1969,1980) found that within the bats low metabolic rates relative to the Brody–Kleiber norm are associated with insectivorous or blood-consuming dietary habits. Referring to Figure 8.2 it is of interest that the metabolic rates of the various natural orders or classes are broadly propor-

Fig. 8.2. The relation between the rate of minimal metabolism and body weight in species for different taxonomic groups showing that, while metabolism increases with $W^{0.75}$, there are considerable differences between groups.

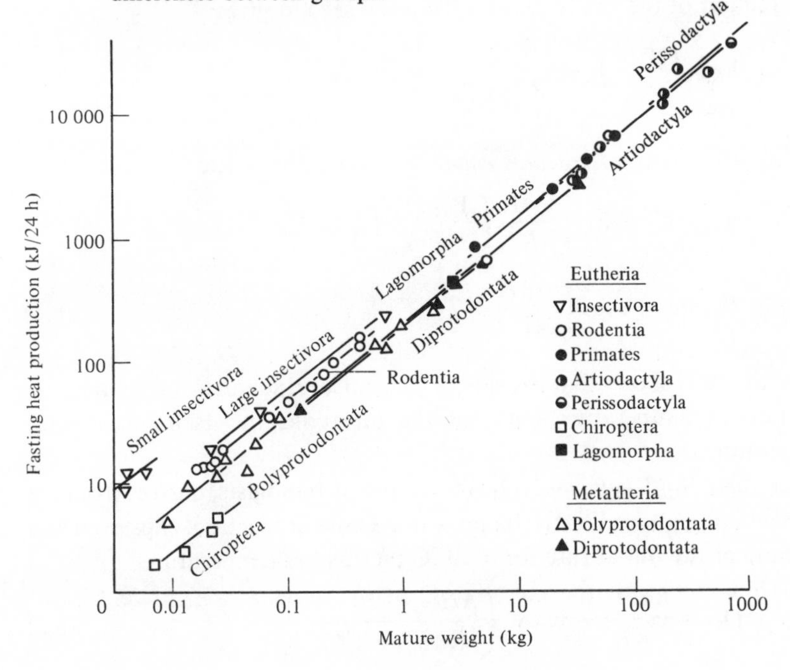

tional to the mean deep-body temperatures of the species concerned. Birds have higher temperatures than mammals and eutherian (placental) mammals have higher body temperatures than marsupials. This relationship between body temperature and metabolic rate has long been known and prompted Robinson, Peters & Zimmermann (1983) to examine the joint effect of body size and temperature on the metabolic rate of organisms.

8.2.2 *The Arrhenius and van't Hoff relationships*

Before dealing with these joint effects on metabolism a discursion is necessary. The second law of thermodynamics and the definition of the equilibrium constant, K_{EQ} can be written:

$$\Delta G = \Delta H - T\Delta S$$

$$\log_e K_{EQ} = -\Delta G/RT$$

Combining these equations gives:

$$\log_e K_{EQ} = -\Delta H/RT + \Delta S/R$$

where R is the gas constant (8.314 J/(K mol)) and T the absolute temperature (K). The effect of temperature on the equilibrium constant can be arrived at by considering two temperatures, T_1 and T_2 in equation (7.3):

$$\log_e K_{T_1} = -\Delta H/RT_1 + \Delta S/R$$

$$\log_e K_{T_2} = -\Delta H/RT_2 + \Delta S/R$$

Subtraction of the first equation from the second gives:

$$\log_e \left(\frac{K_{T_1}}{K_{T_2}}\right) = \frac{\Delta H}{R}\left(\frac{1}{T_1} - \frac{1}{T_2}\right)$$

this equation is the *Arrhenius equation* and can be written:

$$K_{T_2} = K_{T_1} \exp\left\{\frac{\Delta H}{R}\left(\frac{1}{T_1} - \frac{1}{T_2}\right)\right\}$$

or

$$K_{T_2} = K_{T_1} \exp\left\{\frac{\Delta H}{RT_1T_2}(T_2 - T_1)\right\} \quad (8.1)$$

Equation (8.1) shows that the rate of a chemical reaction is an exponential function of temperature but that the rate constant itself varies with temperature.

The *van't Hoff coefficient* (Q_{10}) favoured by biologists, derives from the Arrhenius equation. It is the increase in the rate of reaction, expressed as a multiple of the initial rate for a 10 °C increase in temperature:

$$Q_{10} = \frac{K_{T_1} + 10}{K_{T_1}} = \exp\left\{\frac{\Delta H}{R}\frac{10}{(T_1 + 10)T_1}\right\} \quad (8.2)$$

In most biological applications the value of Q_{10} is about 2.5, which means that a 10 °C increase in temperature increases the rate of reaction 2.5-fold.

8.2.3 *Body temperature and minimal metabolism*

The equation which Robinson and his associates employed to analyse the joint effects of temperature and body weight on metabolism was:

$$O_{2(MIN)}/W = \mathrm{a}W^{\mathrm{b}}\,\mathrm{e}^{\mathrm{c}T} \tag{8.3}$$

where $O_{2(MIN)}$ is the volume of oxygen consumed per hour (l/h); T is body temperature (°C); W is fresh body weight (g); and a, b and c are constants. Using a heat equivalent of oxygen of 20.1 kJ/l (see Section 2.2) the values they obtained for the relationships between metabolism, H_{MIN} (W) and body weight (kg) were, for homeotherms:

$$H_{MIN} = 15.26\ W^{0.79}\ \mathrm{e}^{0.087} \tag{8.4}$$

and for poikilotherms:

$$H_{MIN} = 0.104\ W^{0.76}\ \mathrm{e}^{0.051} \tag{8.5}$$

The powers of weight to which metabolism was proportional did not differ from the standard 0.75 and if body temperature for homeotherms is taken to be 38 °C, then the rate of metabolism is 3.56 W/(kg $W^{0.79}$). If the body temperature of poikilotherms is taken to be 20 °C, their metabolism is 0.202 W/(kg$^{0.76}$). The value for homeotherms is similar to that which Hemmingsen obtained (4.1 W/kg$^{0.75}$) and close to the value given by the original Brody–Kleiber relationship (3.34 W/kg$^{0.75}$). Additionally, the exponential terms in the equations show that the van't Hoff coefficient for homeotherms is 2.4 and that for poikilotherms, 1.7. The value for homeotherms is in accord with expectation, but that for poikilotherms is low, a result attributed to errors of estimation of the coefficient.

These results are of interest since, despite the uncertainty about the temperature effect in poikilotherms, they show that even if their body temperatures were identical, poikilotherms have a considerably lower metabolic rate than homeotherms. This was first shown experimentally by Krogh (1916) when he compared the metabolism of a curarised dog, in which body temperature falls to about 20 °C, with that of poikilotherms; major differences in metabolic rate still existed. This work was confirmed by Benedict (1932). Furthermore, there is evidence that in the primitive Echidnas which have low body temperatures, increase of body temperature to 38 °C does not result in metabolic rates similar to those species which regulate their body temperature at this higher level.

Table 8.2. *The minimal metabolism of some taxonomic groups of animals expressed per kilogram metabolic body size (kg $W^{0.75}$)*

Group	Minimal metabolic rate	
	$W/W^{0.75}$	$kJ/(W^{0.75}\ d)$
Poikilotherms		
Tuatara	0.19	16
Lizards	0.40	35
Snakes	0.48	41
Crocodiles	0.29	25
Turtles	0.15	13
All poikilotherms (Robinson *et al.* 1982)	0.20	17
Homeotherms, mammals		
Echidnas	0.92	79
Polyprotodont marsupials	2.56	221
Diprotodont marsupials	2.26	195
Rodents	3.21	277
'Hot' shrews	8.96	774
'Cold' shrews	3.87	334
Primates	3.36	291
Artiodactyles	3.31	286
Perissodactyles	3.46	299
Edentata (sloths)	2.66	230
Bats	0.87	76
Homeotherms, birds		
Non-passerine	3.87	334
Passerine	7.07	661
Brody–Kleiber relationship	3.34	289
All homeotherms (Robinson *et al.* 1982)	3.56	308

Sources: Derived from various sources including Poczopko (1971); Blaxter (1972); Hulbert (1980); Vogel (1980); Peters (1983).

8.2.4 *The minimal metabolism of different species*

There is little doubt about the validity of using $W^{0.75}$ as a scaling factor when comparing species. There is equally little doubt that there are differences between taxonomic groups or perhaps other groupings of animals in terms of their metabolism/(kg $W^{0.75}$). Table 8.2 summarises a number of estimates of minimal metabolism within taxonomic groups of both poikilotherms and homeotherms, enlarging on the data in Figure 7.2. None of these estimates have been corrected to a standard body temperature. Results for cetaceans and pinnipeds are not included since there is some uncertainty about whether, as was once accepted, these aquatic mammals have a high metabolism (Lavigne *et al.* 1986).

Even such groupings do not encompass all the between-species variation.

Thus Shkolnik (1980) has reported that among the canids the Fennec Fox (*Fennecus zerda*) has a very low metabolism when compared with domestic species of dog and the Bedouin goat a low metabolism compared with the Mediterranean goat.

Figure 8.3 compares the fasting metabolism of domesticated sheep and cattle determined under standard conditions. Sheep have a characteristically low metabolic rate compared with the ungulate mean and cattle a relatively high one. Within cattle breeds, those selected for beef appear to have a lower metabolic rate than those selected for milk production. Racial differences in the metabolism of man are discussed later. These differences in metabolic rate as between different species and strains of animal raise many questions of evolutionary significance. Interesting discussions of them are given by Schmidt-Nielsen (1985).

8.2.5 *Why the three-quarters power of weight?*

The Brody–Kleiber relationship is an empirical one; the fact that metabolism is proportional to the $W^{0.75}$ is an observation owing nothing to theoretical considerations. Understandably, however, attempts have been made to provide a theoretical basis. The surface area law, which suggested that metabolism was proportional to the $W^{0.667}$ was similarly supported by a number of theoretical considerations and these were refuted in a classic review by Kleiber (1947). Most of the explanations suggested for the 0.75 power have been reviewed and similarly criticised (Calder 1981; Economos 1982; Peters 1983). It suffices to state that some explanations propose that metabolic rate is composed of two components, one varying with $W^{0.667}$ and the other directly with weight. A variant of this is the suggestion that one component is mass-related and the other weight-related (that is, dependent on the gravitational field). Some studies suggest that statistical artifact is responsible, due to the mixing of records, or failure to take into account circadian variation in metabolic rate. Other approaches use methods of dimensional analysis to argue that time has a dimension of length and that an operational time-constant should be applied to the argument, based on geometry, that surface is proportional to the two-thirds power of volume (Günther 1975). One explanation, which has received much support, stems from analysis of animal shape and the resistance of the animal to bending and buckling strain (Rashevsky 1960; McMahon 1973). This suggestion states that animals are 'designed' not to be geometrically similar but to be elastically similar, so as to withstand mechanical stress. The dimensional argument in this instance infers that the weight of an animal (assuming constant density) should be proportional to the fourth power of its length rather than to the third power. This could be true if larger animals were

Fig. 8.3. A comparison of the rates of minimal metabolism, in this instance fasting metabolism, of sheep and cattle and their relation to the inter-species norm for artiodactyles shown in Fig. 8.2. Sheep have a low metabolic rate compared with other artiodactyles and cattle a high one. Breeds may also differ: ○ refers to Ayrshire cattle and ● to Aberdeen Angus.

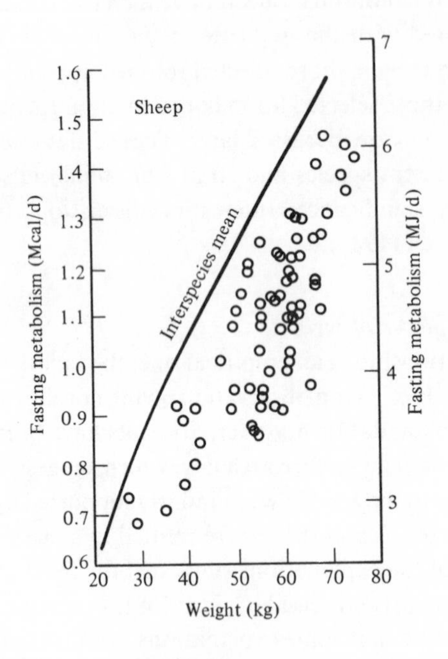

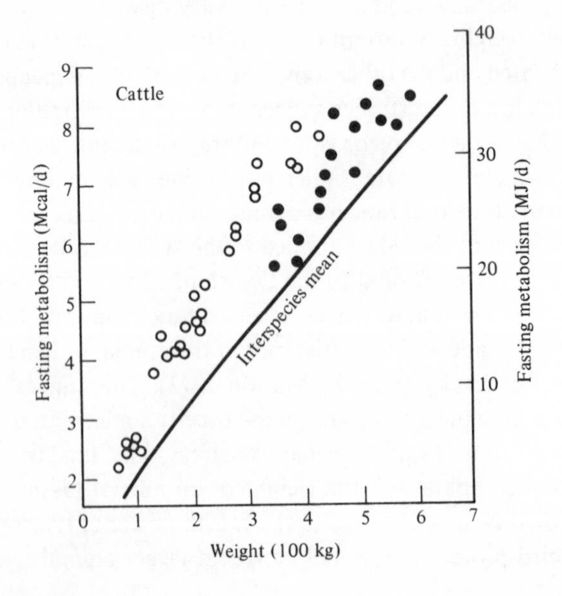

shorter and dumpier than small ones, but statistical analysis of the relation between weight and body length shows insufficient departure from the third power – the value of the exponent of length to which the weight of land mammals is related is 3.23 (Economos 1981).

Whether these explanations are plausible and intellectually satisfying or not, the fact remains that the three-quarters power is firmly based on observation, and that the Brody–Kleiber relationship for adult homeotherms represents a broad description of their metabolism as a function of their size. It could be argued that it is more logical on dimensional grounds to employ $W^{0.667}$ as the first approximation to the effect of size, thus increasing the variation in the coefficient relating metabolism to scaled size (Heusner 1985). From the observed relationship between metabolism and size, however, all that this approach means is that the coefficient then includes a term $W^{(0.75-0.67)}$. Metabolic body size, defined as $W^{0.75}$, provides a basis on which to compare the metabolism of a particular species with that of all other species, or, where the metabolism per unit metabolic size of a particular taxonomic groups is known, with the average and size-independent metabolism of that group.

8.3 Time of day, time of year and minimal metabolism

When Benedict (1938) prescribed the conditions necessary to measure basal metabolism in man – a prescription from which definition of conditions for measuring minimal metabolism in other species has stemmed – he did not mention circadian or circannual variation in metabolism. There is little doubt that there are changes in overall metabolism associated with circadian changes in activity and body temperature. In rodents which are nocturnal, body temperature and metabolism both show a diurnal rhythm with highest values at night. A similar but reverse relationship is seen in most birds (Aschoff & Pohl 1970). These relationships could be due to activity, but Heusner (1956) showed that only a small proportion of the variation arose from this cause. An intrinsic rhythm of body temperature and of the conductance of heat from the body core to the skin (see Section 10.2.4) appear to be the determinants of the circadian changes in minimal metabolism (Aschoff 1981).

In man minimal metabolism occurs in the small hours of the night even in the absence of sleep. An early example of this observation was in a study by Passmore *et al.* (1955) of the metabolism of young men. They were given food sufficient to maintain weight and their last meal at 7.00 pm. Metabolism was measured by mask methods with the subject in bed and awake at 11.00 pm, 3.00 am, and at 7.00 am. The last measurement certainly can be regarded as a measure of basal metabolism. Metabolism measured at 3.00

am, which because of the shorter period of fast might have been expected to be higher, was in fact lower by 12%.

There is also good evidence of a seasonal component affecting the minimal metabolism of many species – quite apart from those which hibernate. Measurements of metabolism made in wild species usually show a reduced metabolism in those captured at times of the year when food supplies are low. In birds, separate equations have been published relating minimum metabolism to body weight during summer and winter (Kendeigh, Dol'nik & Govrilov 1977). Similar seasonal effects have been noted in some mammals which have low metabolic rates in winter. These findings could reflect depression of metabolism by under-nourishment during the winter. However, an analysis of metabolism experiments with sheep, in which the feeding level was at maintenance over several years, showed that in this species there was a seasonal component, well described by a sine-curve with an amplitude of ±14%, with a maximum at mid-summer and a minimum at mid-winter (Blaxter & Boyne 1982). Direct experiments with red deer (*Cervus elaphus*) further showed that metabolism was associated with an imposed light cycle (Argo & Smith 1983). In man there is some early evidence (Gustafson & Benedict 1928) that basal metabolism tends to be lower in winter than in summer, but little more recent information.

8.4 The effect of age

The metabolism of the newborn mammal is characteristically low at birth and increases rapidly thereafter. This postnatal increase in metabolic rate has been noted in rats (Taylor 1960), dogs (Crichton & Pownall 1974), pigs (Mount & Rowell 1960), lambs (Dawes & Mott 1959), cattle (Roy, Huffman & Reincke 1957) and man (J.R. Hill & Rahimtulla 1965). Metabolism immediately after birth is considerably higher than that estimated from the oxygen consumption of the foetus *in utero* immediately before birth. The postnatal increase appears to occur whether or not the newborn animal receives an initial meal. Body weight shows relatively minor change during this period and scaling by body weight does not affect the finding.

Subsequent to this postnatal increase, minimal metabolic rate, expressed per unit metabolic weight, is usually higher than that noted in the same species at maturity. As a generalisation it can be stated that within a species it is common for the relation between minimal metabolism and body weight to vary with a power of weight less than the 0.75 which relates to adults. This arises because the younger and smaller animal has a higher metabolic rate per unit metabolic size. Examples of the relationship between the metabolism of growing animals and their body weight have been fully documented by Brody (1945).

As animals which have reached their mature size increase in age, their metabolism tends to fall. This phenomenon has been most studied in man. Concomitantly there is usually an overall decrease in body weight and certainly in the weight of the non-fatty constituents of the body. When comparisons are made between old people and young people of equal height and stature, men aged more than 60 y are found to have a basal metabolism about 18% less than have those aged 18–30 y (FAO/WHO/UNU/ 1985). In sheep (which reach old age at about 10 y) there is a distinct fall in metabolism per unit metabolic size with advancing age after maturity is reached at about 2 y (Blaxter 1962a). Results with rats also suggest a reduction in metabolism at advanced age.

8.5 Nutritional status and minimal metabolism

The length of fast employed in measures of basal metabolism is that which suffices to obliterate effects of previous food on metabolism. If fasting is continued, heat production continues to fall. So too does body weight, but there is little doubt that metabolism falls at a faster rate. The classic studies of Benedict (1915) on men showed declines in metabolism per unit surface area of about 25% in 30 d of fast. Declines of even greater magnitude have been noted in rats fasted for 15–40 d (Benedict, Horst & Mendel 1932; Kleiber 1961). In calves fasted for up to 4 d, metabolism per unit surface area also declined continuously. There is some evidence that in species which under natural conditions fast for long periods metabolism is not necessarily depressed by continued fasting. This has been noted in the American badger (*Taxidea taxus*) (Harlow 1981) and in the Amazonian manatee (Gallivan & Best 1986).

Continued starvation is the most extreme under-nutrition that can be imposed. Less extreme under-nutrition also depresses the minimal metabolism. In man, metabolic studies to illuminate the consequences of severe food shortages in Europe were undertaken by Benedict and his coworkers in 1919 when the USA entered World War I and by Keys at the University of Minnesota at the end of World War II (Keys *et al.* 1950). Both studies involved giving young men semi-starvation diets continuously for periods of weeks. Both showed a decrease in basal metabolism. In Key's study body weight fell by 24%, and oxygen consumption by 39%. Expressed per m^2 of body surface the fall in metabolic rate was 31%; expressed per kg of weight, the fall was 19%. Apfelbaum (1973) summarised 24 experiments in which human subjects had been given amounts of food which resulted in weight loss; in all cases basal metabolic rate was reduced. Similar falls due to chronic under-nutrition have been found with other animal species. Marston (1948) fed sheep for 10 weeks about half the amount of feed necessary to maintain them such that they lost 2 kg fat and 1 kg protein and became weak and incapable of sustained effort. These animals were then fasted and their metabolism measured on the fourth day of fast. Their metabolic rate was 231 kJ/($W^{0.75}$ d); when the same sheep were studied after having received a maintenance diet for 10 weeks their metabolism was 265 kJ/($W^{0.75}$ d). The fall due to under-nutrition was 13%. With cattle, studies by Benedict & Ritzman (1923,1927) showed a diminution in fasting metabolism after sub-maintenance feeding. Results with rats (Will & McCay 1943) show a comparable depression. The most extreme experimental under-nutrition that has been imposed was by McCance & Mount (1960). Baby pigs were under-nourished for a year; they did not grow at all and instead of weighing 150 kg weighed only 5 kg. Their metabolic rate was 45% of that of normal pigs of equal body weight at the end of the first month and subsequently rose but was always less than that of a normal animal. There is thus a general concordance that under-nutrition for long periods depresses minimal metabolism as determined under standard conditions.

The question arises whether over-nutrition elevates minimal metabolism. Marston (1948) found that sheep which had been given twice the amount of feed necessary to maintain their weight for 10 weeks and then fasted, took considerably longer to reach the post-absorptive state than did sheep fed at the maintenance level. Measuring minimal metabolic rate during a period of over-nutrition may well necessitate a longer than normal period of fasting to remove effects of diet. Even so, Marston found that when a stable level of metabolism was reached it was somewhat higher than that of animals fed a diet sufficient only for maintenance. Garrow's (1978) analysis of many experiments on the effects of overfeeding on the resting metabolism of man failed to reach an unequivocal conclusion but suggested that any increase in minimal metabolism was only apparent following considerable increases in intake. Since resting metabolism was determined after the same length of fast in overfed subjects as in controls, it is possible that any increase observed reflected failure to reach the post-absorptive state. Such a criticism cannot be levied against the experiments of McCracken & McNiven (1983) with rats which had reached stable adult weight and were then force-fed additional food – a procedure that resulted in very considerable gains in weight. The fasting heat production was measured 24–48 h after the last meal. The experiment showed no effect of previous overfeeding on metabolism; fasting heat production was constant at 317 ± 24 kJ/(kg $W^{0.75}$ d). Thus it appears that, provided the measurement is made in the true post-absorptive state, over-nutrition does not significantly increase minimal

Table 8.3. *The effect of 10 weeks of over- or under-nutrition on the fasting metabolism of young pigs*[a]

Breed or strain	Previous nutrition Low	Medium	High
Duroc × Yorkshire			
Weight (kg)	40.8	41.0	40.3
Fasting metabolism (MJ/d)	4.51	5.43	6.36
Selected obese pigs			
Weight (kg)	40.8	41.6	39.6
Fasting metabolism (MJ/d)	5.57	6.29	6.98
Selected lean pigs			
Weight (kg)	40.2	42.0	41.6
Fasting metabolism (MJ/d)	5.26	7.34	8.04

Note:
[a] From Koong *et al.* (1982, 1983). The fasting metabolism in the experiment with Duroc × Yorkshire pigs was estimated by extrapolation. In the remaining experiments the pigs were fasted for 30 h before the measurement. The minimal value for the fasting metabolism of a pig weighing 40 kg is obtained after a 96 h fast and can be expected to be 5.2 MJ/d.

metabolism, suitably scaled by an attribute of body size. This conclusion is further supported by the analysis of fasting metabolism data for sheep; the amount of feed consumed immediately before a fast had a significant effect on fasting metabolism but the amount consumed over the preceding 1–2 months did not (Graham *et al.* 1974). The same conclusions can be drawn from the experiments of Ferrel, Koong & Nienaber (1986).

8.6 Body composition and metabolism

Many of the experiments conducted to ascertain whether under- or over-nutrition affect minimal metabolism are complicated by the fact that underfed animals are smaller and overfed ones larger than the norm. To make comparisons metabolism has to be referred to body weight or some function of it. Koong *et al.* (1982) and Koong, Nienaber & Mersmann (1983) circumvented this problem. They studied young pigs which were all of the same body weight but had reached this weight by different means. One group (MM) had grown continuously, a second (HL) had first been given excess food and then a reduced amount leading to weight loss, and in a third group (LH) this pattern was reversed. The results of their experiments are summarised in Table 8.3 and show that fasting heat production reflected the amount of feed given before the fast. It was least for those which had lost weight in the final phase. It is doubtful whether the fasting values represent metabolism in the post-absorptive state because the fasts were shorter than those found necessary to estimate minimal fasting metabolism in the pig. Even so, Koong *et al.* (1982) point out that the animals which had previously received the highest feed intake and had the highest metabolism also had the highest weights of metabolically-active organs – liver, gut and kidney.

This suggestion that the rate of minimal metabolism reflects the proportional contributions the organs make to the whole body, may be regarded as an extention of the concept that metabolic rate should ideally be related to what has been termed *active body mass* (Miller & Blyth 1953). This term is not precisely defined; it must cater for the continuous gradation of the rates of metabolism of unit weights of tissues. These range from the relatively inert bone, through depot fat and resting muscle, to the highly active liver, gut and kidney. Miller and Blyth recognised the difficulty and suggested that metabolic rate should be referred, not to some power function of body weight, but to lean or fat-free mass an an index of active mass (see Section 5.1.2). The suggestion does not imply that body fat is metabolically inert, but rather that there is a difference between the metabolic rates of fat and non-fat components. In a series of experiments with 48 college students who had body fat contents, estimated from their specific gravity, varying from 3 to 44%, they found better relations of basal oxygen consumption with lean body mass than with body weight or body surface area (Miller & Blyth 1953). N. McC. Graham (1967), working with sheep containing from 7 to 32% fat, also found a close relationship between metabolism and lean body mass. The relationships found for man and sheep were:

$$\left.\begin{array}{ll}\text{Man:} & \dot{H}_{\mathrm{B}}=711+97L \\ \text{Sheep:} & \dot{H}_{\mathrm{F}}=209+105L\end{array}\right\} \tag{8.6}$$

Where $\dot{H}_{\mathrm{B}}$ is basal metabolism (kJ/d); $\dot{H}_{\mathrm{F}}$ is fasting metabolism (kJ/d); and L is the lean body mass (kg). It is of interest that the two coefficients of lean mass are similar for these two species of similar size, but the intercepts differ – possibly reflecting the low metabolic rate of sheep relative to other homeotherms. Within each species there is evidence that differences in metabolism due to sex and age can be predicted reasonably well from the one general relationship, suggesting that simply by allowing for crude differences in body composition much of the differences in metabolism within a species can be accounted for. It is, however, obvious that the assumption of constancy of metabolism per kg of fat-free body mass cannot account for the large between-species differences in metabolic rate in the animal kingdom. Nevertheless it is of interest to examine the contribution made to overall oxygen consumption, measured in the basal state, by individual organs and tissues.

8.7 Organ and tissue metabolism and body size

Several attempts have been made to assess the contribution made by different organs to overall oxygen consumption. Some have been based on measurements of blood flow and arterio-venous differences in oxygen tension, others on the product of organ weight and its *in vitro* oxygen consumption. Virtually all of these assessments do not add up to give total measured oxygen consumption (Grande 1980); it seems highly probable that this reflects technical problems rather than a biological reality. They nevertheless show that the metabolism per unit weight of liver, kidney and alimentary tract is higher than that of the body as a whole, while that of depot fat and the skeleton is much lower. The *in vitro* metabolic rate of depot fat is certainly not zero as might be inferred from some of the arguments adduced to explain differences in metabolic rate (Heusner 1985).

Several attempts have also been made to examine whether the decline in metabolic rate in mature animals of different species with increase in size is reflected in the metabolism of their tissues measured *in vitro*. These are again complicated by the difficulty of measuring the metabolism of tissues *in vitro* under conditions which might be thought to simulate metabolism *in vivo*. The investigations made by Krebs (1950) embraced nine species varying in weight from the 21 g mouse to the 725 kg horse. The fasting metabolism of the mouse, estimated from Brody's general relationship, 295 kJ/(kg $W^{0.75}$ d), is 776 kJ/(kg d) while that of the horse is 56 kJ/(kg d), a range of 14-fold. The range in oxygen consumption per gram of liver in these species was only 5-fold and that of brain 2-fold. Thus there was a tendency for tissue metabolism to decline with increase in body size but not at the same rate as total metabolism. Investigations on the nucleated erythrocytes of birds (Girard & Grima 1980) gave similar results to Krebs; oxygen consumption per litre of cells in birds ranging in size from the 7 g waxbill (*Estrilda* sp.) to the 16 kg turkey (*Melleagris gallopavo*) declined with $W^{-0.10}$, a considerable difference from the power -0.25 linking overall oxygen consumption/unit weight to body weight. In carp (*Cyprinus carpio*) ranging in weight from 0.5 to 1100 g, metabolism of tissue per unit weight declined with $W^{-0.06}$ to $W^{-0.18}$ (Oikawa & Itazaway 1984); that is, with powers higher than the expected -0.25. In this species, however, whole-animal metabolism per unit weight declines with $W^{-0.16}$. Generally, it appears that in accounting for body-size-related variation in metabolic rate, account has to be taken of the proportional contributions of different tissues to total mass as well as to the metabolic rate of the tissues concerned.

There is somewhat better agreement between total metabolism and the concentration of cell organelles and enzymes which are closely concerned with providing energy to the cells of the body. Thus, between species, the concentrations of cytochrome C and cytochrome oxidase correlate closely with metabolic rate (Drabkin 1950; Jansky 1961), and mitochondrial density in muscle correlates closely with maximal metabolic rate (Mathieu *et al.* 1981). In liver the total number of mitochrondria relates to $W^{0.72}$ (Schmidt–Nielsen 1984). In the summated animal tissues, the surface area of the mitochondrial membrane is proportional to $W^{0.76}$ (Else & Hulbert 1985). These relationships are understandable since the mitochondria and their constituent cytochromes are concerned with primary relations involving oxygen consumption.

8.8 Functional explanations of metabolic rate

Baldwin *et al.* (1980), in an analysis of the origin of heat produced in the basal state, regarded the energy expenditure of the whole organism as consisting of two types of function. The first includes 'service functions' necessary for the organism as a whole and exemplified by expenditures by the heart and lungs in providing oxygen, by the kidney in removing waste products and by the central nervous system in integration. The second consists of 'cellular maintenance functions', a term which includes ion transport and the turnover of cellular protein, lipid and other constituents. Service functions accounted for 36–50% and cell maintenance functions 40–56% of total energy expenditure. The components of this model have been discussed by Milligan & Summers (1986); they may not be independent of one another and at present cannot be defined with sufficient precision to allow an explanation of differences in metabolism between species of different body size.

Other types of analytical approach can be employed to explain the origin of heat production in the basal state. One is to separate the 'internal' work done by the body from the heat which arises during metabolism to furnish the free energy to undertake this work. Internal work consists of pressure–volume work including the work of the heart and lung, the synthesis of new chemical compounds from monomers and their subsequent breakdown, i.e. 'turnover', and maintenance of electrochemical gradients between the cell and its extracellular environment and between cell organelles and the intracellular environment. This last work is exemplified by the work done in secretion and by the various 'ion pumps' at cell membranes. These work terms, which are work in the thermodynamic sense, can be calculated. The work of the heart can be estimated from pressure–volume time integrals for both ventricles. The work of synthesising new polymers can be calculated as the free energy of the bonds formed, the free energy of their spatial orientation and the rate of their formation, while the work of maintaining electrochemical gradients can be estimated from ion fluxes and the free energy change estimated from concentration gradients. No work is done on the environment; it is all internal to the body and in the steady state appears as heat and entropy formation. Thus pressure–volume work appears as fluid friction and the work of forming new bonds and configurations appears as heat when the compound is denatured and hydrolysed. Such calculations involve assumptions and extrapolations but suggest that the work done by the body in a fasting resting state is small. In man it probably accounts for 10–15% of the basal heat production (Agricultural Research Council & Medical Research Council 1974). It follows that the major part of the heat arises from the provision of free energy to accomplish this work. The oxidation of the organic constituents of the body to provide ATP and reduced coenzymes, and their participation in the transductions which accomplish work, appear to be the major source of heat. Some of these aspects were discussed in more detail in Chapter 6.

Within such a conceptual framework, between-species differences in metabolism could arise from differences in the amount of work done. There is evidence that this is probable. The work done by the heart and lung in providing oxygen to the tissues obviously relates to tissue metabolism. The minute volume of the heart is proportional to a power of body weight close to 0.75 and mean arterial blood pressure appears to be constant in mammals, irrespective of their size, at about 100 mm Hg. Total work done by the heart is thus proportional to $W^{0.75}$. This simply reflects the fact that the supply of oxygen has to meet metabolic rate; it does not explain why metabolic rate per unit mass is high in small animals.

There is some evidence that the work done in maintaining posture by isometric contraction of the muscles is greater per unit mass in small animals than in large ones (see Section 9.2.3.a). While this could be a factor accounting for the variation in metabolism with size, it can hardly be supposed that it is of major significance in determining minimal metabolism which presupposes muscular repose. Work done by the cells of the body must be of major significance, that is, work done in chemical syntheses and in maintaining electrochemical gradients at cell membranes. The turnover of protein in the bodies of animals (see Section 6.4) of different species is shown in Table 8.4 (Reeds & Harris 1981). It appears that about 15 g protein are synthesised per kg $W^{0.75}$/d, irrespective of size. Thus, as between species, protein synthesis appears to be proportional to the mini-

Table 8.4. *Body protein synthesis in adult mammals and in immature mammals given only sufficient food to maintain body weight*

Species	Weight (kg)	Protein synthesis	
		g/d	g/(kg $W^{0.75}$d)
Rat	0.2	5.6	18.8
	0.35	7.7	16.9
	0.82	11.1	12.8
Rabbit	3.6	33	12.6
Pig	32	268	18.9
Sheep	63	351	15.7
Man	71	328	13.4
Ox	575	1740	14.8

Sources: From Reeds & Harris.

mal metabolic rate. There is little information about species difference in ion fluxes in the whole body. However, Else & Hulbert (1987) have compared the permeability and Na^+,K^+ transport in an Agamid lizard (*Amphiobolurus vitticeps*) and in a laboratory rat. These two species are about the same size, but the overall metabolism of the rat is 5 times that of the lizard. Passive permeability of ^{42}K was measured in tissue slices and in isolated cells showing that the cells of the rat were several-fold more 'leaky' than those of the lizard. The isolated tissues of the rat consumed 2–4 times as much oxygen as those of the lizard and 3–6 times as much of the oxygen was employed for ion transport. Else and Hulbert postulate that the higher metabolism of endotherms than ectotherms is due to 'leakier' membranes which necessitate greater ion pump activity. Whether a similar explanation applies to differences in metabolism related to body size is a matter for further study.

8.9 Variation in minimal metabolism within a species

Provided the measurements are carefully made under standardised conditions, minimal metabolism measured in the same individual on different occasions is highly reproducible as evidenced by coefficients of variation of about ±3%, values which are not much greater than instrumental accuracy. There is a wider variation between individuals of the same species, but here there are often problems of interpretation. Individuals may differ in body mass and the extent of the differences between their metabolic rates then depends on the ways in which corrections are made for size differences. This becomes particularly evident when comparisons are made between

Table 8.5. *Equations relating the basal metabolism (H_B) of men and women (MJ/d) to body weight (kg) for different age classes*

Age group	Sex	Equation
Under 3 years	♂	$H_B = 0.249W - 0.127$
	♀	$H_B = 0.244W - 0.130$
3–10 years	♂	$H_B = 0.095W + 2.110$
	♀	$H_B = 0.085W + 2.033$
10–18 years	♂	$H_B = 0.074W + 2.754$
	♀	$H_B = 0.056W + 2.898$
18–30 years	♂	$H_B = 0.063W + 2.896$
	♀	$H_B = 0.062W + 2.036$
30–60 years	♂	$H_B = 0.048W + 3.653$
	♀	$H_B = 0.034W + 3.538$
Over 60 years	♂	$H_B = 0.049W + 2.459$
	♀	$H_B = 0.038W + 2.755$

Source: Calculated from published data by W.N. Schofield (1985).

obese and normal people. The general conclusion reached by Mitchell (1962) from consideration of many studies in different species is that variation from individual-to-individual within a species is about $\pm 6\%$. His conclusion, as far as man is concerned, is supported by more recent work (Garby, Lammert & Nielsen 1986).

8.9.1 *Prediction of the basal metabolism of man*

The number of clinical determinations of the basal metabolism of man must be many hundreds of thousands. At intervals during the last sixty or more years these have been summarised to provide standards whereby the normality of individuals can be judged. Many of these earlier standards were expressed in terms of metabolism per unit surface area (m^2) per unit time and the source material for them has been compiled by C. Schofield (1985). The most recent analysis has been made by W.N. Schofield (1985) on behalf of a FAO/WHO/UNU committee concerned with human energy requirements. Their use by this committee has been reviewed by Waterlow (1986). W.N. Schofield used over 7000 determinations in his analysis and divided the data into six age groups, separating the two sexes. He then used linear regression methods to relate basal metabolism to body weight. The equations are given in Table 8.5 and the values finally used by the FAO/WHO/UNU committee are summarised in Table 8.6. It is of interest that Schofield found no increase in precision of

Table 8.6. *The basal metabolism (H_B, MJ/d) of adult men and women, classified by age and height*[a]

Height (m)	Weight (kg)	Age in years		
		18–30	30–60	Over 60
Men				
1.5	49.5	6.03	6.07	4.81
1.6	56.5	6.44	6.40	5.23
1.7	63.5	6.90	6.78	5.65
1.8	71.5	7.41	7.15	6.07
1.9	79.5	7.91	7.53	6.53
2.0	88.0	8.49	7.95	6.99
Women				
1.4	41.0	4.60	4.98	4.31
1.5	47.0	4.98	5.19	4.56
1.6	54.0	5.40	5.44	4.85
1.7	61.0	5.82	5.69	5.15
1.8	68.0	6.28	5.94	5.48

Note:
[a] The body weights are the median acceptable ones for the given height. These were arrived at by stating that the acceptable ratio of weight to $(height)^2$ (Quetelet's index) is 22 for men and 21 for women.
Sources: FAO/WHO/UNU (1985).

estimation by including body height in the analysis, suggesting that whether a person of a given age is slim or rotund, his or her metabolism is solely determined by body weight. It was found, however, that both male and female Indian subjects had lower metabolic rates per unit weight than had Europeans. The differences were not small – about 10%. The data used to establish this racial difference are the results of numerous investigations conducted many years ago. Recent work (Shetty 1984) shows that poor Indian labourers do indeed have a basal metabolism which is 17% less than the FAO/WHO/UNU standards, and other studies (McNeill *et al.* 1987) suggest a lowering by 12%. Whether the low basal metabolism in Indians is a true racial characteristic, is due to climatic factors or is the result of depression of metabolism by long-term under-nutrition is not known.

8.9.2 *Minimal metabolism in other species*

No investigations have been made of the metabolism of any animal species on a scale anywhere approaching those relating to the basal metabolism of man. For most species the number of observations is limited and for herbivorous animals these have been summarised by Hudson & Christopherson (1985) in terms of the rate per (kg $W^{0.75}$).

Graham and his coworkers in Australia (Graham *et al.* 1974), however, have undertaken an analysis of a fairly extensive series of measurements of fasting metabolism in sheep, separating effects of body weight, age, and the amount and type of feed given before the fast. They found that metabolism varied $W^{0.78 \pm 0.02}$, indistinguishable from the inter-species power 0.75, and that this term alone accounted for 89% of the variation in metabolism. Age accounted for a further 6% of the variation and the level of feeding immediately before the fast but not the level during the preceding 1–2 months also had an effect. The final equation which they obtained for ruminating (as distinct from suckling) sheep, was:

$$\dot{H}_F = 257\ W^{0.75}\ e^{-0.083A} + 46I \qquad (8.7)$$

where $\dot{H}_F$ is the fasting heat production; W body weight of the shorn fasted animal, A is age in years; and I the pre-fasting intake of metabolisable energy in MJ/d. The residual standard deviation from this prediction equation was 7–8% of the mean.

8.10 Theories and conjectures

It was noted at the beginning of this chapter that to compare the metabolisms of individuals within a species, or make between-species comparisons, the conditions under which metabolism is measured must be standardised. Measurement of a minimal metabolism under a set of reproducible conditions suffices for such purposes. Minimal metabolism is not the only measure that can be used; indeed Prothero (1979) has compared species with respect to the maximal metabolic rate they can achieve over very short periods of time. He found that between species, maximal oxygen consumption varied with $W^{0.75}$ – the same power of weight to which minimal metabolism appears to be proportional.

However, such comparisons must be qualified. Most inter-species analyses employ information drawn from different sources, and it is not certain whether the measurements of minimal – or maximal – metabolic rate are truly comparable from species-to-species. Most of the observations on large domesticated animals represent 24-h determinations of fasting metabolism, and these, because of body movement and circadian variation in metabolic rate, are certainly higher than a measurement of minimal metabolism made using Benedict's conditions. Even so, the concordance of results does suggest that in endotherms, over a range of body mass of 100 000-fold – from mice to elephants – metabolic rate is proportional to $W^{0.75}$. There are deviations from this broad relationship. Some may well be due to differences in measurement techniques. Most are, however, probably due to the physiological, biochemical and behavioural strategies that individual species adopt to deal with the consequences of being large or small. Furthermore, body weight is hardly a precise description of an animal; a mouse and a bat may weigh the same but their body form is very different.

A number of theories accounting for the empirical relationship of Brody and Kleiber, between weight and metabolism have already been discussed. They are not particularly convincing and are not readily extended to explain why some species deviate from the general relationship. In seeking explanations it is perhaps useful to consider the one attribute of animals in which they have close similarity, namely their body temperature. Animals attempt to maintain body temperature, and particularly the temperature of their central nervous systems, within a range of 33–39 °C. Temperatures below an optimum within this range lead to a reduction in the integrative action of the central nervous system, torpor, and a reduction in reaction time to environmental hazard. Temperatures above the optimum have equally adverse effects and the generalisation is true of both endotherms and ectotherms. To maintain stability of body temperature, heat production is set at such a level that heat loss can be facilitated and metabolic acceleration, through the Arrhenius–van't Hoff relationship, avoided. This suggests that the metabolic rates of different species (that collectively lead to the Brody–Kleiber relationship) represent the end results of a series of different solutions to the problem of maintaining temperature stability in a physical environment which, during evolution, has probably been fairly constant in terms of its effects on avenues of heat loss. It should be appreciated that the temperature and radiation limits for sustained plant and animal life are really quite small. Different anatomical and physiological strategies, some of which are dictated by dimensional constraints, have been adopted by different species in reaching such solutions. The real question is perhaps not so much why metabolic rate should appear to vary with $W^{0.75}$, but why central temperature should be set within such relatively narrow limits.

9
Muscular work

The mass of the voluntary muscles accounts for 40–45% of the total mass of the bodies of vertebrates, irrespective of their body size. The musculature is capable of considerable metabolic activity; muscular exertion can, for short periods, increase heat production to values which are more than 10-fold the minimal metabolism. In 1842, at the same time that the first law of thermodynamics was being propounded, Justus von Liebig wrote 'The contraction of muscles produces heat; but the force necessary for the contraction has manifested itself through the organs of motion in which it has been excited by chemical changes. The ultimate cause of the heat produced is therefore to be found in the chemical changes'. This was a prescient remark, and during the ensuing 150 years, a continuous search has taken place to elucidate these chemical changes and the nature of the generation of force by muscle. The fascinating history of these many investigations and the theories of contraction, has been recorded by D.M. Needham (1971) and the search is not complete. While this chapter largely deals with the energetics of muscular contraction in the whole animal, it is essential to summarise the basic physiological and biochemical information which has accumulated from studies on isolated preparations. A full account is given by Bagshaw (1982) and a useful summary by Perry (1985).

9.1 The physiology and biochemistry of muscle

9.1.1 *The structure of muscle*

Skeletal muscle is made up of a number of parallel muscle fibres which run virtually the length of the muscle. The fibres have a surface membrane, the *sarcolemma*, under which lies a sparse number of nuclei. The membrane encloses arrays of *myofibrils* within a *sarcoplasm*. The myofibrils have a characteristic, repeating pattern of crossbanding. Mitochondria are present within the fibre, and in addition there is a sarcoplasmic reticulum of microtubular structure.

Muscle fibres can be classified according to their dependence on different metabolic pathways for the production of ATP. The classification is undertaken by histochemical staining methods to reveal myosin-ATPase activity and the dependence on anaerobic glycolysis. There are two main fibre types. The first consists of the *slow twitch* or *type I* fibres. These have a high oxidative capacity, a low glycolytic capacity, a slow speed of contraction and are slow to fatigue. The second class consists of the *fast twitch* or *type II* fibres. These have a higher glycolytic capacity, a lower oxidative capacity and a fast speed of contraction; they are quicker to fatigue than are type I fibres. The type II fibres are usually subclassified; *type IIA* are fibres in which ATPase activity is poor and greater reliance is placed on oxidative metabolism to furnish ATP. *Type IIB* have a much higher glycolytic activity and their ATPase activity is not reduced significantly at low pH. A single type of fibre can make up whole muscles in some species. In fish, for example, cruising is sustained by recruitment of superficial muscles containing slow type I fibres and the deeper fast fibre muscles are only recruited at faster swimming speeds (Johnston & Goldspink 1973). In land vertebrates there can be similar distinctions but in most muscles the fibre types are mixed. Because of this mixed energy transduction system in muscles, patterns of recruitment of fibres is complex.

The distinction of fibre type is relevant to the function of the muscle. Two types of contraction of muscle can be distinguished. In the first, *isometric contraction*, there is no change in length of the fibre but force is generated. In the second, *isotonic contraction*, the muscle shortens under a constant load and work is done, estimated as force × distance. Maintenance of posture largely involves isometric contraction while movement involves isotonic contraction.

9.1.2 *The contraction process*

The contractile system of vertebrate striated muscle consists of four proteins; myosin, actin, tropomyosin and troponin. These are composed into two filaments within the myofibril, the *A*, or *thick filament*, which consists almost entirely of myosin and the *I*, or *thin filament*, which while largely composed of actin, also contains tropomyosin and troponin. These latter proteins regulate the activity of the whole complex. In smooth muscle this two-filament system is not so precisely organised as in striated or skeletal muscle. In 1954 A.F. Huxley and R. Niedergerke in Cambridge, England and H.E. Huxley and J. Hanson in Cambridge, Massachusetts, proposed the sliding filament model to account for the process of muscular contraction (see H.E. Huxley 1969; A.F. Huxley 1980). The model states that the two filaments slide over each other during contraction through the formation of cross bridges, and that during this process ATP is hydrolysed to ADP and inorganic phosphate. The model has been generally accepted, although there is a number of modifications of it and the precise nature of the conformational changes that lead to the relative movement of the two filaments is still not understood.

In more detail, and commencing with the neural stimulus, the sequence of events in a cycle of contraction and subsequent relaxation is as follows. First the neural stimulus depolarises the sarcolemmal membrane and this leads to depolarisation of the tubular system of the sarcoplasmic reticulum. Calcium ions are released which diffuse to the thin filaments where they bind to the troponin component to make a troponin–Ca complex. This binding has the effect of unblocking the actin sites on the thin filament by tropomyosin. The actin filament can then form cross bridges with the heads of the myosin molecules of the thick filament. These heads, which protrude from the thick filament, have a high ATPase activity and ATP is hydrolysed when the cross bridge 'swivels'. During the relaxation stage the calcium ions diffuse back to the sarcoplasmic reticulum, calcium is released from the troponin–Ca complex and the tropomyosin can thus return to its position blocking the actin sites. The myosin–actin cross bridges break and the ATP-myosin complexes on the heads of the myosin molecules return to the resting state.

9.1.3 *Energy transduction in muscle*

There is now no doubt that the source of energy for muscular work is from the hydrolysis of ATP. Muscle contains very little ATP despite the fact that it is continuously hydrolysed. Its concentration is maintained from the immediate reserve of phosphocreatine, through the action of the enzyme creatine kinase. A secondary source of ATP arises from the action of AMP kinase which catalyses the reaction whereby two moles of ADP form one mole of ATP and one of AMP. This latter reaction takes place when muscles are working near to their physiological limit and when the supply of phosphocreatine is nearly exhausted. As phosphocreatine is depleted it is replaced through the oxidation of the body fuels notably in the first instance the glycogen of the muscles and, as exercise proceeds, the fatty acids of the blood. The creatine liberated when phosphocreatine is dephosphorylated by the contracting myofibril moves to the mitochondrion and is phosphorylated there. This process is termed the creatine–creatine phosphate shuttle (Bessman 1985). When the oxygen supply to the muscle mass is sufficient, the sources of ATP for the phosphorylation of creatine through the reverse of the creatine kinase reaction are aerobic glycolysis and the tricarboxylic acid cycle. When oxygen supply is insufficient, then anaerobic glycolysis is the source and lactic acid accumulates.

It is the exergonic process of ATP formation which is linked to the endergonic ones of muscle contraction. The latter consist firstly of those associated with the actomyosin ATPase of the mechanical event itself and secondly with those related to the Ca^{2+}–ATPase and the Na^+,K^+–ATPase activities involved in the conduction of the neural signal. These components relate to the isolated muscle; in the whole animal undergoing exercise there are additional endergonic reactions required to furnish the energy for the transport of oxygen and energy-yielding substrates to the muscle.

9.1.4 *Muscle and the first law of thermodynamics*

Contracting muscle produces heat and undertakes work. Expressing the energy change in terms of enthalpy, that is, stating that pressure is invariant, the relation between the heat produced by a contracting muscle, the work that is done by it and the enthalpy of the chemical events which take place can be expressed as:

$$\text{heat} + \text{work} = \sum_{0}^{i} \varphi \Delta H_i \qquad (9.1)$$

where the summation is over all 'i' reactions; φ is their extent; and ΔH_i is the enthalpy of each reaction. Whether this equality can be demonstrated experimentally constitutes a 'proof' of the first law. It is similar to the comparison of measurements of the heat produced by the whole animal, estimated from direct calorimetric study, with indirect estimation from knowledge of the substrates oxidised, as discussed in Section 5.4.

The heat produced by a muscle was studied in detail by A.V. Hill & Hartree (1920) who distinguished four phases during a contraction. Their nomenclature has been somewhat superceded. The phases now accepted are;

activation heat, which is thought to be the net result of movement of calcium ions and the activation of the troponin and tropomyosin of the thin filament.

maintenance heat, which is the steady state rate of heat production during an isometric tetanus. This is thought to represent the net heat arising from the actomyosin interactions, the steady state rate of turnover of calcium and its uptake by the sarcoplasmic reticulum.

shortening heat, which is found when a muscle shortens under load, that is, contracts isotonically. This is higher than the maintenance heat measured with an isometric preparation and the difference is thought to be a by-product of movement within the sarcomeres.

recovery heat, which represents the heat arising from active metabolic processes concerned with restoration of the initial state of chemical intermediaries.

Careful experiments in the late 1960s (Carlsson, Hardy & Wilkie 1967; Wilkie 1968) appeared to show that the equality shown in equation (9.1) did apply. Frog muscle kept at 0 °C was poisoned with iodoacetate to prevent lactic acid formation and oxygen was excluded to prevent aerobic generation of ATP. The only source of ATP for contraction could thus be assumed to be from the splitting of phosphocreatine. The heat + work term was initially found to be −42 and later −46 kJ/mol phosphocreatine. This latter value agreed with the earlier determination by Meyerhof & Schulz (1935) of the enthalpy of phosphocreatine breakdown. This was also −46 kJ/mol. It thus seemed that the first law, as exemplified by equation (8.1), applied to isolated muscle. However, Curtin & Woledge (1978) showed that Meyerhof's calculation of the enthalpy change was incorrect and that the true value was −34 kJ/mol phosphocreatine. Subsequent critical work has shown that in experiments in which attempts have been made to isolate the energy source as phosphocreatine, the heat + work term of equation (9.1) exceeds the enthalpy change which can be explained chemically. This does not mean that there is a divergence from the first law; it implies, however, that the summation of chemical changes taking place is so far incomplete and that there are important chemical processes which have not been taken into account. These have been considered in detail by Homsher (1987), particularly the enthalpy changes associated with Ca^{2+} binding in the sarcoplasm. His studies confirm that the discrepancy occurs early in the contraction process and is associated with the excitation–contraction coupling (Kushmerick 1983).

9.2 Muscular work in the whole animal

In summarising his work on muscle, which spanned more than 50 years, A.V. Hill (1965) wrote,

> Work on hard muscular exercise in man provided the same opportunity of getting accurate and reproducible results as is found in experiments on isolated muscle. Indeed in some ways man was the better experimental object, as other people have learnt; when trained he can repeat the same performance again and again. And it remained a matter always of satisfaction, sometimes even of excitement, as the work evolved, to find how experiments on man and those on isolated muscle confirmed and threw light on one another.

These remarks emphasise the importance of cooperation of the subject and of familiarity with the techniques imposed if reproducible results are to be obtained. They must be borne in mind when results with animals are considered. It follows that it is much easier to undertake studies on muscular exercise in man than it is in other species. Understandably there is much more information on man than on other animals.

9.2.1 *Methods of measurement and expression*

Most of the estimates which have been made of the energy expended in muscular work have been obtained from measurements of oxygen consumption or of oxygen consumption and carbon dioxide production. When, in a resting animal, work at a constant rate is undertaken, oxygen consumption does not increase immediately. After a period of 1–2 min, however, a steady state of oxygen consumption is realised which is usually but not invariably a good measure of the metabolism during work. How good a measure it is depends on the extent to which the muscles oxidise anaerobically. Thus if work is severe, at its cessation oxygen consumption is considerably elevated, declining slowly over a period of many minutes to the pre-exercise level. This lag is termed the *oxygen debt* and reflects the recovery process in the muscle in which lactic acid, accumulated in anaerobic muscular activity, is oxidised and creatine is phosphorylated. In these circumstances the energy cost of undertaking work is the sum of that expended during the exercise period itself and the oxygen consumed above the pre-exercise level during the recovery period. A perspective on the magnitude of the oxygen debt is given by an example for man. In an athletic man the respiratory and circulatory system can supply oxygen at the rate of about 5 litres/min. The potential oxygen debt incurred by an athlete is about 15 litres. The maximal amount of oxygen available to the man in 1 min is thus 5 + 15 = 20 litres and in 12 s it is 1 + 15 = 16 litres. In the second case the maximal rate of oxygen supply is 80 litres/min. Very severe exercise can only be undertaken for short periods and at the expense of an oxygen debt.

Whether the measurement of oxygen consumption and hence of heat production is made in the steady state or by integrating observations over a period of work and subsequent recovery, care has to be taken in the interpretation of the results. The overall measurement includes both the heat produced by the work undertaken and that produced in the absence of the work, that is, heat produced in the resting state. A common procedure is to estimate the net cost of muscular activity by deducting from the measured heat production the heat produced in an initial period when no work is undertaken. The energy cost of standing is thus estimated as an increment of heat above that measured when the subject is lying, and the energy cost of forward movement in a land animal is usually measured as an increment above the metabolism when it is standing still. In a flying animal or bird such a device is not possible; the heat produced during hovering flight – comparable to a land animal standing still – is considerably greater than that in forward flight. Some investigators report their results simply in terms of the primary observations on the heat produced during exercise. For example, in the definitive summary of human energy expenditure during exercise by Passmore & Durnin (1967), results are expressed in terms of kJ/min. The values thus include the heat produced when no work is done. The device of deducting or estimating the heat produced in the absence of work to arrive at a net cost of work can produce curious results. For example, it is shown later (Section 9.2.3) that in small species a separation of the heat production when no work is done from the energy cost of moving the body on a horizontal plane gives results which are somewhat difficult to understand.

9.2.2 *Standing*

In man the total heat produced when sitting comfortably with the back supported is only marginally greater than that noted when the subject is lying, and sitting upright on a stool increases heat production by only 3–5%. Standing is more expensive and also more variable. An average value as an increment above lying is 15 kJ/(kg d) and the standard deviation about this mean approaches 45%. The reason for the variation relates to the fact that different postures are adopted by people when standing. The most economical posture is that in which the feet are separated, the knees locked, the vertebral column flexed forward in the lumbar and cervical regions, the arms loose at the sides and the head relaxed forward. This is the posture adopted in fatigue. The least economical stance is that of standing to attention with constant use of muscles to maintain balance. During fatigue, upright posture is maintained by ligamentous bracing and movement of the centre of gravity of the body is kept minimal.

In farm animals most measurements of the incremental cost of standing have been based on regressions of heat production on time spent lying. These have been summarised for sheep and cattle (Agricultural Research Council 1980) to produce a mean value of 10 kJ/(kg d), again with considerable variation – 6 to 12 kJ/(kg d) – between different studies. In the horse Winchester (1943) found no increase in heat production on standing. Horses have powerful suspensory ligaments and can sleep while standing. Work by Toutain *et al.* (1977) throws light on the variation in the energy cost of standing. As shown in Table 9.1, the baseline of rest varied with the state of vigilance. In addition, they showed that the increase in heat

Table 9.1. *The heat production of mature sheep when standing, lying awake, lying drowsily, in slow-wave sleep or in paradoxical sleep*

	Watts/kg		kJ/(kg d)	
Activity	Actual	Increment over lying	Actual	Increment over lying
Lying awake	3.45	—	107.9	—
Standing overall	4.23	0.78	132.4	24.5
in first 5 min	4.45	1.00	139.1	31.2
after 60 min	3.83	0.38	119.9	12.0
Lying drowsily	3.15	−0.30	98.5	−9.4
Asleep, slow wave	3.00	−0.45	93.8	−14.1
Paradoxical sleep	3.00	−0.45	93.8	−14.1

Source: Calculated from Toutain *et al.* (1977).

production on standing included considerable initial components, one due to the act of rising to the feet and others with those muscular activities that take place immediately on rising, such as urination, defaecation, stretching, grooming, scratching and investigatory behaviour. The equilibrium value for standing, discounting these initial activities, was about 12 kJ/(kg d), similar to that found in the ARC (1980) compilation.

Similar variation has been noted when attempts have been made to measure the increase in heat production for wild species. Fancy & White (1985), reviewing such studies, tentatively concluded that the increases in heat production in wild herbivores when standing were probably greater than in domesticated ones. However they also found that it was extremely difficult to isolate a component specifically due to the act of standing since standing was inevitably associated with greater alertness and with minor activities.

When the mean energy cost of standing in man is compared with that in a quadruped of about the same body mass, it does, however, appear that standing in man is the more expensive. There is some evidence from indirect studies (see Section 9.2.3 dealing with walking and running) that the net cost of maintaining posture is greater in smaller species than in large ones. This probably reflects the fact that small animals stand on bent legs and thus cannot economise by 'locking' their joints.

9.2.3 *Walking and running*

The commonly adopted method for estimating the heat production which accompanies running and walking uses a motorised treadmill.

The subject stays in the same place and the apparatus for measuring the gaseous exchange is permanent and static. A variation is to mount the measuring equipment on a motor vehicle which accompanies the animal during walking. Expired air is passed by flexible tubing to the equipment mounted on the vehicle. This approach has been used successfully with horses (Pagan & Hintz 1986). Alternatively, the animal or man can run on a suitable terrain carrying an apparatus which either enables the ventilating air to be measured and samples taken for analysis or is designed to provide a direct record of oxygen consumption.

When animals walk on a treadmill belt, there is usually an attachment between them and the framework of the apparatus. This may be a tether or the suspension of the mask or hood which the animal wears. It could be a backboard behind the animal, which in the instance of some devices for measuring metabolism in small animals, would be the rear wall of a chamber sliding on the moving belt of the treadmill (C.R. Taylor, Heglund & Maloiy 1982). These various attachments can all provide assistance to the animal during walking or running; the animal can drag on its tether or receive a push from the backboard. In most of the published work on locomotion in animals scant mention is made of the possibility of such assistance. Assistance can however be large and real. Boyne *et al.* (1981) measured its extent in sheep and cattle trained to walk on a treadmill. With oxen walking on a gradient of 5° at a speed of 59 m/min, the energy expenditure of walking 1 m, was 6.0 J/(kg m) above that of standing when there was no tension in the tether. Some animals voluntarily developed as much as 150 g tension/kg body weight in the tether. The apparent energy cost of walking fell in these animals to 2.7 J/(kg m), or 45% of the true value. In studies with sheep which were tethered through their respiratory hoods, mean tensions in their tethers increased with both gradient and speed and were present at negative as well as positive gradients. In man, sheep and oxen the effects of tension by connections to the area outside the treadmill belt on energy expenditure when walking are similar. They vary with speed and gradient up to maxima of about 0.04 J/(m g) of tension or negative load. These observations suggest that large errors could arise from such unsuspected assistance and caution must be exercised in the interpretation of experiments in which there is the possibility of such aid.

Total heat production during walking or running increases with speed (expressed here in m/s). In man, many observations have shown that the relationship is curvilinear, there being a disproportionate increase in expenditure at higher speeds. Figure 9.1, adapted from Passmore & Durnin (1967), summarises many observations made during the last half-century on men weighing 70 kg. From the classic study by Benedict &

Murschhauser (1915) on the energy cost of walking above that of standing, the energy cost per kg body weight per metre moved can be calculated to be 2.13 J/(kg m) at a speed of 1.2 m/s (2.6 mph), 2.49 J/(kg m) at a speed of 1.8 m/s (4.0 mph) and 3.93 J/(kg m) at a speed of 2.4 m/s (5.4 mph). At this latter speed, which is very difficult for the untrained person to attain in walking, changing gait from walking to running decreased the energy cost of horizontal motion to 3.40 J/(kg m).

Studies with horses trained not to change their gait when reaching the speed at which they would normally do so, shows that for each of the gaits – walking, trotting, cantering and galloping – the slope of the relationship between heat production and speed is also curvilinear. This is shown, using the data of Hoyt & Taylor (1981), in Figure 9.2. When, however, cognisance is taken of the fact that animals appear to change their gait to minimise the cost of progression at a particular speed, the overall relationship between heat production and speed is almost linear. Observations on migrating African animals, for example, show that they only employ a particular gait for a restricted range of speeds (Pennycuick 1975a). It is this relationship between speed and change of gait that justifies simple linear analysis of the relationship between heat production and speed over a considerable range of speeds.

Fig. 9.1. The rate of heat production of man when walking on the level at different speeds. The symbols refer to results of several investigators as summarised by Passmore & Durnin (1967).

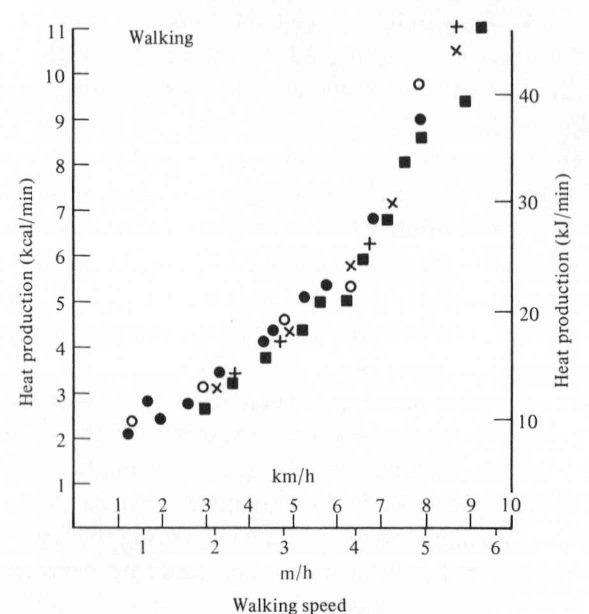

a. Generalisations about walking and running

Until about twenty years ago there was a considerable paucity of information about the energy cost of locomotion in animals other than man or those of agricultural interest. These increases in heat production above the resting, standing position are summarised in Table 9.2. Largely as a result of much investigation by researchers at Harvard University, there is now considerable information about the energy expended in walking and running for a wide variety of species. Seven species were studied by Taylor, Schmidt-Nielsen & Raab (1970); later Taylor *et al.* (1982) included in their analysis 65 species of mammal, ranging from mice to hedgehogs to lions and

Fig. 9.2. The oxygen required by a horse to move 1 m on the level in relation to gait and speed. The horse was trained not to change gait when it would normally do so and each point is total oxygen consumption/min divided by speed. Note that the minimal oxygen consumption/m travelled was virtually the same whether the horse was walking, trotting or galloping. The frequency distributions at the foot of the figure show the gaits chosen by the horse when not constrained to a particular gait. The data are those of Hoyt & Taylor (1981).

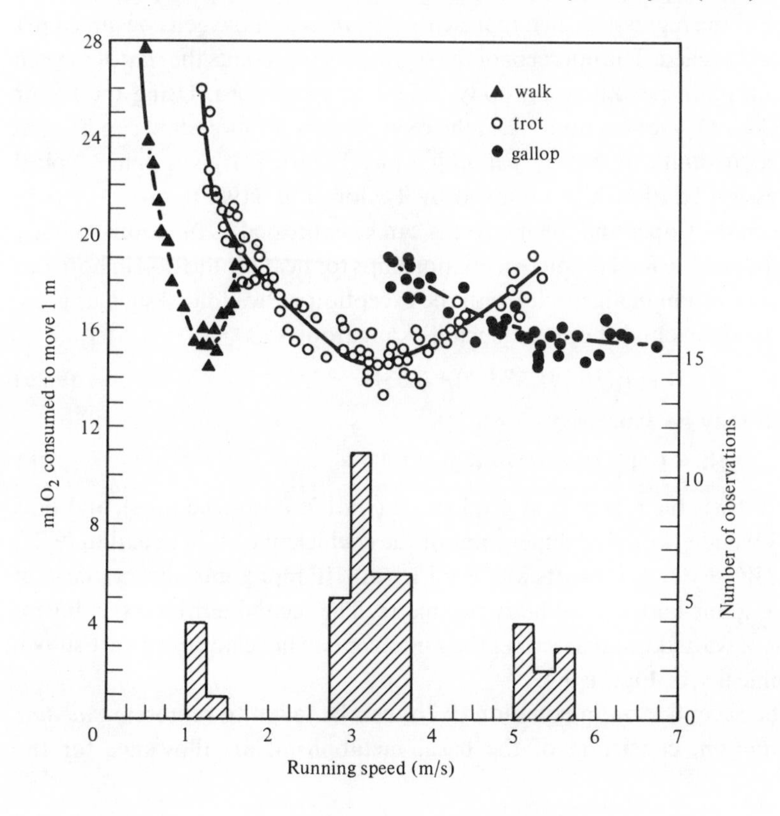

Table 9.2. *The increase in heat production during walking measured as the increase over standing metabolism (J/(kg m))*

	Increase in heat production per kilogram body weight		
	Per metre of horizontal travel		
Species	Separate investigations	Predicted from equation (9.2b)	Per metre travelled vertically
Sheep	2.9	3.1	32
Ox	2.0–2.1	1.6	27
Horse	1.5–1.6	1.4	28
Man	2.4–2.8	2.8	30

Sources: The data were abstracted from Benedict & Murchauser (1915); Hall & Brody (1934); Clapperton (1964a); Hoffmann, Klippel & Schiemann (1967); and Farrell, Leng & Corbett (1972).

gazelles, and 14 species of bird, including ostrich and quail (Taylor & Heglund 1982; Taylor *et al.* 1982).

The approach adopted by these workers was to regress the rate of oxygen consumption per kilogram body weight against speed for each species. The slope of the regression then represents the *amount* of oxygen consumed per metre travelled. The intercept of the regression represents the *rate* of oxygen consumption per kilogram body weight at speed zero. Using the factor 20.1 kJ/l O_2 (see Section 2.2.2) the oxygen consumption data can be converted to units of energy per unit time. Figure 9.3 gives some typical regression relationships obtained by Taylor *et al.* (1982).

Both the slopes and the intercepts can be expressed as functions of body weight and the final predictive relationships for heat production in horizontal locomotion in all species with the exception of 'waddlers' such as geese and penguins and 'hoppers' such as kangaroos was:

$$\dot{H}_P/W = V(10.7W^{-0.316}) + 6.03W^{-0.303} \quad (9.2a)$$

which may be expressed:

$$\dot{H}_P = V(10.7W^{0.684}) + 6.03W^{0.697} \quad (9.2b)$$

where $\dot{H}_P$ is the rate of heat production (watts); V is speed (m/s); and W is body weight (kg). The dimensions of the coefficient of V in equation (9.2a) ($10.7W^{-0.326}$) are (watts/kg)/V = J/kg m). It represents the net cost of moving one kilogram of body one metre. The second term has the dimensions of watts/kg and describes their intercepts. The relationships are shown graphically in Figure 9.4.

The second term in equation (9.2b) can be taken to represent *standing metabolism*, consisting of the basal metabolism, an allowance for the

Fig. 9.3. The relation between the rate of oxygen consumption/kg of body weight and speed of locomotion in some artiodactyls. Note that the relationship for each of the four species appears to be linear.

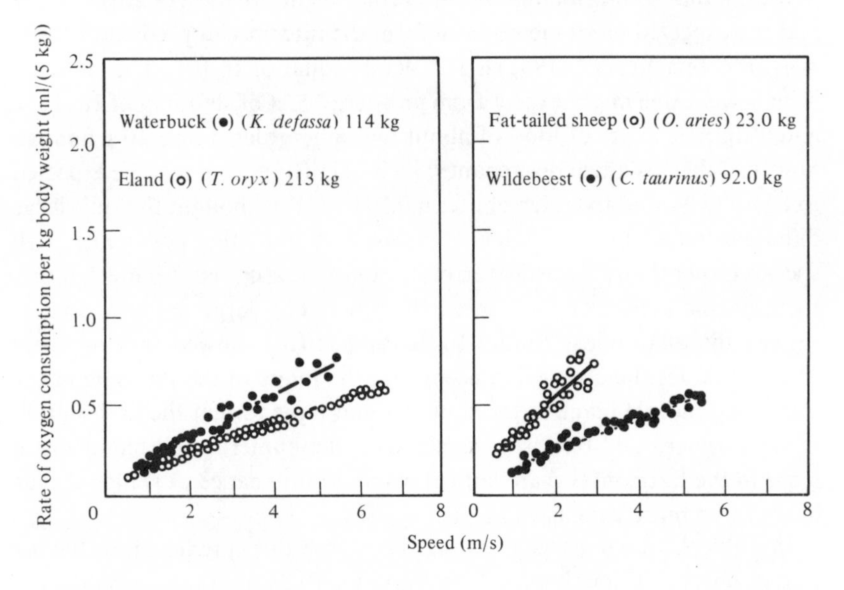

Fig. 9.4. The rate of heat production during horizontal locomotion in animals, computed from equation (9.2b) in the text.

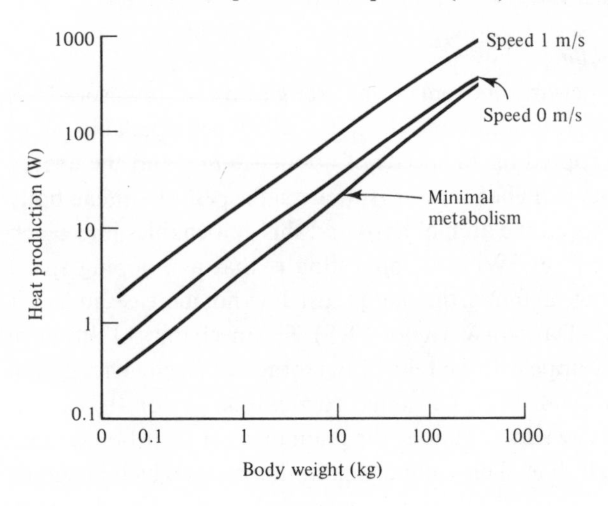

energy cost of standing, and assuming that the animals were fed and not in the post-absorptive state, an amount of heat due to the thermogenic effect of food. For a 100 kg animal the intercept term in equation (9.2b) is 149 W and the expected basal metabolism from the interspecies prediction equation of Robinson *et al.* (1982) is 113 W (see equation (8.10)). The intercept term is what one might expect from an energy cost of standing of 10–15% and a thermic effect of food of about the same order. For a 10 g animal, however, there is a real discrepancy; its basal metabolism may be expected to be 0.113 W and the intercept term 0.243 W. It is thought that this large difference reflects a very high energy cost of maintaining posture in small species rather than a statistical error; it cannot be ascribed to a much larger thermogenic action of food in small rather than large animals. In this respect the anatomical studies by Jenkins (1971) showed that in many smaller species the classical concept that the joints of the limbs lie in the same longitudinal plane as the shoulder and hip and that the limb simply swings forward and backward is not true; the humerus and femur move more in the horizontal than vertical plane. Maintenance of stance is thus likely to be more expensive in such species.

In Table 9.2 a comparison is made between values predicted for the net cost of horizontal locomotion from equation (9.2a) and those observed in extensive experiments with large farm animals and man. Agreement is good over this range of body weight.

b. Hopping and waddling

Kangaroos hop. Forward movement by kangaroos at low speeds is achieved by using the four limbs and the tail – called 'pentapedal locomotion'. This gait is adopted up to speeds of about 1.6 m/s and the energy expended per (kg m) is higher than it is in a quadruped of similar body weight. Above this speed the animal hops and this gait enables it to reach speeds of 15 m/s or more. What is interesting is that as hopping speed increases over the range 2–6 m/s, the energy cost does not increase, indeed it declines very slightly (Dawson & Taylor 1973). The mechanics of this have been examined; they appear to be related to storage of elastic energy (see R.McN. Alexander 1968, 1977; Cavagna, Heglund & Taylor 1977).

Penguins and geese waddle. That is, the main mass of their bodies does not move in a straight line. This can be seen in the tracks which penguins make in snow, for their tails draw a sinusoidal path. Waddling is a slow gait, and for speed on a down-gradient penguins 'toboggan'. Pinshaw, Fedak & Schmidt-Nielsen (1977) made a study of Emperor, Adelie and white-flippered penguins (*Aptenodytes forsteri, Pygoscelsis adeliae, Eudyptula albosignata*) and showed that the oxygen consumed per (kg m) was considerably higher than that of a mammal or bird of similar weight using the normal upright gait. The same high cost is true of geese walking at low speeds. To increase speed while walking these birds use their wings in unison. It may be remarked in passing that there appears to be no difference in the energy cost of locomotion between walking on two legs or on four.

c. The effect of terrain on the energy cost of locomotion

Most studies of walking and running have been under laboratory conditions where the terrain is the smooth surface of the belt of a treadmill or running track. The energy cost is considerably greater if the surface is rough or uneven. In man the energy cost, above that of standing, of walking over a ploughed field is about 50% greater than that of walking on a smooth surface. It may be that the effect of type of terrain on the energy cost of locomotion is somewhat greater in man with his bipedal gait than in quadrupeds. There is evidence, for example, that the increment of heat production noted when a man walks over tundra compared with an asphalt road is proportionately greater than it is for a reindeer (White & Yousef 1978). This can in part be ascribed to the low foot-loading of reindeer and caribou relative to man – man sinks more at each step. Walking in snow is even more energetically expensive than walking on a rough, uneven surface and again the extent of foot-loading is critical. In ungulates, the energy cost of forward locomotion in snow increases exponentially as the depth to which the animal sinks approaches the height of its chest. Thus in white-tailed deer the total energy expenditure in moving a given distance doubles if the sinking depth is 25% of the height of the brisket, triples when the height is 50% and is five times greater when it approaches 70% of brisket height (Fancy & White 1985). Not all animals in a herd incur such large energy costs. One animal usually acts as a 'trail blazer' with its companions following in its tracks.

d. The mechanics of locomotion and its energetic efficiency

The mechanics of locomotion have been studied intensively and excellent accounts are given in works edited by R.McN. Alexander & Goldspink (1977), Pedley (1977) and Elder & Trueman (1980). Briefly, the external work done by an animal moving along the ground can be divided into four distinct components. The first consists of the sum of the kinetic and gravitational potential energy imparted to the animal's centre of mass; the second is the kinetic energy of the limbs and other parts of the body as they accelerate and decelerate during each stride relative to the centre of mass; the third is the elastic potential energy stored during part of each stride and released later; the last is work done against external friction, notably, and on terrains on which the feet do not slip, friction due to air resistance. The first three components are not entirely independent of one another. For example, when the foot is placed on the ground during a stride the decrease in energy due to the slowing of movement of the centre of the body mass can be used to accelerate the limbs in a forward direction, and part of the kinetic and potential energy imparted during a stride is certainly stored in elastic strain energy in the tendons. In treadmill walking, since the animal stays in the same place the component due to air resistance is negligible.

The magnitude of the work terms due to acceleration and deceleration of the centre of mass and the limbs have been determined by high speed photography of animals. These, together with determinations of the masses of the body and its components, allow estimates to be made of the external mechanical work done (Heglund Cavagna & Taylor 1982). The studies by the Harvard group (Taylor & Heglund 1982) showed that the mechanical work done by animals of different sizes could be described by the equation:

$$w_{\mathrm{mechan}}/W = 0.478V^{1.53} + 0.685V + 0.072 \quad (9.3)$$

where w_{mechan} is the work done in kinetic and gravitational terms (watts); W is body weight (kg); and V is speed of locomotion (m/s). The elastic component of the work done is not included in the mechanical work nor is the air friction term, which in experiments with treadmills must be negligible since the subject does not move. It is immediately evident that the mechanical work done in moving the body is directly proportional to its mass and increases curvilinearly with increasing speed.

This equation can be compared with those determined in the same laboratory for the oxygen consumption and hence heat production associated with locomotion (equations (9.2a) & (9.2b)). The first term in equation (9.2a) gives the net cost of moving the body J/(kg m). The ratio of equation (9.3) to this first term is an estimate of the efficiency with which energy derived by chemical oxidation of nutrients during muscular activity is transduced to mechanical work. As pointed out by Taylor & Heglund (1982) the efficiencies so calculated for any particular velocity increase with body mass, and for an animal weighing 100 kg travelling at high velocity, given impossible values of more than 100%. At low body weights the efficiencies are minute and in no way in agreement with studies made on the efficiency of isolated muscle preparations. The very high efficiences found in the larger animals could be due to the storage of kinetic and gravitational energy as elastic energy in tendons and muscles during part of a stride and its release later in the cycle of the stride. This contention would agree with a conclusion reached by R.McN. Alexander (1980) from purely theoretical considerations. No explanation is forthcoming for the incredibly low mechanical efficiencies found in the smaller animals. These studies have forced Taylor & Heglund (1982) to conclude that the rate at which animals oxidise nutrients during locomotion is not determined by the external mechanical work which they can be calculated to do. They point out that this external work can only result from shortening of muscles – that is by isotonic contraction – and that the stabilisation of joints which entails isometric contraction and resistance to stretching is a considerable component of the total muscular activity accompanying walking. The question arises whether more isometric contraction occurs in body movement by a small mammal, and this is not readily answered. Taylor & Heglund (1982) have further suggested that the metabolic cost per unit time of generating force by muscle, irrespective of whether mechanical work is performed, is the determinant of the rate at which heat production increases during locomotion.

e. Walking and running on gradients

When animals walk uphill, work has to be done to move the body vertically against gravity as well as horizontally. Movement on gradients is conventionally expressed as the difference between the observed heat production during ascent and that noted when walking on the level. Expressed as the energy expended per kilogram body weight moved vertically one metre, this is virtually invariant between species at 27 J/(kg m). The work done by the animal in this vertical movement is mass × gravitational acceleration × distance, that is 9.81 J/(kg m). The efficiency with which work of ascent is undertaken is thus 9.81/27 or 36%, and for most purposes can be regarded as invariant with size. This calculation of efficiency by separation of the vertical component from the horizontal component of movement can be criticised. Some of the work done by raising the body's centre of gravity at each step during horizontal locomotion is ascribed to the vertical component by this method of calculation. Indeed, careful study shows that the efficiency changes with gradient. It is, however, apparent that movement up a hill produces less metabolic strain on a small animal than on a large one since the metabolic rate of a small animal is so much greater per unit weight than it is for a large one.

When movement is down a gradient there is assistance from acquired potential energy. By measuring the heat production in animals walking downhill the proportional recovery of this potential energy can be calculated. It amounts to about 50%, with variation occurring as a result of steepness of the gradient because muscular work has to be done to prevent falling. If the animal does fall, the whole of the potential energy is recovered as the heat of impact!

Table 9.3. *The effect of wind on the energy cost of walking and running by man*

Speed of walking or running (m/s)	Wind velocity (v) (m/s)	Actual O_2 consumption (ml/s)	Heat production (Watts)	Increase in energy cost with square of wind velocity (v) (Watts/v^2)
Walking				
1.25	0.0	14.63	294	
1.25	10.0	19.86	399	1.05
1.25	14.1	25.09	504	
2.08	0.0	27.48	552	
2.08	10.0	37.01	744	1.91
2.08	14.1	46.54	935	
Running				
3.75	0.0	47.27	950	
3.75	10.0	54.05	1086	1.36
3.75	14.1	61.83	1243	
4.47	0.0	50.17	1008	
4.47	10.0	61.83	1243	2.52
4.47	14.1	75.41	1516	

Source: Calculated from the results of Pugh (1971).

f. Wind resistance and movement

A number of studies have been made of the energy cost of running and walking against wind. All have been made with man. Such information is highly relevant to the assessment of athletic performance and to the occurrence of exhaustion and hypothermia in those walking in high mountains and moorland where the effort of walking into gale force winds is a contributory factor. Similar work with animals has not yet been undertaken.

The additional work done under such circumstances is the product of the force of the wind (which varies with the square of its velocity and the profile area presented by the subject to the wind) and the speed with which the subject moves. Pugh (1971) approached the problem of how to measure the effect of wind by having his subjects walk or run on a treadmill in a wind tunnel and comparing the increase in oxygen consumption due to wind with that noted when, in the absence of wind, they walked on inclines. Oxygen consumption due to wind could thus be equated with that due to work of ascent, to give an 'equivalent vertical force'. Table 9.3 summarises these experiments and shows that heat production in walking or running is considerably augmented by the presence of wind. It also shows the equivalence of wind expressed as work of ascent. In more simple terms, an athlete running at marathon speeds (4.45 m/s) in minimal wind would consume oxygen at the rate of 3.0 l/min (heat production of 60.3 kJ/min or 1 kW). Running into a wind of 18.5 m/s would increase oxygen consumption to 5.0 l/min, equivalent to a rate of heat production of 1.68 kW. The energy cost of overcoming air resistance in running in still air is probably about 8% of the total for middle-distance runners and up to 15% for sprinters.

9.2.4 *Load carrying and draught*

Much of the interest in the energy cost of carrying loads in man has been for military reasons – how much a soldier can carry and how best the load should be dispersed on his person. The first studies of this nature were carried out by Zuntz & Schumberg (1901) who showed that the increase in heat production was proportional to the total weight of subject plus load. Minor differences were found reflecting how the load was carried and much study has been devoted to designing military packs to ensure an optimal disposition of the load. Similar work has been done with horses and oxen, leading to the same conclusion; disposition of the load results in minor deviation from a direct proportionality to the weight carried. This general relationship was amply confirmed by Taylor *et al.* (1980a) with dogs, rats, men and horses. Their results are shown in Figure 9.5. Oxygen consumption increased in direct proportion to the mass of the load carried, irrespective of speed or size of animal. Load-carrying is not of course limited to beasts of burden and man. Predators remove prey from sites of kills, food is taken by adults to the young in the nest and some hibernators store considerable food in caches for the winter. The European dormouse (*Glis glis*), for example, each year stores 10–15 kg of food for the winter.

Work done by traction in pulling a load can be measured as the product of the force recorded by a dynamometer and the distance travelled. The energy cost of undertaking this work can be estimated by simultaneous measurement of the respiratory exchange. There are problems, however, in making the measurement of force since in many applications in which animals, even when they are well trained, pull loads, the draught force varies considerably, mainly as a result of the alternate forward thrusts of the animal's legs. With poorly-trained animals an even more jerky draught force is observed. Estimate of the work done when periods of observation are short thus show considerable variation. Lawrence & Pearson (1985), for example, found that the draught force exerted by oxen in ploughing in the same field using the same plough varied from 589 Newtons (N) to 2160 N. Oxygen consumption does not exhibit such rapid oscillation.

The earliest work on the energy cost of draught was carried out by Zuntz

& Hagemann (1898). They determined the efficiency with which well-trained horses undertook tractive work. This efficiency is the ratio of the work done measured as the product of dynamometer force times distance divided by the energy cost of undertaking the work. The latter is measured as the difference between heat production walking without load and that when walking pulling the load. The efficiency was 33%. In experiments with man, Lloyd & Zacks (1972) found an efficiency of 36% and Boyne *et al.* (1981) one of about 33%, declining with both increasing load and when the work was performed on an uphill gradient. Knowing the tractive force and distance travelled, the increase in heat production due to pulling loads can be calculated with relatively little error using an efficiency factor of 33%. It may be noted in passing that the efficiency of the performance of work in pulling a load is very similar to that observed in overcoming wind resistance (Pugh 1971; Davies 1980).

9.2.5 *Hovering, flying, gliding and soaring*

Birds and flying mammals exhibit a wide range of modes of flight and an equally wide range of body forms – suggesting evolutionary adapta-

Fig. 9.5. The ratio of the rate of oxygen consumption of animals when carrying a load to that when they were unloaded, irrespective of the speed of locomotion. The results are those of Taylor *et al.* (1980a).

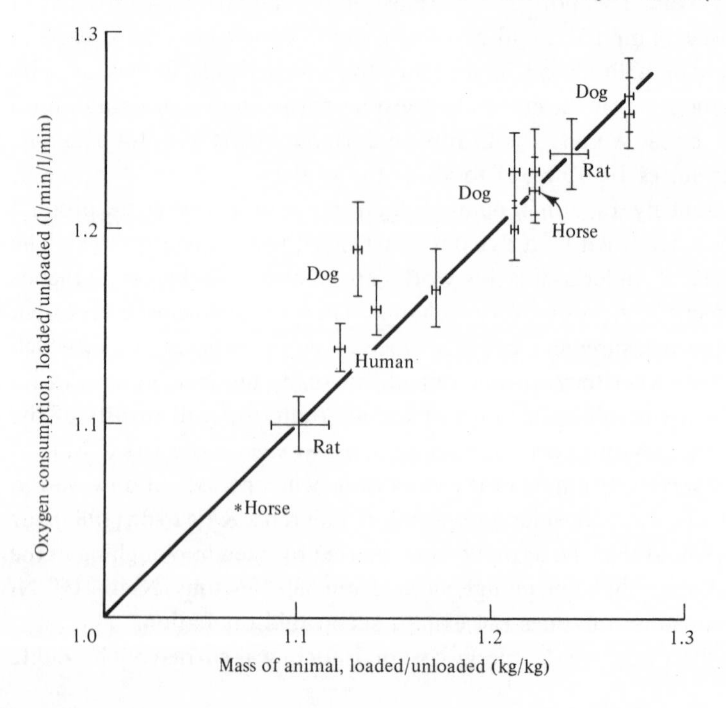

tion. All birds exhibit active flapping flight. The extreme of this mode is *hovering* when rapid wing movement provides no forward thrust and the bird is stationary relative to the surrounding air. Hummingbirds are the best examples of birds adopting this mode of flight, which is virtually confined to birds weighing 50 g or less (Weis-Fogh 1977). Raptors and petrels are often said to hover, but for the most part they fly very slowly into the wind. In *gliding* the bird descends during the forward movement of its body and the wings do not move to generate thrust. This mode of flight is evident in large birds, notably gulls. *Soaring*, which is seen in vultures and storks (Pennycuick 1972), entails the bird making use of 'thermals' – regions of convective updraught – to maintain horizontal motion as well as ascent by circling. Some birds combine flapping flight and gliding. Two types of this combination have been distinguished. In *bounding* flight, which is seen in a number of small birds with short broad wings (such as the chaffinch (*Fringilla coelebs*)), the flapping phase is followed by a glide phase in which the wings are folded close against the body to give the bird a streamlined shape. In *undulating* flight the wings are outstretched throughout and the bird gains height for the glide phase by making a few wing strokes. This mode of flight is seen in gulls, crows and most raptors.

Estimation of the heat produced by birds during flight is technically difficult. The respiratory exchange during hovering flight has been determined by confining hummingbirds in a respiration chamber (Pearson 1950), and a limited number of studies have been made in which birds have been trained to wear masks and fly in wind tunnels. Tucker (1968), using budgerigars, made a mask from a cellulose acetate centrifuge tube. This was connected by light vinyl tubing to analysing equipment. The results obtained in these pioneering studies are shown in in Figure 9.6. They show that as speed increases the total oxygen consumption of the bird passes through a minimum. Measurements of rate of heat production from the respiratory exchange in birds adopting different modes of flight show considerable differences. These are summarised in Table 9.4 which shows that hovering is particularly expensive and gliding the least.

Other methods of estimating the energy cost of flight have been based on studies of free-living birds. In migrating birds determination of the loss of body fat allows an approximate estimate to be made of the energy cost of flying from one point to another. Evidence from laboratory experiments shows that during flight the respiratory quotient falls to close to 0.7 indicating that fat is the major fuel, but speed and actual distance travelled have to be assumed and it is not necessarily known whether the bird has fed or rested during flight. The same criticisms can be made of the estimation of carbon dioxide production of free-flying birds and mammals by the doubly-labelled water technique.

It is perhaps because of these problems of measuring energy expended in flight, that considerable effort has been exerted to derive theoretical models. These predict the work done in flight, when, by assuming an efficiency with which the flight muscles work the metabolic cost can be ascertained. Adding this to the minimal metabolism should result in an estimate of the total energy cost of flight which can then be compared with estimates determined from the gaseous exchange. The components of these models derive from aerodynamic theory. Thus in steady flapping flight the components of the power consumption are; work against body drag (called the *parasite power*), work against form and friction drag on the wings (called the *profile power*) and thirdly work to generate lift and thrust (called *induced power*). The major models in use are those of Pennycuick (1969), Tucker (1973) and Rayner (1979). A full discussion of these and other theoretical treatments is given in Pedley (1977) and Rayner (1982). What perhaps is surprising is that

Fig. 9.6. The mean rate of oxygen consumption/unit body weight of two budgerigars, trained to wear masks and fly in a wind tunnel at different speeds (Tucker 1978). Curve A relates to level flight, curve B to ascending flight at an angle of 5° from the horizontal and curve C to descending flight at an angle of −5°.

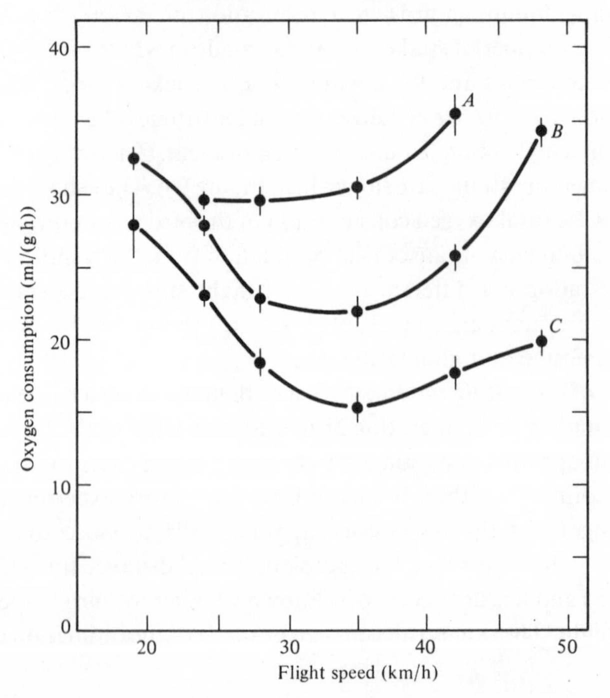

Table 9.4. *The energy cost of different modes of flight*

Species	Mode of flight	Velocity (m/s)	Heat production (w/kg)
Hummingbird (*Archilocus* sp.)	Hovering	0	240
Budgerigar (*Melopsittacus undulatus*)	Level flapping	10	120
	Undulating	7, 10	125
Fish crow (*Corvus ossifragus*)	Flapping	10	80
Laughing gull (*Larus atricilla*)	Flapping	10	60
Herring gull (*Larus argentatus*)	Gliding	—[a]	15

Note:
[a] No value available.
Source: From a summary by Goldspink (1981).

these various models of flapping flight agree fairly well, particularly in relation to the variation of rate of metabolism with speed, with the limited amount of information on the rate of heat production obtained by measuring the gaseous exchange.

In hovering, the aerodynamic analogy is with helicopter rotors rather than with fixed wing aircraft (Hainsworth & Wolf 1975), and estimates from models agree reasonably with the limited respiration data. In gliding the wings are held rigidly and the pectoral muscles maintain this posture by isometric contraction. The energy cost of this type of flight has been estimated by Pennycuick (1972) to be equal in magnitude to the minimal metabolism of the bird.

What is abundantly clear is that it is extremely difficult to ascribe a unique metabolic cost to flight applicable in all circumstances. Flight modes vary, different strategies of flight are adopted to meet different circumstances, and there are differences in wing form, size and loading which result in species differences in the cost of movement.

9.2.6 *Swimming*

Some birds swim as well as fly. Ducks use their feet as paddles, and studies of swimming ducks show that their rate of oxygen consumption is constant at speeds from 0.2 to 0.5 m/s and rises sharply above this level (Prange & Schmidt–Nielsen 1970). The explanation for this curious phenomenon is that the duck produces both bow and stern waves and, as speed increases, the wave length of the bow wave reaches the length of the body in contact with the water. The duck is thus trapped in a trough between the two waves and considerable power has to be exerted to overcome the increase in resistance.

The energetics of fish is outside the scope of this book but nevertheless a brief account is pertinent since aquatic mammals and some diving birds move in similar ways to fish. An excellent account of many aspects of the energy exchanges of fish is given by Tyler & Calder (1985). As with bird flight, different types of locomotion have to be considered, ranging from the almost flapping mode adopted by some rays with their flattened body form and highly muscular fins, the undulatory mode seen in cylindrical fish such as eels and the classical swimming mode in which the tail acts as a hydrofoil as it moves from side to side. This latter mode of swimming is seen in a number of aquatic mammals. Again, as with bird flight, a considerable body of theory of considerable sophistication has been built up to analyse the power requirements of swimming. Briefly, a fish moving through water is subject to an opposing force termed *drag*. Drag varies with the maximal cross-sectional area of the fish in the line of its motion and with the square of the velocity of movement. This component force would be present even if the fish was towed. An additional force is involved to impart motion and this is achieved by movement of the tail acting as a hydrofoil. This provides 'lift', a term which is used to imply that the force is at right angles to the line of motion rather than that the fish is moved vertically. Lift is also proportional to the square of velocity and varies with the area of the hydrofoil component. The tail, acting as a hydrofoil, also contributes to drag, introducing a further term into an overall force expression (see Lighthill 1975; Alexander 1981; Blake 1983).

Power, the rate of energy use per unit time, is the product of force × speed. The theoretical studies briefly outlined above suggest that since the major terms involved in calculating the forces vary with the square of velocity, the power required in swimming will vary approximately with the cube of the fish's speed. If the efficiency with which the muscles of fish work is known, then this power requirement divided by the efficiency should be a good indication of the increase in metabolism of a fish above its resting rate. There is every reason to believe that these theoretical calculations agree with estimates of rates of metabolism in swimming fish. As speed increases the metabolic cost of movement, in addition to that of the resting animal, increases enormously.

The technical problems of measuring oxygen consumption by swimming fish are less daunting than those encountered with birds. The fish is confined in a tank, which is an analogue of a treadmill for land animals. The laminar flow of water is varied, the fish swims at the speed of this flow, and effectively stays in the same place. The tank system is such that the consumption of oxygen can be estimated from changes in the concentration of dissolved oxygen in the water. The work by Brett (1965) on swimming in the sockeye salmon (*Oncorhynchus nerka*) is one of the most complete of such investigations. He distinguished three levels of swimming performance; a speed which could be sustained indefinitely, one which could be achieved and sustained for one or more hours but which led to fatigue and burst speed which could only be maintained for about 30 s. In passing it might be mentioned that fish incur oxygen debts just as do land animals but recovery times are probably somewhat longer – 4 to 5 h in some species.

Ware (1978) undertook a statistical analysis of records obtained for swimming fish to obtain the relationship:

$$\frac{\dot{H}_{SWIM}}{W}=\frac{\dot{H}_{REST}}{W}+1.17WV^{2.42} \qquad (9.4)$$

where $\dot{H}_{SWIM}$ is the total metabolism of the fish (W); W is body weight (kg); and V velocity (m/s). $\dot{H}_{REST}/W$ is the resting metabolism of the fish per unit weight as estimated by Winberg (1960) by the relationship:

$$\frac{\dot{H}_{REST}}{W}=0.285W^{-0.19} \qquad (9.5)$$

The fact that the second term in equation (9.4) includes $V^{2.42}$ is in broad agreement with theory. A similar relationship has been described by Beamish (1978). To compare the energy expenditures of different species of fish, a choice of swimming speeds must be made. Schmidt–Nielsen (1984), using data from both Brett and Beamish, showed that in fish swimming at a speed which is three-quarters of that which they can sustain for an hour, energy expenditure, expressed per unit weight transported unit distance, declines with increasing weight of fish. The logarithmic rate of decline is -0.25 to -0.30, values which suggest a parallelism with resting metabolic rate/weight.

In man, swimming produces the same massive increase in energy cost with speed. Energy cost also varies with stroke. In descending order of energy cost for the same speed are crawl, back stroke, breast stroke and side stroke. Furthermore there appear to be much wider differences between people in the energy cost of swimming than in walking or running. It is not uncommon to find five-fold differences in energy cost between un-trained and well-trained subjects.

9.2.7 *Other activities*

Many animals live part of their lives underground. The amounts of soil moved by burrowing species is considerable. The volume of the nest of a mole is about 2000 cm^3 approached by 6 m of tunnel implying a removal of 0.08 m^3 of soil and it has been observed that a mole will move about 0.01 m^3 of soil per day. The yellow-necked field mouse (*Apodemus flavicollis*) can dislodge 50–150 times its body weight in a day (Golley, Ryszkowski &

Sokur 1975). Vleck (1979) measured the energy expended by the pocket gopher (*Thomomys bottae*) in burrowing by building a respiration chamber and filling it with soil. He found the energy expended during burrowing was on average 4.1-times the resting metabolic rate, the cost varying with the nature of the soil he used, and being maximal for clay soil. He remarked that burrowing for 1 m is up to 3600-times as expensive as moving 1 m on the surface.

Standing, walking, running, flying, swimming and burrowing are activities common to many species; they are usually major in terms of their contribution to total heat production. There are of course many other activities which animals undertake, and if muscular movement is involved then heat is produced. Preening, grooming, licking, stretching, eating, drinking, playing, jumping and a host of other activities all increase heat production. The metabolic rate during these activities has been measured in many species and in man a considerable number of observations has been made of heat production when complex tasks are undertaken. These are exemplified by the definitive summary by Passmore & Durnin (1967) of activities associated with both work and leisure.

In addition to these recognisable and defined tasks involving muscular movement there are many minor activities which are less easily classified and which are perhaps best described as 'fidgeting'. Studies with man show that small muscular movements made when a particular posture is adopted increase metabolism by about 25% (Garrow *et al.* 1977). This minor activity, difficult though it is to quantify, probably accounts for a large part of the inter-individual variation in metabolism (Dauncey 1987).

9.3 Interpretation of metabolic rates during exercise

The rates of metabolism during complex tasks such as those listed by Passmore & Durnin (1967) for man, by Buttemer *et al.* (1986) for budgerigars and by others for other species, were invariably determined on normally fed animals. The total metabolic rate as measured can thus be regarded as consisting of two components; a rate in the absence of exercise and the net cost of the activity above this base line. The rate in the absence of exercise may be regarded as the basal or minimal metabolism plus heat arising from the thermogenic effects of food; the net cost of the activity is thus that measured above a base line in which the subject is in the state of complete muscular repose required for a basal metabolism determination (see Chapter 8). In the compilation of metabolic rates of man made by the FAO/WHO/UNU Expert Consultation (1985) the energy cost of activity is expressed as a multiple of the basal metabolic rate (BMR). For example, the energy cost of office work is assessed as $1.3 \times$ BMR, bricklaying as

3.3 × BMR and pedalling a laden rickshaw as 8.5 × BMR. This approach implies that the energy cost ascribed to the activity includes all the heat produced above the base line of basal metabolism, including any thermogenic effect of the diet and any thermogenesis induced by the climatic environment. The effects of the environment on heat production are discussed in Chapter 10. In most of the studies on the effect of different activities on heat production it can be assumed that the physical environment elicits no metabolic response and that the measurement of the heat produced in excess of the basal metabolism represents the net cost of the exercise above lying and the thermogenic effect of the food consumed.

Exercise does not affect the thermogenic effect of food. Clapperton (1964b) made sheep walk 12.6 km daily up a gradient of 1 in 17 on a treadmill within a respiration chamber. Each sheep was fasted and then given different amounts of food. The 24 h-increment of heat per joule of metabolisable energy consumed, measured from fasting to maintenance, was the same in exercised as in unexercised animals. Jequier & Schutz (1983) used a somewhat different approach in experiments with man. A radar system based on the Doppler effect was employed to obtain a continuous record of the activity of their subjects. They found that the slopes of the relationships between heat production and activity were virtually the same as between the subject when fasted and the same subject when fed. The intercept (corresponding to zero activity as measured by radar) for the fasting subjects was in agreement with the basal metabolism and in the fed subjects was in agreement with the resting metabolism. The energy cost of activity was thus independent of the amount of food consumed.

Conversely, there is no evidence that exercise affects the magnitude of the increase in heat production which occurs in response to food. It had been suggested (Miller *et al.* 1967b) that the increase in heat production due to a meal consumed during exercise was double that found when a meal was consumed in the resting state. Careful experimentation has refuted this claim; there is no significant interaction between exercise and the thermic effect of food, either when muscular work is performed or over the whole 24-h period in which exercise occurs (Dallosso & James 1984).

9.3.1 *The contribution of muscular activity to total heat production*

The relative importance of muscular work in contributing to the total metabolism of an animal can be approximated as the ratio of total heat production to minimal metabolism. The ratio is only an approximation to the amount of muscular work relative to the minimal metabolism for two reasons. The first has already been given; total heat production in an animal consuming food to meet its maintenance needs includes the thermogenic

Table 9.5. *Time–energy budget of a Scottish coal miner*

Activity	Time spent min	Time spent (proportion of day, a)	Heat production relative to basal metabolism (b)	Product $a \times b$
In bed	501	0.348	1.00	0.348
Non-occupational				
Sitting	331	0.230	1.51	0.347
Standing	19	0.013	1.71	0.022
Walking	129	0.089	4.67	0.416
Washing and dressing	43	0.030	3.14	0.094
Gardening	17	0.012	4.76	0.057
Cycling	21	0.015	6.28	0.094
Occupational				
Sitting	130	0.090	1.60	0.144
Standing	18	0.012	1.71	0.021
Walking	58	0.040	6.38	0.255
Hewing	11	0.007	6.38	0.045
Timbering	59	0.041	5.43	0.223
Loading	104	0.072	6.00	0.432
Totals	1440	1.000	—	2.497

Source: Recalculated from the data of Garry *et al.* (1955).

effect of that food and any heat arising from thermoregulation. In the interpretation of the ratios a rough guide is that in animals at the maintenance level of nutrition the thermogenic effect of food constitutes about 15% of total heat production in an omnivorous species and about 30% in one in which extensive fermentation of food takes place. The second reason is that some increase in the deposition of fat and protein may well take place, indeed is likely to do so in growing animals. Such depositions are associated with food intakes which are higher than maintenance ones and with an additional production of heat. Limitation of studies to adult animals and to those kept in thermoneutral environments minimises these uncertainties of interpretation.

Total daily heat production can be calculated from a 'time–energy budget', that is, a summation over 24 h of the times spent in particular activities, each multipled by the total rate of heat production of the activity concerned. Table 9.5 gives an example of such a calculation for a coal miner working at the coal face in a mine in Fife, Scotland during the early 1950s before massive mechanisation of the mining industry had taken place. The table was recalculated from information presented by Garry *et al.* (1955) based on measurements of the heat production during each of the activities listed. Averaged over one week, the ratio of daily rate of heat production to the basal metabolic rate for this physically demanding life-style was 2.5. During working hours heat production was 4.3 × BMR. These are very high values. The report by FAO/WHO/UNU (1985) suggest that the daily heat production of adult men varies from 1.55 × BMR for occupational work which can be classified as light, through 1.78 × BMR for moderate occupational work to 2.10 × BMR for heavy occupational work. An office clerk is estimated to have a heat production of 1.54 × BMR, a retired, healthy, elderly man 1.51 × BMR, and a housewife in an affluent society of 1.52 × BMR. Their calculations for a subsistence farmer in a developing country show that total metabolism is about 1.8 × BMR and the ratio is much the same for a rural woman in a developing country who works both in the house and in the fields. Only those who undertake exceptionally heavy work approach the values found for the Fife coal miner 30 years ago. Indeed a value of 1.5 × BMR may well be too high for modern sedentary people. Studies with women in the United Kingdom (Prentice *et al.* 1985) suggest a value of 1.38 × BMR or 1.33 × BMR (Dauncey (1987)).

Making allowance for the thermogenic effect of food, which is included in the ratio, it appears that the *increase* in the heat production of modern, affluent men and women above the basal level which is due to muscular activity is relatively small (< 40%). It is interesting to compare this small increase over minimal metabolism with the increases found in wild species of animal and bird.

9.3.2 *Field metabolic rate as an index of muscular work*

Time–energy budgets can be prepared for wild species in the same fashion as they are computed for man. Observations on activity patterns under free-ranging conditions have been made for many species and these can be combined with measurements of the heat production associated with particular activities in captive ones to estimate the 24 h-rate of metabolism. Division by the 'minimal' rate of metabolism measured in captive animals provides a ratio comparable with that derived for man. Whether the minimal rate or the rate associated with activity, as measured in captive animals in the laboratory, are true reflections of rates of metabolism in the wild, is uncertain (see Weathers *et al.* 1984).

A more acceptable approach is to use the doubly-labelled water method (see Section 4.2.2) of determining carbon dioxide production in free-living animals and to estimate the ratio of rate of heat production, calculated from carbon dioxide production, to minimal metabolic rate measured in the laboratory. Such estimates of 24 h-metabolism in the field are usually termed *field metabolic rates*. A major problem in using this technique is the recapture of the animal given the doubly-labelled water in order to deter-

Table 9.6. *Field metabolic rates of mammals and birds in relation to body weight (kg)*

Group	Number of observations	Field metabolic rate (kJ/d)	Basal metabolism (kJ/($W^{0.75}$ d))	Ratio field/basal when $W=1.0$
All eutherians	46	918 $W^{0.813}$	300	3.1
All marsupials	28	631 $W^{0.576}$	208	3.0
Passerine birds	26	1570 $W^{0.749}$	661	2.4
Non-passerine birds	24	848 $W^{0.749}$	334	2.5

Note:
From the compilation of Nagy (1987) and the basal metabolism data in Table 8.2. It will be noted that with birds the ratio of field to basal metabolism does not change with body weight; with eutherian mammals there appears to be a slight increase with weight; and, with marsupials a decline.

mine the final concentrations of ^{18}O and ^{2}H in its body water. This is particularly true of birds, which seem to learn how to avoid being captured more quickly than other vertebrates. Most values for birds are thus made during the breeding season when they return repeatedly to their nests.

Nagy (1987) has summarised estimates of field metabolic rates made using the doubly-labelled water technique for 23 species of eutherian mammals, 13 species of marsupials and 25 species of bird. In Table 9.6 his regression estimates of field metabolic rate are given together with estimates of minimal metabolism derived from Table 8.2. Table 9.6 shows that for wild mammals under natural conditions, the ratio of heat production to minimal metabolism is about 3.0, and for birds about 2.5. Making allowances for the thermogenic effects of diet, it seems that muscular activity in wild species is probably more than four-times greater than it is in man [(3.0 − 1)/(1.4 − 1)]. The low values for the field metabolic rate in birds may reflect the fact that breeding birds were mainly used. Thus in Wilson's storm petrel (*Oceanites oceanicus*), the smallest of the Antarctic endotherms, weighing only 40 g, the ratio field/minimal metabolism was 2.2 when on the nest and 4.2 when off the nest (Orst, Nagy & Ricklefs 1987). This bird forages on open water and is continuously on the wing using the flapping–gliding mode of flight. There is also considerable inter-species variation in the ratio. The wandering albatross (*Domedea exulans*), an oceanic soaring bird has a ratio of 1.83, flying seabirds usually have values of about 3.5 and swimming birds such as penguins values around 3.0 (Adams, Brown & Nagy 1986). The same considerable variation is also true of mammals. The sugar glider (*Petaurus breviceps*), a small (121 g) marsupial, shows a ratio of field metabolic rate to basal metabolism of 3.8 while

Table 9.7. *Comparison of the minimal and field rates of metabolism in koalas (Phascolarctos cinereus) and three-toed sloths (Bradypus tridactylus)*

Attribute	Koala	Three-toed sloth
Minimal metabolic rate (kJ/(kg$^{0.75}$ d))	151	119
Field metabolic rate (kJ/(kg$^{0.75}$ d))	391	209
Field metabolic rate: minimal metabolic rate	2.59	1.75
Body temperature (°C)	36	30–38 (cycles)

Sources: From data of Nagy & Martin (1985); Nagy & Montgomery (1980).

Leadbeater's possum, a marsupial of similar size, exhibits a ratio of 5.8 (Nagy & Martin 1985). Table 9.7 compares the minimal metabolisms and field metabolic rates of the koala with those of the three-toed sloth. In terms of the ratio of field to minimal rate, the sloth appears less slothful than is modern man!

9.3.3 *Maximal metabolism by sustained muscular work*

Obviously modern, sedentary man is capable of undertaking far more muscular work than he usually performs. Questions arise about the limit to heat production when exercise is at the physiological limit. *Maximal aerobic capacity* is that level of oxygen consumption which can be achieved without incurring an oxygen debt, and as mentioned earlier, in well-trained athletes is about 5 l/min. In terms of heat production this is about 100 kJ/min. The calculated value for 24 h is 144 MJ – a heat production which is hardly likely to be sustained! Basal metabolism in a young athlete is probably about 7 MJ/d (see Table 8.6) and thus maximal aerobic capacity in a athlete is about 20 × BMR. In an ordinary man a value of 12 × BMR would be more usual (Durnin 1985). C.R. Taylor *et al.* (1980b) made studies with 19 wild African mammals and 8 domesticated ones in which they were run on a treadmill until lactate began to accumulate in the blood. Oxygen consumption was measured and related to body weight. The result obtained for the wild species, expressed in terms of heat production (20.1 kJ/l O_2) was:

$$\dot{H}_{MAX} = 2428\ W^{0.79} \qquad (9.6)$$

where $\dot{H}_{MAX}$ is maximal aerobic capacity (kJ/d); and W is body weight (kg). Since the minimal metabolism of mammalian species is about 289 kJ/(kg$^{0.75}$ d), aerobic capacity in a 1 kg animal is about 8.4 × minimal, rising to 10.4 × minimal in a 200 kg animal. Taylor *et al.* (1980a) point out that within domesticated and laboratory species the range of the ratio of maximal aerobic capacity to minimal metabolism is very considerable; for dogs and horses aerobic capacity is 3.5 times greater than it is in sheep or goats (which are about the same size as the dog) or it is in cattle (which are about the same size as horses). Relative to all species, cattle and sheep have low aerobic capacities while horses and dogs have high ones. The generalisation that aerobic capacity is simply 10 × minimal metabolism is hardly applicable to all species anymore than it applies to all individuals within a species. Even so, the generalisation does permit the statement to be made that muscular activity associated with normal daily existence in all species studied to date, does not amount to more than 30% of maximal aerobic capacity and in most species it is considerably less. In modern, sedentary man daily muscular exercise amounts to less than 5% of aerobic capacity.

9.4 The fuel for muscular exercise

Early studies of the fuel used in muscular exercise were based on examination of the respiratory quotient; more recent studies have employed biopsy techniques to measure glycogen depletion and have monitored the rates of glucose and fatty acid uptake. In man the older and recent studies agree. When prolonged exercise is undertaken at about 70% of aerobic capacity (such as running a marathon), muscle glycogen is the major source of energy during the first 20 min. Liver glycogen is also hydrolysed to provide glucose and by about 60 min increased reliance is placed on hepatic gluconeogenesis for glucose and about 40% of the energy is derived from free fatty acids. As the exercise continues glucose production in the liver declines and a greater proportion of the energy is obtained from free fatty acids, some of which derive from lipid in the muscle but most of which reach the muscle in the blood. Muscle glycogen is never completely depleted and endurance is proportional to the muscle glycogen content before the exercise. After exercise it is common to find that muscle glycogen concentrations rise to levels which are higher than those noted prior to exercise (Leblanc & Labrie 1981). This may account for the increase in post-prandial oxygen consumption seen post-exercise (Bahr & Maehlum 1986).

There has been considerable discussion for a century about whether protein is oxidised during exercise and serves as a source of energy for muscular contraction. Amino acids do not appear to serve as a direct fuel, but there is a considerable release of alanine from exercising muscle as part of the alanine shuttle, and the alanine uptake by the liver results in gluconeogenesis. The contribution of alanine in continuing exercise is probably less than 3% of the total energy. Nitrogen balance studies have usually shown no change in N-excretion due to exercise. Exceptions are when the dietary food intake is fixed and in the absence of exercise, growth involving the net deposition of protein occurs. Then exercise increases nitrogen excretion and reduces deposition since exercise reduces the energy available for the support of growth. An example of this is seen in the studies of Clapperton (1964a) in which young sheep were exercised when given different amounts of food. It was only when fat and protein deposition were taking place in the unexercised animal that the imposition of exercise augmented nitrogen excretion.

10

Metabolic effects of the physical environment

10.1 Heat production in defined environments

Many thousands of experiments have been undertaken to assess the effects of environmental warmth on the metabolism and heat production of different species of animal and of man. Chapter 7 showed that the warmth or coldness of an environment is not specified solely by air temperature; air movement and incoming long and short wave radiation have also to be considered. Many of the early (and indeed some of the more recent) experiments, however, have simply described the physical environment in terms of the ambient temperature alone. An early exception was that of Herrington (1940) with laboratory species. The walls of the animal chamber he used were made of materials of high thermal conductivity and the room housing the chamber was kept at the same temperature as the air inside the chamber. Radiant temperature was thus close to air temperature. The ventilation rate of the chamber was constant and the air velocity within it may be calculated to have been 0.07–0.10 m/s. There was of course no solar radiation. Mice, rats and guinea pigs were all fasted for lengths of time sufficient to obliterate previous effects of feeding – the results therefore apply to a standard nutritional status. His results are shown in Figures 10.1, 10.2 and 10.3. All three species showed a decline in heat production as temperature increased, until a minimum value was attained. With the rat, there was a clear increase in heat production after this minimum, but with mice and guinea pigs this was not so easily discerned. With the rat, the zone of thermoneutrality – the environment in which temperature had no effect on heat production – was extremely small; Herrington estimated it to be between 28–29 °C and minimal heat production was 2.9 MJ/(m^2 d).

Hey (1969) conducted similar experiments with babies. The chamber was of Perspex, which is opaque to infra-red radiation (Hey & Mount 1967), and was water-jacketed so that its temperature was close to that of the air within

Fig. 10.1. The rate of heat production of mice in relation to environmental temperature. The precise definitions of the environment and the nutritional status of the animals are discussed in the text. The data are those of Herrington (1940).

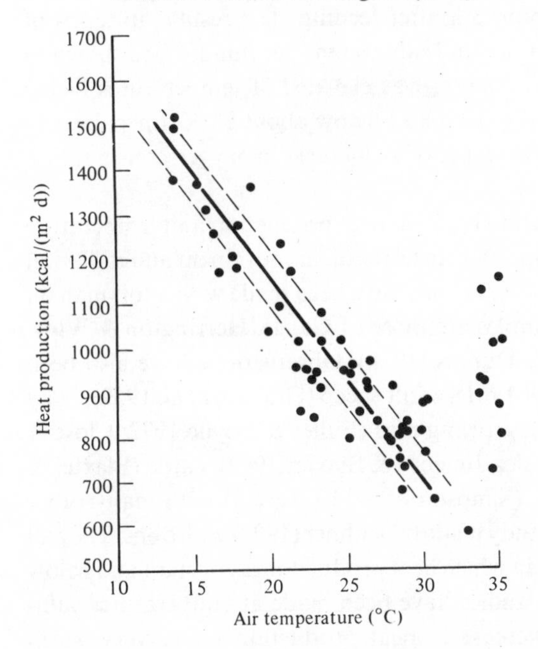

Fig. 10.2. The rate of heat production of rats in relation to environmental temperature. The precise definitions of the environment and the nutritional status of the animals are discussed in the text. The data are those of Herrington (1940).

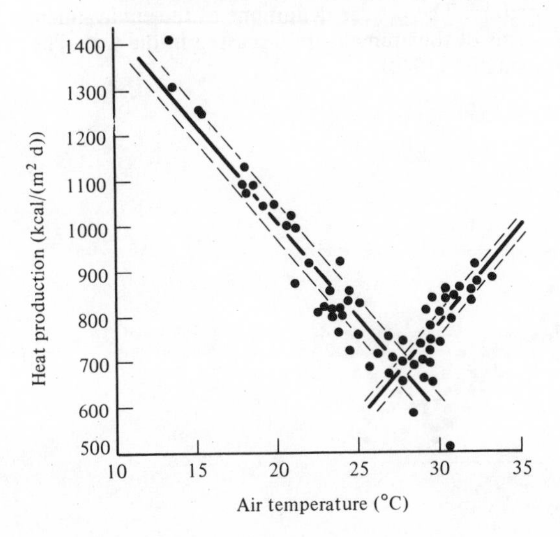

it. Air and wall temperatures were thus almost the same. The babies also conducted heat to the floor on which they lay. Air velocity was low and constant. The babies aged 4–12 h had not been fed; those aged 9–11 d were subject to measurement about 3 h after feeding. The results, in terms of oxygen consumption per kilogram body weight per minute, are shown in Figure 10.4. Again heat production was elevated at temperatures below about 35 °C for babies aged 4–12 h and below about 33 °C for babies 9–11 d of age. There was little evidence of an increase in oxygen consumption at higher temperatures.

Such examples of the relation between heat production and temperature in environments which are specified in terms of air movement and radiation can be multiplied. Similar observations have been made with adult man on many occasions. Early examples are those of Gagge, Herrington & Winslow (1937) and of Hardy & Dubois (1938). Observations have also been made with lambs (Alexander 1961), adult sheep (Graham *et al.* 1959), baby pigs (Close & Stanier 1984), growing pigs (Fuller & Boyne 1972; Close & Mount 1978), calves (Gonzalez-Jimenez & Blaxter 1962), cattle (Blaxter & Wainman 1964b), red deer (Simpson *et al.* 1978), and with many other species, notably the classic study made by Rubner (1902) with dogs. They all show a zone of temperature in which heat production is constant and below which it increases. Not all studies have been made at temperatures sufficiently high to show an increase in heat production under very warm conditions.

Fig. 10.3. The rate of heat production of guinea pigs in relation to environmental temperature. The precise definitions of the environment and the nutritional status of the animals are discussed in the text. The data are those of Herrington (1940).

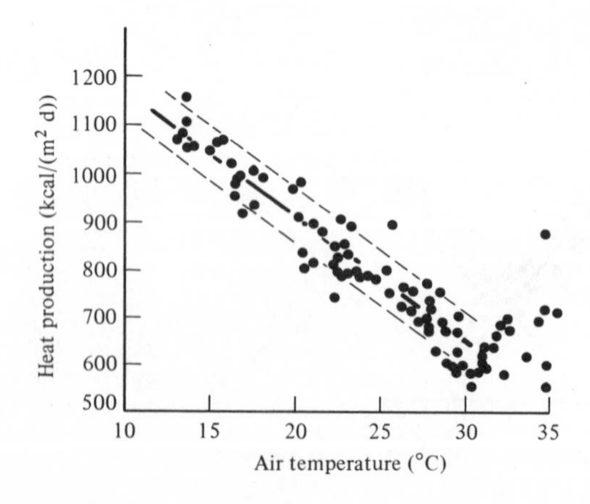

10.1.1 *Nutrition and the avenues of heat loss*

In some of the above studies the amount of food given to the animals was the same at each environmental temperature studied, or, as in Herrington's and Hey's studies, no food was given. Figure 10.5 shows the heat production of the same adult sheep given three different amounts of food for several days and kept at different temperatures. The sheep was closely shorn at short intervals of time to ensure that its covering of fleece was always the same. At low temperatures heat production was independent of the amount of feed given, but as environmental temperature was increased clear differences in heat production due to the amount of food emerged. As temperature was further increased, heat production increased markedly when 1800 g of feed was given, slightly when 1200 g was consumed and not at all when 600 g of feed was given.

In this experiment the losses of heat by non-evaporative and evaporative pathways were separated (Fig. 10.6). A small additional component of heat loss due to the warming of ingested feed and drinking water to body temperature is included in Figure 10.6. At all environmental temperatures non-evaporative heat loss was independent of the amount of feed given.

Fig. 10.4. Rate of oxygen consumption of human babies/unit body weight at two ages in relation to environmental temperature. The closed symbols indicate babies who were underweight for their period of gestation. The precise definition of the thermal environment and the nutritional status of the babies are discussed in the text. The data are those of Hey (1969).

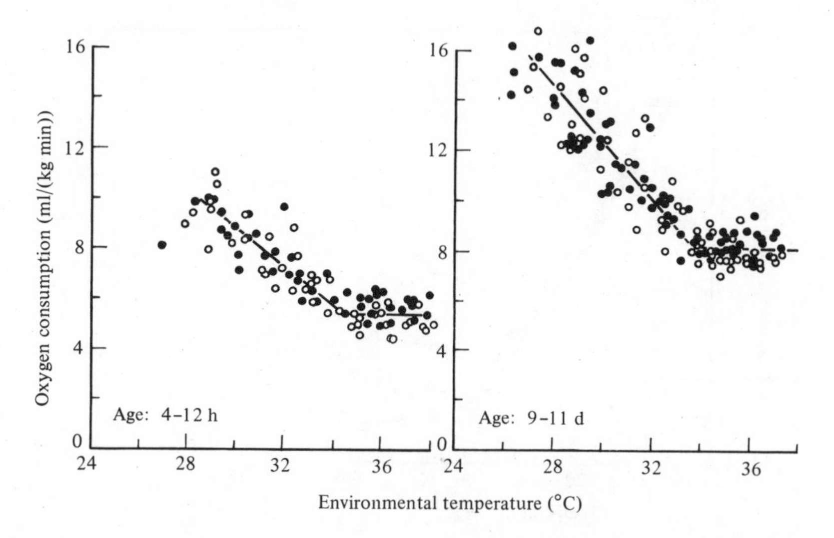

Evaporative loss, however, was only invariant with the amount of feed supplied at very low temperatures and a clear effect of nutrition was apparent at higher ones; the increase in the loss of heat by vaporisation of moisture occurred at a lower temperature the higher the amount of feed given. Similar results to these have been found in the pig (Close 1971); although some results with pigs suggest that small effects of diet on the sensible loss of heat, with higher heat production at the higher feeding levels, can occur. These results may well reflect differences in behavioural patterns, or, because animals given more food have larger gut contents and higher weights, simply reflect the choice of function used to calculate body surface area from body weight (Fuller & Boyne 1972; Bruce & Clark 1979).

These observations have been systematised, commencing with the explanation and terminology used by Rubner in 1894 (see Rubner 1902). The terminology used to describe them was broadly agreed at an international conference on temperature regulation, and Figure 10.7 summarises current views (see Mount 1974). The *critical temperature* (sometimes called the lower critical temperature) is defined as that environmental temperature at which heat production first begins to rise when environmental temperature falls. The *temperature of hyperthermal rise* (sometimes called the upper critical temperature) is defined as that environmental temperature at which heat production increases when temperature is further increased. In be-

Fig. 10.5. The effects of giving different amounts of food on the rate of heat production of a closely shorn sheep kept at different environmental temperatures.

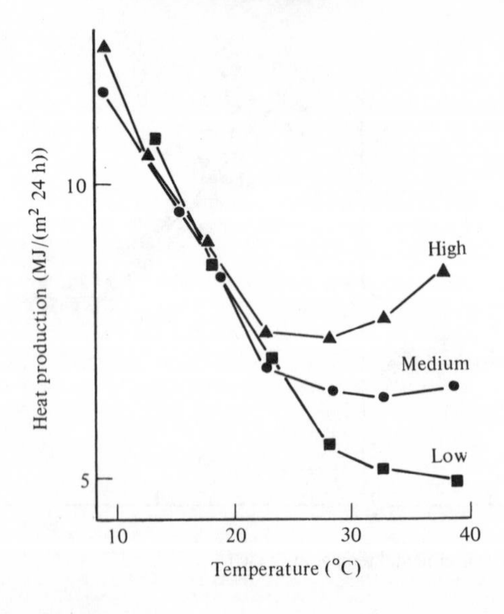

tween these two limits is the *zone of minimal metabolism* or the *thermoneutral zone*. Below the critical temperature heat production continues to rise as the environment becomes cooler until a maximum is reached, termed the *summit metabolism*. Below the environmental temperature at which summit metabolism occurs, heat production then falls coincidental with body cooling and hypothermia. Figure 10.7 also shows that in environments below the critical temperature the loss of heat by the vaporisation of moisture is low and constant but that loss of heat by non-evaporative pathways is directly proportional to the environmental gradient of temperature. Above the critical temperature the non-evaporative loss remains proportional to the environmental gradient (though the slope may change

Fig. 10.6. The partition of the heat loss of the sheep depicted in Fig. 10.5 between sensible heat (the sum of radiation and convective loss), evaporative loss from respiratory passages and skin, and from warming food and drinking water to body temperature when the animal was given different amounts of food and kept at different environmental temperatures.

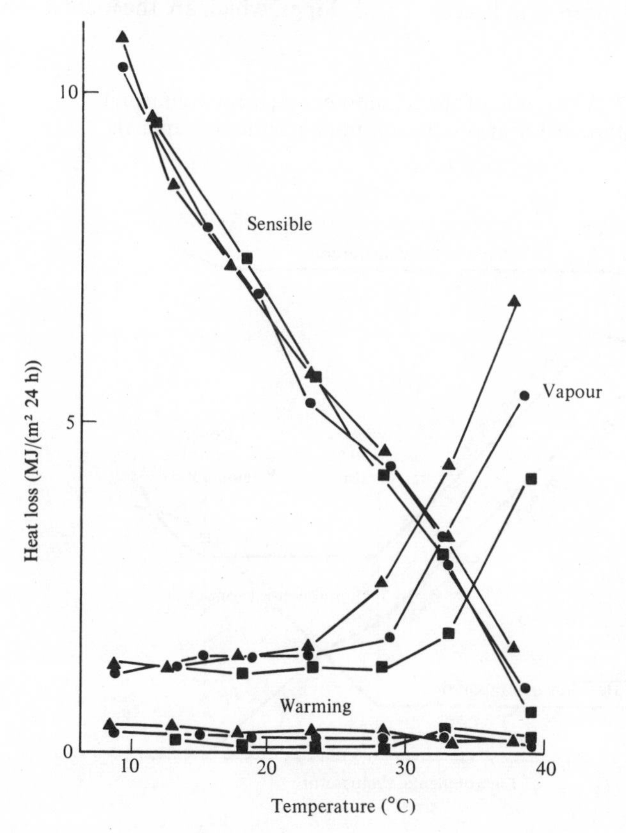

from that noted below the critical temperature) but the evaporative component increases. The constancy of heat production in the thermoneutral zone is attained by an increase in evaporation which balances the decrease in the non-evaporative loss of heat as environmental temperature rises.

The same model can be used to explain the effects of different amounts of feed which were shown in Figure 10.6. The critical temperature is clearly affected by the amount of feed supplied. It coincides with that environmental temperature at which the sum of the minimal loss of heat by vaporising water and that lost by non-evaporative pathways is equal to the heat produced in the thermoneutral zone.

10.1.2 *Non-evaporative heat loss*

It can be seen from Figure 10.7 that the slope of the relation between heat production and environmental temperature, when the animal is under cold conditions below its critical temperature, is the same as the slope of the non-evaporative heat loss curve. In Table 10.1 the slopes for the animal species shown in Figures 10.1–10.5 are given, together with additional slopes for sheep with fleeces. These slopes, which are the increases in

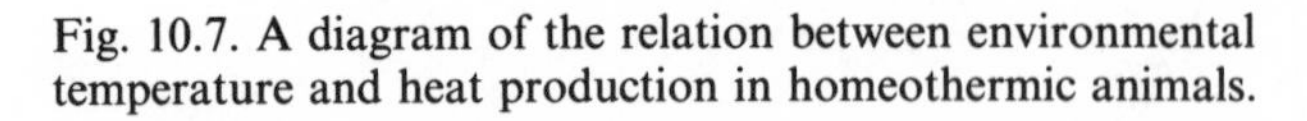

Fig. 10.7. A diagram of the relation between environmental temperature and heat production in homeothermic animals.

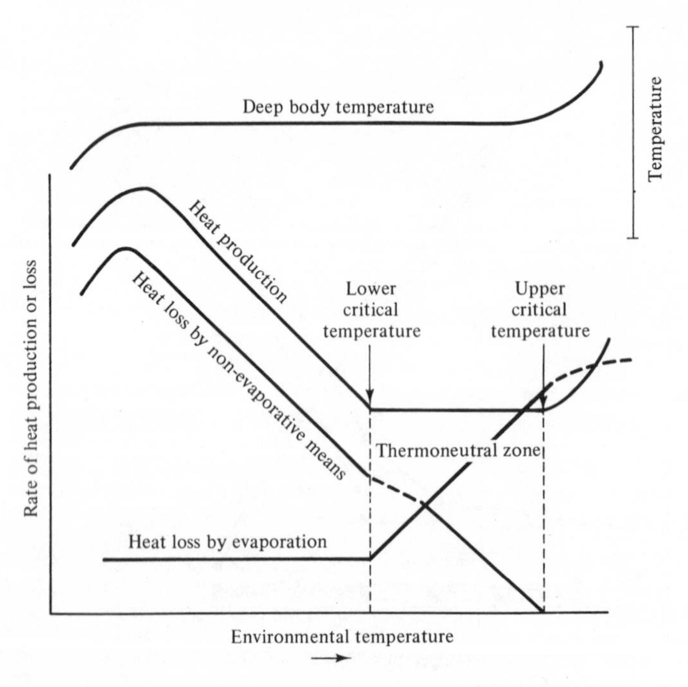

Table 10.1. *Increases in heat production for every 1 °C fall in air temperature (total conductance) at environmental temperatures below the critical*

Species	Total conductance measured below the critical temperature	
	W/(m² °C)	kJ/(m² d °C)
Mouse	2.24	194
Rat	2.09	181
Guinea pig	1.26	109
Human baby	5.17	447
Sheep		
0.1 cm fleece	3.96	343
2.5 cm fleece	2.66	230
4.5 cm fleece	1.73	150
10 cm fleece	1.27	110

Note:
Calculated from the data in Figure 10.1, 10.2, 10.3, 10.4, and 10.5 with additional data for sheep.

heat production (or non-evaporative heat loss), per degree Centigrade fall in environmental temperature are called *whole body conductances*. The table shows that they vary considerably, despite the fact that in the experiments chosen, air velocity and radiant temperatures were very similar. The largest whole body conductance was noted for the human baby and the lowest for the guinea pig. The results for sheep show that whole body conductance falls as fleece length increases – as the animal becomes more insulated from its climatic environment.

10.2 Thermal insulation

To facilitate the analysis of factors which determine the non-evaporative loss of heat it is more convenient and informative to use the reciprocals of conductance terms. The reciprocal quantity is termed *thermal insulation* and represents the resistance to heat flow, that is, the temperature gradient associated with unit heat flow per unit area. Its units are (°C m²)/W or (°C m² d)/MJ. An alternative unit of measure favoured by environmental physicists is a resistance which is simply the insulation expressed in °C m²/W multiplied by the volumetric specific heat of air at a particular standard temperature and pressure. This unit has the dimensions m/s and its advantage is that its use facilitates simultaneous study of heat, mass and momentum transfer (Cena & Clark 1978). Its derivation from measured quantities is not immediately evident from its dimensions and the more complex expression of insulation is preferred.

The total insulation, I_{TOTAL}, can be regarded as consisting of three terms in series. The first, tissue insulation, I_T, represents the thermal resistance of the pathway between the deep parts of the body where heat is generated and the skin surface. The second, the fur, coat or clothing insulation, I_F, represents the resistance to heat flow provided by the fur, coat or garments. The last component is termed the insulation of the air, I_A. This represents the resistance to the flow of heat from the surface of the fur or clothing which is determined by the extent of radiant and convective heat losses.

10.2.1 *Air insulation*

All insulations can be expressed as:

$$I = \frac{\text{temperature gradient}}{\text{sensible heat flux/unit surface}} \quad (10.1)$$

so that, measured in the absence of solar radiation, the insulation of the air can be expressed as:

$$I_A = \frac{T_F - T_A}{(\dot{C} + \dot{R}_I)/A_F} \quad \text{(10.2)el-7}$$

where T_F is the temperature of the coat surface; T_A is the temperature of the air; $\dot{C}$ and $\dot{R}_I$ are the heat losses by convection and infra-red radiation, respectively; and A_F is the surface area of the coat. Alternatively – and equivalently – air insulation can be defined as the reciprocal of the sum of the coefficient for convective heat transfer (h_C) and the first power radiation heat transfer coefficient, (h_R) (see Section 7.1.5).

Table 10.2 gives the convective conductance and insulation of the air for the naked human subject in different air velocities. The convective heat transfer coefficient (h_C) was computed from the relationship found by Kerslake (1972) after reviewing many experimental determinations:

$$h_C = 8.3\ V^{0.5} \quad (10.3)$$

where V is air velocity (m/s). The first power radiative exchange factor (h_R) used was 5.5 W/(m² °C) (see Table 7.9).

The surface of a naked man which exchanges heat with the surroundings is his skin surface (A_S) and this is identical with the surface area of the coat A_F used in the discussion above. When man is clothed the area for the exchange of heat increases and the extent of the increase depends on how much clothing he wears. Because of the variability in the amount of clothing

Table 10.2. *The insulation of the skin–air interface for naked man*

Wind velocity (m/s)	Convective heat transfer coefficient W/(m² °C)	Air insulation	
		(°C m²)/W	(°C m² d)/MJ
0.1	2.6	0.123	1.43
0.2	3.7	0.109	1.26
0.3	4.5	0.100	1.16
0.4	5.2	0.093	1.08
0.6	6.4	0.084	0.97
0.8	7.4	0.077	0.90
1.0	8.3	0.072	0.84
1.2	9.1	0.068	0.79
1.4	9.8	0.065	0.76
1.6	10.5	0.062	0.72
1.8	11.1	0.060	0.70
2.0	11.7	0.058	0.67
2.5	13.1	0.054	0.62
3.0	14.4	0.050	0.58
3.5	15.5	0.048	0.55
4.0	16.6	0.045	0.52
4.5	17.6	0.043	0.50
5.0	18.6	0.041	0.48

worn and the constancy of the skin surface area, it is convenient to refer the insulation of the interface at the coat or clothing surface to the skin surface area. This is a matter of direct proportionality; the sum of the convective and radiative heat transfer coefficients computed from first principles and referring to the outer surface are multiplied by the ratio A_F/A_S. Conversely the insulation terms are multiplied by the ratio A_S/A_F.

In a similar way, if the convective and radiative heat transfer coefficients are different for different parts of the body – as, for example, the legs and trunk – then the total heat loss from the body by these pathways is obtained by weighting them according to their proportions of the total body surface. This has been done for man who is regarded as consisting of six elements (legs, arms, head and trunk) and for animals who are regarded as consisting of three elements (legs, head and trunk) (see McArthur 1981).

10.2.2 *Insulation of hair coats and clothing*

In Section 7.3 it was stated that heat flow through hair coats and clothing could be dealt with as a problem in conduction in which allowance was made for the radial flow of heat. Accordingly, the insulation of a hair coat per unit thickness, that is, the reciprocal of its thermal conductivity can be expressed as:

$$1/k = I_F/x = \frac{(T_F - T_A)}{x(\dot{C} + \dot{R}_I)/A_S} \tag{10.4}$$

where k is the effective thermal conductivity of the coat; x is the effective thickness of the coat, that is, actual thickness corrected for the effects of radial heat flow as described in Chapter 7.

The idea that the insulation of hair coats could be regarded as reflecting the thermal conductivity of the air which is trapped by the coat received support from observations that the insulation per millimetre of thickness remained constant when an animal's coat or a blanket were compressed to change their thicknesses. Hammel's (1955) observations that the insulation of a hair coat could be massively changed by replacing the air trapped by the coat with the refrigerating gas, Freon (which has a higher thermal conductivity than air), supported this in that it suggested that conduction and convection alone were the sole processes involved. The heat transfer is certainly not due to conduction alone since the values calculated for the thermal conductivity of animal coats are usually more than double the conductivity of still air, which at 20 °C is 25 mW/(m °C). Cena & Monteith (1975a,b) have established that radiation, conduction and convection are all involved. Conduction of heat along the fibres of a hair coat is negligible but conduction through the still air trapped in the coat is of considerable magnitude. Radiation from fibre-to-fibre takes place and the thermal gradients established within the coat result in free convection. Obviously in winds there is a further effect on convection, particularly at the boundary of the coat. (Cena & Monteith 1975a,b; McArthur & Monteith 1980a,b). With human clothing another factor comes into play. The juxtaposition of clothing to the skin surface is rarely close and bodily movement sets up considerable local air movements between the skin and the first layer of clothing and indeed between successive layers, producing a forced convective component of heat loss within the overall insulating layer. Nevertheless, the insulation provided by coats and clothing can be expressed in terms of the equation above; it has to be remembered, however, that to do so does not imply that the heat transfer is purely one of molecular conduction.

10.2.3 *External insulation*

The sheep is an excellent animal to use in experiments to explore the insulating effect of coats since the thickness of the sheep's fleece can vary from 2 mm to 12 cm. By measuring skin surface temperatures, air temperature and the non-evaporative loss of heat, the external insulation, I_X can be determined. This is defined as the sum of the insulation provided by the air interface and that of the fleece:

$$I_X = I_A + I_F \tag{10.5}$$

A typical series of insulation experiments with sheep is shown in Figure 10.8 for different air velocities when the animals were exposed at right angles to different winds (Joyce & Blaxter 1964b). External insulation can be regressed on the measured fleece length as shown in Table 10.3. The intercepts of the equations represent the insulation of the air interface and the slopes represent the insulation of the coat per millimetre of its actual thickness. Both the intercepts and the slopes decline as air velocity increases. The table also includes data for cattle (Blaxter & Wainman 1964b; Webster 1974). The intercept term for cattle is greater than for sheep. This finding is understandable; the greater ratio of surface area to mass means that convective heat losses are greater in smaller animals. Analysis of intercepts and slopes from these and other experiments showed that the insulation of the fleece was related to air velocity (V, m/s) and to effective thickness of the fleece by the relationship

$$I_F = x(0.0122 - 0.0028\,V^{0.5}) \tag{10.6}$$

where x, the effective fleece length, derives from considerations of radial heat flow, and in this instance is the quantity:

$$x = r_S \log_e(r_F/r_S) \tag{10.7}$$

Fig. 10.8. The effect of length of the fleece coat on external insulation, $I_X = I_A + I_P$, of sheep exposed to three wind speeds. From data of Joyce & Blaxter (1964b).

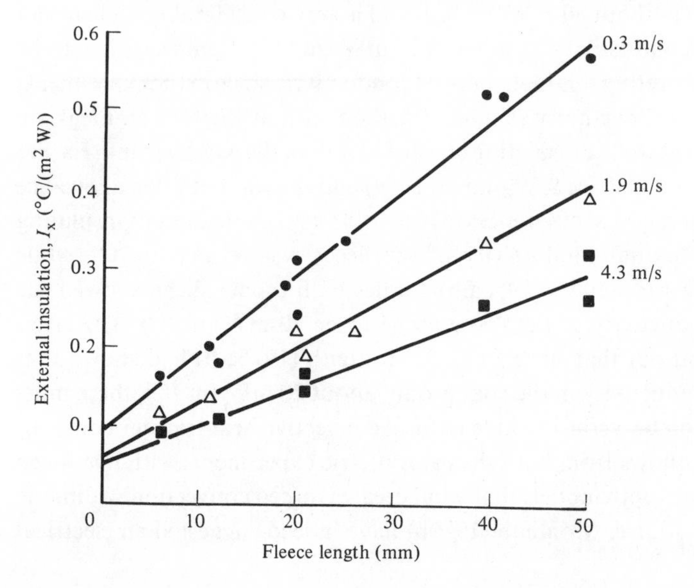

Table 10.3. *The external insulation,* I_X, *of sheep and cattle in relation to coat length, (x, mm), and wind velocity*

Wind velocity (m/s)	External insulation	
	(°C m²)/W	(°C m² d)/MJ
Sheep		
0.3	$I_X=0.0989+0.0096x$	$I_X=1.14+0.110x$
1.1	$I_X=0.0836+0.0082x$	$I_X=0.97+0.095x$
1.9	$I_X=0.0656+0.0068x$	$I_X=0.76+0.079x$
3.4	$I_X=0.0603+0.0072x$	$I_X=0.72+0.083x$
4.3	$I_X=0.0446+0.0057x$	$I_X=0.52+0.065x$
Cattle		
0.2	$I_X=0.122+0.0076x$	$I_X=1.41+0.090x$
0.7	$I_X=0.080+0.0071x$	$I_X=0.93+0.082x$

Note:

$$\text{External insulation} = \frac{(T_A - T_S)A}{(\dot{C}+\dot{R}_I)} = I_F + I_A$$

Source: Blaxter (1977).

where r_F and r_S are the radii of the sheep at the fleece surface and skin surface, respectively. The equation shows that the insulation provided by the fleece at low air speeds is about 0.012 (°C m²) for every millimetre of thickness. The reciprocal of this is the thermal conductivity of the fleece in still air. This is about 80 mW/(m °C), and is very considerably higher than the thermal conductivity of still air (25 mW/(m °C)). Some care has to be taken in interpreting this high thermal conductivity since in the experiments concerned no allowance was made for the fact that the face and legs are covered with a smaller insulating layer of hair than the woolly trunk. Earlier work (Blaxter, Graham & Wainman 1959) had shown that when allowance was made for regional differences in heat loss and the thickness of insulating layers, the thermal conductivity of the fleece was 65 mW(m °C), while McArthur & Monteith's (1980a,b) studies with model sheep showed the thermal conductivity of the fleece alone to be 70 mW/(m °C). The latter authors point out that since Cena & Monteith (1975a,b) had shown that radiative conductivity in the coat is only about 20 mW/(m °C), there must by implication be a considerable natural convective heat transfer component. It also follows from both the calorimetric experiments with live sheep and the studies with models that wind creates forced convection within the fleece. McArthur & Monteith (1980b) have indeed suggested an electrical analogue of the fleece showing the separate components of resistance to the flow of heat (Fig. 10.9).

What applies to the fleece equally applies to other animal coats and to human clothing. The attributes of a hair coat which impart thermal insulation are firstly its thickness, or rather its effective thickness as defined above, secondly its density, that is, the extent to which the proximity of the fibres which make up the coat or garment trap air and prevent free convection and thirdly the nature of its surface with respect to wind penetration. This last attribute alters the coat's behaviour in winds through prevention of forced convection. Table 10.4 summarises observations on the insulation provided by different hair coats and garments.

The data in Table 10.4 refer to insulations per unit thickness – they are 'apparent' thermal conductivities. Piloerection in mammals or ruffling of feathers in birds in response to cold results in an increase in the thickness of the coat but a dimunition in its effectiveness per unit thickness. This is seen in the coat of the calf. When piloerected the coat is almost twice as thick as it is when flat. The insulation added by piloerection is, however, only 40%. With denser coats the effect of piloerection on total coat insulation is greater since the fall in the apparent thermal conductivity is less.

A further factor which affects coat insulation relates to water vapour. It is

Fig. 10.9. (*a*) An electrical analogue for heat transfer through an animal coat in wind, distinguishing conduction, radiation, forced convection and free convection. (*b*) A situation in which there is penetration of the coat by wind to depth *t*. From Cena & Monteith (1975b).

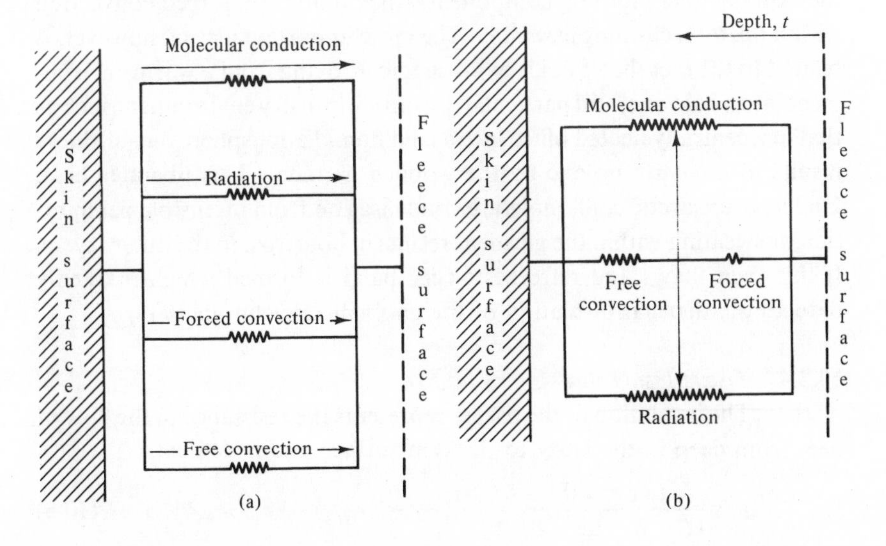

Table 10.4. *The insulation provided by animal coats and fabrics per centimetre of their effective thickness ((°C m²)/W per centimetre)*

Coat or material	Insulation per centimetre
Thermal conductivity of still air	0.36
Ayrshire cattle, coat flat	0.10
Ayrshire cattle, piloerected	0.07
Beef cattle, piloerected	0.08
Merino sheep	0.24
Down sheep	0.16
Scottish Blackfaced sheep	0.13
Husky dog	0.24
Pig	0.04
Closely felted wool	0.29
Human clothing (average value)	0.25

Note:
The values for insulation may be converted into resistances by multiplying by the product of the density of air and its specific heat which at 20 °C is 11.92.

common experience that a woollen garment feels warmer than one made of a single filament fibre, despite the fact that their thicknesses are the same and indeed that the two garments have the same insulation measured in the laboratory. One reason for this relates to the contact the wool makes with the skin or other clothing component which minimises forced convection within the total clothing assembly. The more important aspect, however, is related to the fact that wool is hygroscopic. Moving from a warm environment into a cold one, and particularly from a warm dry environment such as that in a centrally heated office into a cold humid atmosphere out-of-doors results in moisture uptake by the woollen garment. This liberates heat. Under severe arctic cold, moisture condensation from insensible perspiration or sweating within the garment results in hoarfrost in the outer layers. If this then thaws and refreezes an ice band is formed which markedly reduces the thermal insulation of the garment (Newburgh 1949).

10.2.4 *Tissue insulation*

The insulation of the tissues represents the resistance to the flow of heat from deep in the body to the skin surface and is defined:

$$I_T = \frac{(T_R - T_S)}{(\dot{C} + \dot{R}_I + \dot{E}_S)/A_S} \qquad (10.8)$$

where I_T is tissue insulation; A_S is skin surface area; T_R is deep body temperature or rectal temperature; T_S is skin temperature; $\dot{C}$ is convective heat loss; $\dot{R}_I$ is radiative heat loss; and $\dot{E}_S$ is the loss of heat by vaporising moisture from the skin surface. The last term is also equal to the total evaporative loss ($\dot{E}$) minus the loss of heat from the lungs and respiratory passages ($\dot{E}_L$). In a number of experiments in which I_T has been determined, however, the divisor has not been the sensible heat plus the skin evaporation alone but the total heat produced. Values obtained in this way are obviously low, and at high temperature in an animal relying on panting rather than sweating, could be considerably in error.

Tissue insulation has been determined in many species and the results show that it varies over a four- to eight-fold range, being low and relatively constant at temperatures below the critical and increasing markedly as ambient temperature increases above the critical. These differences reflect changes that take place in the convection of heat by the circulating blood to the skin. In hot environments, vessels and capillaries dilate and more blood passes through per unit time. In cold environments, the vessels constrict, limiting the flow of heat to the body surface. The distribution of vessels in the skin varies from species-to-species and so too does the amount of subcutaneous fat. Thus, in sheep, the Down breeds have a lower tissue insulation in the cold than does the Cheviot. This reflects the virtual absence of mid-dermal vessels in Cheviots. Representative values of tissue insulation for different species are listed in Table 10.5.

In a classic paper, which did much to clarify thought about the relationships between the heat production of animals and the flow of heat from the body, Burton (1934) devised a 'circulation index' which is in effect the ratio of the external insulation (I_X) to tissue insulation (I_T) with a small correction for evaporative heat loss. His work with man showed clear differences in skin temperature and hence the circulation index for different parts of the body. The same is true of other species and many experiments have been undertaken in which surface temperatures and deep body temperatures of different regions of the body have been measured in relation to environmental temperature. In a calf the ears, tail and feet all show precipitous falls in temperature as the environment cools (Fig. 10.10). The critical temperature of this animal was about 14 °C. The uncorrected circulation indices for the various regions are summarised in Table 10.6. Thermal indices on the trunk remained fairly constant at environmental temperatures of 35, 10, and 5 °C, despite the erection of the hair coat and thus a change in external insulation, but those of the lower limbs, tail and ears fell precipitously. The extremities thus play a considerable role in heat conservation.

The extremities equally have a role to play in heat dissipation. Direct

Table 10.5. *The thermal insulation of the tissues, I_T, at environmental temperatures below the critical and at high temperatures. These insulations approximate the maximal and minimal values for tissue insulation*

Species	Insulation in cold (°C m²)/W	Insulation in heat (°C m²)/W
Adult man	0.10	0.03
Human baby	0.05	0.015
Adult ox (European)	0.14	0.04
Adult Brahmin ox	—[a]	0.015
Calf	0.09	0.03
Adult sheep	0.08	0.02
Growing pig	0.08	0.03
Adult hen	0.09	0.025

Note:
[a] No value available.
Sources: The values were compiled or calculated from the results of Hardy & Soderstrom (1938); Hardy & Milhorat (1941); Hardy (1961); Hey *et al.* (1970); Blaxter (1977); Gonzalez-Jimenez & Blaxter (1962); Richards (1976, 1977); and Finch (1985).

Table 10.6. *Thermal circulation index of different body regions at different environmental temperatures in a young calf. The critical temperature of this animal determined by indirect calorimetry was 14 °C*

Body region	Environmental temperature (°C)		
	35	10	5
Trunk	2.52	2.45	2.50
Upper leg	2.38	2.10	2.03
Shin	1.75	0.93	0.38
Foot	1.31	0.35	0.13
Tail	1.75	0.49	0.25
Ear	3.25	0.14	0.13

Source: From Gonzalez-Jiminez & Blaxter (1962).

measurements of the blood flow in the tail of the rat show that flow increases over eight-fold when environmental temperature increases from the critical temperature of 28 °C to 33 °C. Gradient layer calorimetry of the tail shows that the tail dissipated 17% of the *total* heat production of the animal although it comprised only 5% of the total surface area (Rand, Burton & Ing 1965). In heavily fleeced sheep the hair-covered head, feet and ears can dissipate over 50% of the heat lost by radiation and convection by the whole animal under warm conditions.

Whether the extremities vasoconstrict and thus increase their insulation depends on the overall insulation of the animal. Thus, in closely clipped sheep, the ears vasoconstrict at environmental temperatures of 18 °C, but in

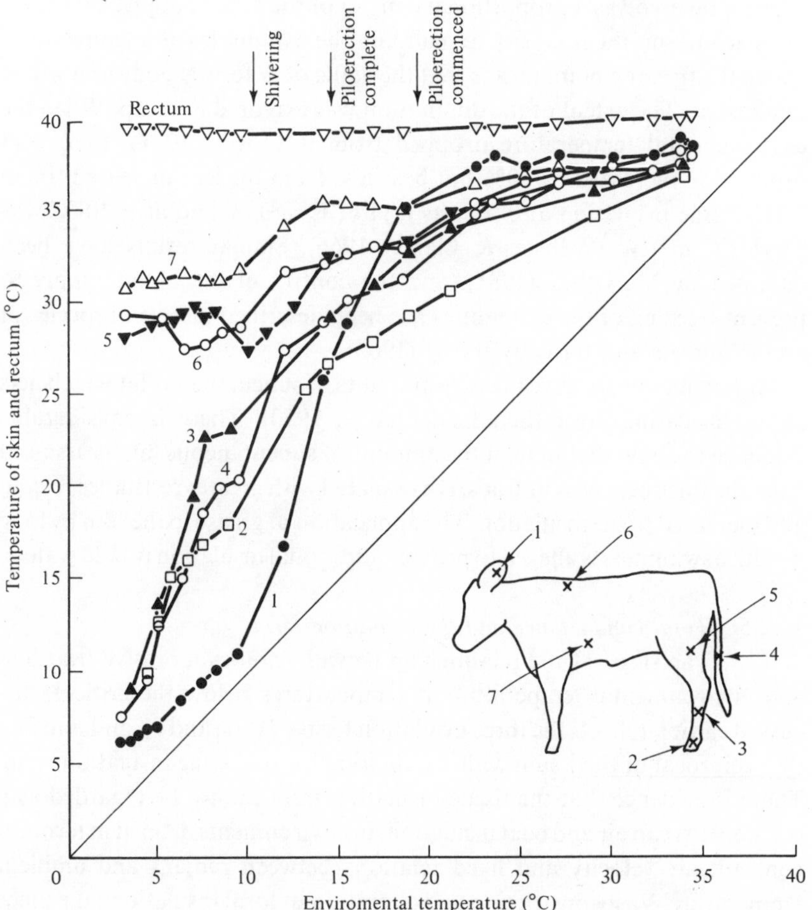

Fig. 10.10. Mean temperatures of the rectum and of skin surfaces of a calf during experimental cooling. From Gonzalez-Jimenez & Blaxter (1962).

heavily fleeced and well-fed sheep the ears have surface temperatures of 34.5 °C even at environmental temperatures close to 0 °C. Once the ears or other extremities are vasoconstricted, their temperatures remain but a few degrees above that of the environment. The question arises about what occurs when an animal with vasoconstricted extremities encounters freezing temperatures. The ear may be taken as an example. When the temperature of the ear reaches near the freezing point, vasodilation occurs, typically by a sudden opening of the capillary network admitting blood which heats the tissue. Subsequently the ear cools, and when its temperature again approaches freezing vasodilation again takes place. This 'hunting reaction', first noted by Lewis (1930), has been found in the ears of a number of species, including man, rabbit, pig, sheep and ox. It has also been found to occur in the foot of the cat but not in that of the artic fox (*Alopex lagopus*) or the wolf (*Canis lupus*) or indeed the sheep where protection against freezing apparently involves a proportional control of the blood supply. Whatever the mechanism, the necessity to maintain the extremities at a temperature above the freezing point means that the value of I_T for the body as a whole diminishes. The extent of the diminution was explored in sheep. When the environmental temperature dropped from 0 °C to −10 °C, total heat production increased by 60% but heat loss from the legs increased three-fold. Tissue insulation at 0 °C was 0.114 (°C m^2)/W and at −10 °C was 0.095 (°C m^2)/W (Webster & Blaxter 1966). Similar results have been obtained by McArthur (1981), and the control of the blood supply to prevent freezing of the extremities has been described in several species of arctic animals and birds by Irving (1964).

Marine mammals have massive amounts of subcutaneous fat which acts as an insulating layer (Scholander *et al.* 1950). There is considerable evidence to show that in man the amount of subcutaneous fat, as assessed from the thickness of skin-folds, is associated with greater resistance to cold and increased tissue insulation. The application of grease to the skin by long distance swimmers is alleged to provide additional insulation in cold waters.

10.2.5 *Total conductance and the insulation terms*

The slope of the relationship between non-evaporative heat loss and environmental temperature at temperatures below the critical, discussed earlier, reflects the three insulation terms, I_A, I_F, and I_T, and is in fact the reciprocal of their sum with a weighting for the tissue insulation term. There is evidence that the tissue insulation term cannot be regarded as a constant, as can air and coat insulation, in environments defined in terms of constant air velocity and fixed relations between radiant and ambient temperature. Vasoconstriction and alteration of local insulation take place over a range of temperatures at and below the critical level. If air temperature falls below freezing point insulation can decrease. It can be questioned why the overall conductance curve appears linear when one of the insulation terms is non-linear. The reason probably relates to the relative magnitude of the terms involved. I_T is usually about 0.1 (°C m²)/W; I_A at low air velocities also has a value of about 0.1 (°C m²)/W, and is higher in smaller species. The value of I_F is about 0.2 (°C m²/W) for animal coats and clothing 1 cm in depth. Proportionally weighting the tissue insulation term to take into account the low amount of water vaporised from the skin at low temperatures, shows that although tissue insulation does change it does not affect conductance greatly until vasodilation occurs at the extremities.

a. Species differences in conductance

Table 10.1 listed total thermal conductances for rats, mice, guinea pigs, human babies and sheep. Conductances, defined as the slope of the heat production curve at temperatures below the critical, have also been measured in many wild species, particularly those weighing less than 5 kg and their relation to body weight has been systematically examined (Herreid & Kessel, 1967; Bradley & Deavers, 1980). The latter obtained data for 192 species of mammal and regressed specific thermal conductance, C^*, expressed as ml O_2/(g h °C), on body weight, W, in grams. They found an overall relationship:

$$C^* = 0.760 W^{-0.426} \tag{10.9}$$

If the surface areas of animals are proportional to the two-thirds power of their body weight, their specific surface area (A/W) will vary with $W^{-0.33}$. Conductance per unit surface area will therefore vary with $W^{-0.093}$, and its inverse, total insulation, will increase with body weight. Schmidt-Nielsen (1984), using the results of Bradley & Deavers (1980) and also those of Herreid & Kessel (1967) and Aschoff (1981), arrived at an exponent of weight in a similar equation relating conductance to W of -0.50. The total insulation per unit surface thus increases on an inter-species basis with $W^{0.17}$. From the previous discussion this difference could be explained by differences in air insulation at the same air temperature as a result of greater convective losses for the smaller animal. It could also arise from differences in the amount of thermal insulation provided by the hair coat. The thickness of the fur coat tends to increase with size of animal, at least up to body weights of 10 kg. Schmidt-Nielsen's analysis suggests that coat insulation increases with $W^{0.15}$ and this factor may be responsible. Very little is known about tissue insulation in the smaller mammals but in view of the relatively smaller amount of subcutaneous lipid this could also be a factor.

10.3 Responses to cold environments

10.3.1 *Estimation of critical temperature*

At the critical temperature, the animal has made the circulatory adjustments to maximise its tissue insulation and has reduced its cutaneous heat loss by vaporising water to a minimum. The heat produced by metabolism then precisely balances what may be termed the environmental demand for heat. The critical temperature is thus that air temperature at which heat production (as measured in the thermoneutral zone) is equal to this environmental demand. The derivation of the expression for estimating critical temperature is as follows: Generally

$$\dot{H}_P/A = (\dot{C} + \dot{R}_I + \dot{E}_S + \dot{E}_L)/A \tag{10.10}$$

where $\dot{H}_P$ is the heat produced and $\dot{C}$, $\dot{R}$, are the corresponding heat losses by convection and infra-red radiation; $\dot{E}_L$ and $\dot{E}_S$ are the corresponding losses of water vapour from the lungs and skin, respectively.

Knowing the insulation terms and the heat emitted, temperature gradients can be expressed in terms of heat production

$$\left.\begin{aligned} (T_R - T_S) &= I_T(\dot{H}_P - \dot{E}_L)/A \\ (T_S - T_F) &= I_F(\dot{H}_P - \dot{E}_L - \dot{E}_S)/A \\ (T_F - T_A) &= I_A(\dot{H}_P - \dot{E}_L - \dot{E}_S)\mathrm{A} \end{aligned}\right\} \tag{10.11}$$

where I is insulation, with subscripts T for tissues, F for the coat or clothing and A for air. Temperatures have subscripts to indicate deep body R, skin surface S, coat surface, F and air A.

Denote $\dot{H}_{P(TZ)}$ as the value of $\dot{H}_P$ in the thermoneutral zone, $\dot{E}_{L(C)}$ and $\dot{E}_{S(C)}$ as the minimal losses of heat by evaporation in the cold and $\dot{I}_{T(C)}$, $\dot{I}_{F(C)}$ and $\dot{I}_{A(C)}$ as the maximal insulations in the cold. The three equations (10.11) can then be added to eliminate T_S and T_A to give the air temperature $T_{A(CRIT)}$ at which heat emission from fully constricted skin and from skin losing minimal amounts of water vapour is precisely the same as heat production in the thermoneutral zone:

$$T_{A(CRIT)} = T_R - (I_{T(C)} + I_{F(C)} + I_{A(C)})\frac{(\dot{H}_{P(TZ)} - \dot{E}_{L(C)})}{A} + (I_{F(C)} + I_{A(C)})\frac{\dot{E}_{S(C)}}{A} \tag{10.12}$$

This relationship has the solar radiation term omitted. It has been used to estimate the critical temperatures of domesticated animals and man under conditions of minimal air movement. Table 10.7, adapted from a compila-

Table 10.7. *Some approximate critical temperatures of farm animals and man under conditions of minimal air movement*

Species and conditions	Critical temperature (°C)
Newborn	
Calf	14
Lamb	29
Piglet	32
Chick	34
Human baby	35
Adults	
Ox, maintenance feeding	7
fully fed	−30
Sheep, maintenance feeding[a]	−3
fully fed[a]	−40
Pig, maintenance feeding	23
fully fed	14
Hen, normal feeding	18
Man[a]	26

Notes:
[a] Critical temperature is dependent, not only on the amount of heat produced and the characteristics of the thermal environment, but also on the insulation provided by the fur or clothing. In the sheep this was a fleece 10 cm thick and man was naked.

tion made by the National Research Council (1981), shows the critical temperatures of farmed livestock together with data for man. Bruce & Clark (1979) have derived a more sophisticated equation to estimate the critical temperatures of pigs of different sizes, in which an additional term for heat loss by conduction to the ground has been included.

The critical temperatures of a number of wild animal species have been determined or estimated from total body conductances (the reciprocal of total insulation). These have been generalised (Bartholomew 1977) to give the relationship:

$$T_{A(CRIT)} = T_R - 3.8W^{-0.23} \qquad (10.13)$$

where T_R is deep body temperature; and W is body mass (g). Assuming the deep body temperature of all species is 38 °C, then an animal weighing 10 g would have a critical temperature of 31.5 °C and one weighing 50 kg, −7.8 °C. These are reasonable estimates; the small shrew, *Sorex vagrans*, has a critical temperature of 32.2 °C (Tomasi 1985), and depending on coat length, a sheep has a critical temperature well below the freezing point. Larger species thus have lower critical temperatures.

Similar studies have been made with birds (Kendeigh *et al.* 1977). Compilations of studies with 74 non-passerine birds gave the relationship:

$$T_{A(CRIT)} = 47.17W^{-0.181} \quad (10.14)$$

and for 43 passerines, measured in summer:

$$T_{A(CRIT)} = 40.73W^{-0.184} \quad (10.15)$$

Again, critical temperature fell with increasing body size.

10.3.2 *Climate and species distribution*

The above between-species generalisations have many exceptions; it has long been known that many arctic mammals are small and yet have low critical temperatures (Scholander *et al.* 1950). However these generalisations are very similar to others which have been accorded the status of 'climatic rules'. Bergmann's rule, devised in 1847, states that smaller-sized geographic races of a species are found in the warmer regions occupied by that species, Allen's rule, formulated in 1877, states that protruding parts of animals, such as tails, ears, bills and legs, are relatively shorter in the cooler parts of a species' range. Gloger's rule states that there are fewer skin melanins in species which occupy cold climates. These same climatic rules have been applied to man and they all relate to putative devices to conserve heat in cold environments and to their evolutionary significance.

These rules have largely been discounted by showing that within a species or genus, body size and shape do not change with latitude and by calculating that such adaptive changes could only be of minor significance in conserving body heat and thus contributing to survival. Irving (1964) indeed remarked 'The arguments to sustain the value of climatic rules as indicators of economy of heat have become so illogical that their refutation by physiologists is not profitable'. Even so, there is little doubt that the experimental imposition of a cold environment alters the morphology of animals. Rats and mice kept in the cold have shorter tails than those kept in the warm (Ogle 1934). Studies with pigs (Weaver & Ingram 1969) and sheep (McBride & Christopherson 1984) show that the young, when kept in the cold, have smaller ears, shorter legs and reduced trunk length compared with litter-mates kept in warmth. In pigs, initially aged 14 d, these considerable morphological changes were apparent within 30 d of placing them in an environment of 10 °C (Dauncey *et al.* 1983) and were independent of concomitant nutritional effects. The mechanism involved appears to be vasoconstriction which limits the supply of nutrients and oxygen to the tissues and thus limits their growth. Thus Heath & Ingram (1981) insulated one ear of a pig kept in the cold and observed that subsequently, this ear was larger than the other.

10.3.3 *Exercise to keep warm*

The formula for calculating the critical temperature given in equation (10.12) equates heat production by metabolism in the thermoneutral zone, $\dot{H}_{P(TZ)}$, with a situation in which no work is done and there is no heat storage in the body. When exercise is undertaken both these terms have to be considered, for when an animal moves, work is done on the environment and not all the heat is liberated in the tissues. As was well established by Nielsen (1938) with man and subsequently confirmed with other species, body temperature rises during exercise irrespective of environmental temperature. In man, for example, the increase in deep body temperature is linearly related to the work rate relative to the maximal aerobic capacity by the expression (Kerslake 1983)

$$T_R = 36.5 + 3.0(\dot{v}_{O_2}/\dot{v}_{O_2max}) \qquad (10.16)$$

where $\dot{v}_{O_2}$ and $\dot{v}_{O_2max}$ are actual oxygen consumption during work and maximal possible oxygen consumption in work, respectively.

A further factor to be considered is that bodily movement in exercise increases convective heat loss and conversely reduces total insulation. In clothed man exercise also augments the ventilation within the clothing and no doubt there is a reduction in the insulation of the hair coat of animals as they move. In addition the increase in energy expenditure augments the volume of air expired thus increasing heat loss from the respiratory passages. Thus not all of the heat produced, as measured by the gaseous exchange during exercise, is available to keep the body warm at temperatures at or below the critical temperature; heat loss is increased, vasodilation occurs, insulation of the tissues and of the air and coat or clothing diminishes and not all the heat produced is emitted by the body. In laboratory rodents, heat produced by activity does not substitute completely for what may be called thermoregulatory heat production (see Hart 1971). In other species, including man, there is usually a greater proportional substitution of heat produced in metabolism during exercise for thermoregulatory heat production. It has been suggested by Jansky (1973) that whether or not activity increases the rate of metabolism in animals kept at temperatures below the critical temperature depends on the extent to which non-shivering thermogenesis occurs. Species which rely on shivering have similar responses to cold in environments below and above the critical temperature.

10.3.4 *Heat production below the critical temperature and the effect of food*

The rate of increase in heat production for every Centigrade degree fall of environmental temperature below the critical is the *total body conductance*. The small obligatory loss of heat from the lungs which parallels the increase in heat production is necessarily included in this term. Heat production increases until the summit metabolism is reached, when the animal or man is unable to maintain deep body temperature and hypothermia ensues. Total body conductance (multiplied by the surface area of the animal) is thus a measure of the amount of metabolisable energy as food required to offset the increase in heat production per °C of cold.

Animals subjected to cold increase their food intake. Cattle kept at −20 °C have voluntary intakes of feed which are 35% greater than they are at 20–30 °C (McDowell *et al.* 1969); in pigs increases are of the same order over a similar range (Versteghen, Brasscamp & van der Hel 1978). In poultry Sykes (1977) summarised the results of nine separate experiments to show that over a range of environmental temperature from about 5 to 35 °C, the voluntary intake of metabolisable energy per bird increased according to the relationship,

$$\text{intake per bird (kJ/d)} = 1690 - 20.1T_A \qquad (10.17)$$

where T_A is in °C. The slope of the relationship between feed intake and environmental temperature is the same as that for the relationship between heat production and environmental temperature. In man there are few comparable data, no doubt because man is seldom exposed to continued cold for long periods. Results obtained in World War II on servicemen show that there is an association between the amount of food consumed per man and the local mean temperature of the environment (Johnson & Kark, 1947; Fig. 10.11). As Burton & Edholm (1955) pointed out, the results are difficult to explain since the men concerned were clothed to deal with their environments and thus the superficial explanation that the higher intake of food in the cold was the consequence of a greater climatic demand for heat hardly seems tenable.

There is now considerable evidence from animal experiments that digestibility and hence metabolisability declines in the cold. The National Research Council (1981) has summarised many experiments carried out with farm species and some relevant results are summarised in Table 10.8. They show that the digestibility of the energy supplied by feed falls by about 0.25% for every degree fall in environmental temperature. This effect is independent of the amount of feed consumed. The effect of cold on the

Table 10.8. *The effect of environmental temperature on the digestibility of the energy-yielding constituents of the diet by cattle, sheep and pigs. Change in the digestibility coefficient (J digested/100 J consumed) per degree Centigrade rise in environmental temperature*

Animal species	Temperature range studied (°C)	Change in the digestibility coefficient/°C
Cattle	−4–17	0.31
	−11–23	0.03
	20–32	0.25
Sheep	8–38	0.20
	1–18	0.21
Pigs	6–20	0.27
	5–23	0.12

Notes:
The values above are taken from a summary of published work compiled by the National Research Council (1981).

Fig. 10.11. The voluntary intake of energy as food by soldiers stationed in climates of different temperatures. From Johnson & Kark (1947).

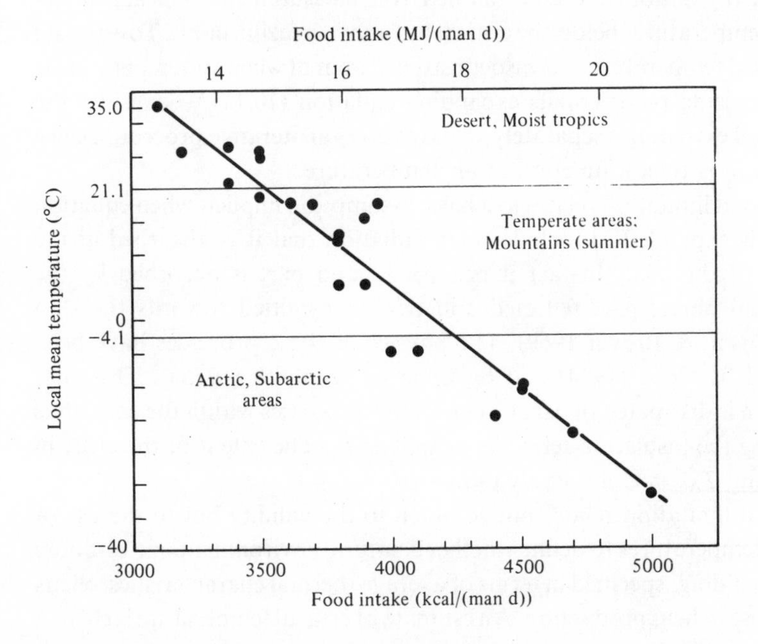

metabolisability of feed has also been noted in domestic poultry (Swain & Farrell 1975) and in small wild birds (see Kendeigh *et al.* 1977). It has been observed in fish but has not been investigated in man.

The mechanism of the reduction in the energy value of unit diet as the result of cold is not completely understood. Part is certainly due to changes in the rate of passage of digesta. Large changes in gastric emptying time have been noted in dogs exposed to cold (Sleeth & Van Liere 1937) and changes in motility of the ruminant stomach have been noted by Kennedy, Christopherson & Milligan (1976, 1982). There is also evidence that environmental cold affects the lifespan of enterocytes. In the pig the time taken by enterocytes to migrate from the crypts to the tip of villus was about double in animals kept at 10 °C rather than 35 °C (Ingram 1986).

10.3.5 *Interpretation of critical temperatures*

The use of critical temperatures and of estimates of the rate of heat loss below the critical temperature have limitations. Firstly the estimation of the critical temperature ignores the fact that when the animal is fully vasoconstricted, the insulation of the extremities falls at temperatures below freezing point. Some of the very low values for the critical temperatures of cattle ignore this and thus are too low. The same is true of man. The results of Erickson *et al.* (1956), for example, show that at a temperature of −6 °C, heat production was about 25% higher than that estimated from linear extrapolation of results obtained from measurements of heat production at temperatures below the critical but above freezing point. To estimate the critical temperature of a vasoconstricted animal when air temperature is below freezing point entails expanding equation (10.11) to consider the trunk and extremities separately. This involves an iterative procedure since $I_{\mathrm{T(extremities)}}$ is then a function of air temperature.

A second limitation relates to a basic assumption implicit when equation (10.12) is expanded to include solar radiation that it is absorbed at the surface of the coat. In fact it penetrates, and part is absorbed by the individual fibres, part reflected and part transmitted towards the skin (Hutchinson & Brown 1969). The physics of these processes have been analysed by Cena (1974) and by Cena & Monteith (1975a). The solar radiation is dissipated in part by convective processes within the coat, thus involving the insulation term, I_{F}, as well as I_{A}. The extent of the error in estimating $T_{\mathrm{A(crit)}}$ is not easily estimated.

A final limitation relates not so much to the validity but to the use of critical temperatures to define whether a natural environment out-of-doors or in a building, specified in terms of average thermal characteristics, elicits a response in heat production. An estimate of critical temperature is correct only for those insulations and heat productions specified, and these change during the day – The external insulation ($I_F + I_A$) can vary by a factor of at least three due to changes in air velocity and incoming long wave radiation. Behavioural reactions of the animal occur to minimise heat loss. Thus animals orient to wind to reduce their profile. They seek shelter, huddle and change their posture so as to change the surface area they present. The same devices are used by man. Additionally, out-of-doors, solar radiation varies continuously during the daylight hours depending on the amount of cloud. Nor is the amount of heat produced in the thermoneutral zone constant. It varies with time of day, with the time elapsed since the previous meal and with the amount of physical activity. Much of this short-term variation in both heat production and the environmental demand for heat is accommodated by changes in body heat content. The response to a change in the environmental demand for heat is not immediate and is characterised by a time constant which in larger species is measured in terms of many minutes. The rat responds to cold in 3 min, the hamster in 12 min but in man and sheep the time is 30–40 min. The studies made by McLean *et al.* (1983a,b 1984) with cattle show that at least 4 h are required to establish a new equilibrium following a large change in calorimeter temperature, and that, although the potential heat storage capacity is only equivalent to about 40 min of metabolism, heat storage acts as an effective buffer when environmental temperature changes suddenly or cyclically. It is probably only rarely that the equilibrium conditions of constant and invariant heat production and insulation, implicit in the use of equation (10.12) are attained in the real environment. Care must therefore be taken in the interpretation and application of critical temperatures determined under laboratory conditions.

10.3.6 *Integrating models*

In assessing the 'coldness' of an environment the best estimate must surely be the animal's reaction to it as judged by its heat production. Continuous measurement of heat production would provide an estimate of both the effects of short-term changes in physical conditions and also the behavioural responses which the animal makes. The doubly-labelled water method of estimating metabolism (see Section 4.2.2) enables this to be done. Another approach measures the extent and duration of deviations of mean skin temperature from the critical temperature. A last approach applicable to man is simply to ask whether he or she feels cold!

Model animals have been used to measure the radiant heat exchange. Lindstedt (1980) placed heated metal models in vacuum chambers and determined their cooling rates. Model animals with an internal heat source

(Bakken *et al.* 1981) have also been used to measure the mesoclimates in which small animals and birds live, while with large animals heated models have been used to measure the coldness of the environment over longer periods of time (Blaxter & Joyce 1963a; Webster 1971; Burnett & Bruce 1978). These devices measure the external insulation of the trunk since they are all cylindrical without extremities. They could be made more sophisticated – as indeed is that of McArthur & Monteith (1980a) – so as to measure effects on separate regions of the body, but they suffer from the limitation that they can hardly have behavioural attributes. An indication of both the variability of the outdoor environment during a 24-h period and the ability of cylindrical models to integrate its components is given in Figure 10.12. The components and output of the model were summated for each hour and show that total losses of heat by convection and radiation varied from a minimum of 102 W/m^2 to a maximum of 138 W/m^2.

Fig. 10.12. Heat losses from two cylindrical models simulating oxen when exposed to a natural outdoor environment (in Aberdeenshire, Scotland, 6 February, 1982). During the course of the day, there was no rain, air temperature ranged from 1 to 8 °C, wind speed from 1 to 6 m/s, and net radiation from −65 to 80 W/m^2. Predicted values were from measurements of air temperature, velocity and net radiation. Records kindly provided by Dr Bruce, Scottish Farm Buildings Investigation Unit.

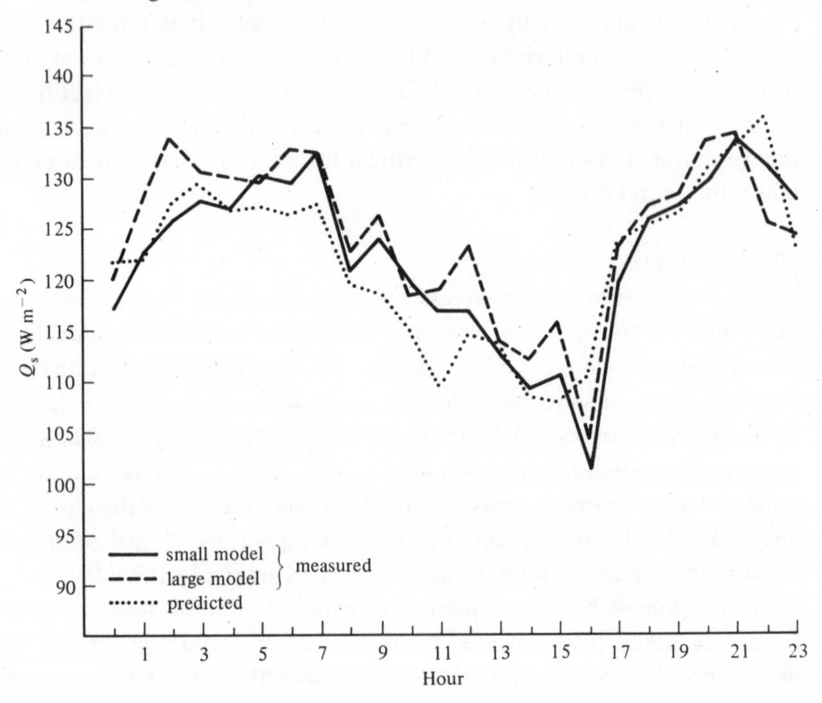

Table 10.9. *Summit metabolism in different species of animals and in man*

Species	Summit metabolism (Watts/kg $W^{0.75}$)	(MJ/kg $W^{0.75}$ d)
Laboratory mouse	23	2.0
White-footed mouse	33	2.8
Red-backed vole	32	2.8
Laboratory rat	22	1.9
Snowshoe hare	26	2.2
Dog	25	2.2
Lamb	28	2.4
Sheep	25	2.2
Baby pig	21	1.8
Man	21	1.8

Sources: Compiled or calculated from results of Giaja & Gelineo (1933); Alexander (1962); Bennett (1972); Wickler (1980); Feist & Rosenmann (1975); Rosenmann, Morrison & Feist (1975); Behnke & Yaglou (1951); Mount & Stephens (1970).

The alternative to using living animals or physical analogues of them is to measure the attributes of the environment and describe and quantify more precisely the physiological and behavioural reactions of animals which modify heat loss in the cold. The two could then be integrated over short time intervals.

10.3.7 *Summit metabolism*

Summit metabolism is rigorously defined as the maximal rate of metabolism which occurs in response to cold without decline in body temperature. It represents a steady state of metabolism. Animals can achieve metabolic rates in the cold higher than this summit for short periods but these are not stable and body temperature falls. Many of the estimates of the maximal metabolic response to cold lie between these two extremes and those that are closest to the definition of summit metabolism (in that they can be sustained for an hour) are summarised in Table 10.9. A compilation of 12 additional values for rodents is given by Hart (1971).

In Table 10.9 summit metabolism has been expressed per unit of metabolic body size (kg $W^{0.75}$) and while there might be a small decline with increasing body weight, the data suggest that summit metabolism is about 25 W/(kg $W^{0.75}$). This amounts to about seven times the fasting rate of metabolism or about 2–3 times the field rate of energy expenditure. A similar estimate of a summit metabolism amounting to five times the resting metabolic rate was noted in rodents by Hart. Marsupials have a lower fasting metabolic rate and lower resting or habitual metabolism than eutherian mammals. Dawson & Dawson (1982) found, somewhat unexpectedly, when comparing Australian marsupials with Australian rodents that, despite their lower metabolic rate, marsupials had a greater response to cold, their summit metabolism being 8–9 times their resting one. In the Australian rodents the ratio was 3–5 times resting metabolism.

Estimates of summit metabolism can be employed to estimate an analogue of the critical temperature, that is, the air temperature below which body temperature cannot be maintained and death from hypothermia results. All that is necessary is to insert in the equation for estimating critical temperature the summit metabolism rather than the thermoneutral metabolism. When this is done for sheep with thick fleeces (10 cm) in environments with little air movement, it appears that sheep could withstand temperatures below absolute zero (Alexander 1974)! This impossible conclusion simply reflects the problems of extrapolation (see Section 10.2.4a). For newly-born lambs with wet birthcoats in windy environments, and in newly-shorn sheep, starved and exposed to rain, however, the estimates of this limiting environmental temperature accord well with practical observations on mortality.

In environments that are colder than those which elicit a summit metabolism, body temperature and heat production both fall. In some species, notably the sheep, the relation between body temperature and metabolic rate is linear (Bennett 1972) but in others such as the dog (Spurr, Hutt & Horvath 1954) the relationship is logarithmic suggesting a Q_{10} of 3.0–3.4 (see Section 8.2.2).

10.3.8 *Shivering*

Shivering is the most obvious sign of increased heat production when temperatures are below the critical. Shivering in this sense is the visible movement of the subcutaneous muscles. There has long been controversy about whether the whole of the increased heat production in man is due to muscular movement because an elevated heat production can be found without apparent shivering. Dauncey (1981), for example, exposed nine women dressed in thin cotton trousersuits to environmental temperatures of either 28 °C – a temperature above the critical – or one of 22 °C. Mean heat production in the former environment was 7716 kJ/24 h and in the latter 8258 kJ/24 h – a highly significant increase showing that 22 °C was below the critical temperature – yet no shivering was observed. Much earlier, Hill & Campbell (1922) had suggested that increases in metabolism of over 40% could occur in the absence of shivering in man exposed to cold outdoor environments. Swift's (1931) studies showed that muscle tension to simulate the initial effects of cold could increase metabolism and concluded that 'the increase in heat production is proportional to the amount and intensity of shivering and any increase not accounted for by shivering may be justly ascribed to muscular tension'. Work with dogs using skin electrodes (Hemingway & Stuart 1963) showed, however, that there was a small component (about 7% of the increase) which could not be attributed to shivering as detected by the skin recording which revealed an abrupt change when shivering commenced. Electromyographic recording in which action currents from single fibres in the muscle were recorded, showed that in the cat prior to synchronous discharge of muscle units during cold exposure, there is an increase in the frequency of discharge of the units and these coincide with the first appearance of a metabolic response to cold (A.C. Burton & Bronk 1937).

There is thus little doubt that in adult man, dog and cat an increase in muscle tone followed by synchronous movement of the muscle, closely parallels the increase in heat production that occurs below the critical temperature. Shivering characteristically takes place in bouts, commencing with small tremors which increase in frequency and severity. The adjustment of heat production to the thermal demand of the environment appears to be accomplished by changes both in severity and duration of these bouts.

No external work is done during shivering, or at least very little, since there is no overt movement of the body. Both agonist and antagonist muscles increase in tension. The whole of the heat which is produced during the muscular events is thus available to combat cold. This may be contrasted with the effect of muscular exercise, discussed in Section 10.2.3.

10.3.9 *Non-shivering thermogenesis*

While there is considerable evidence to show that the major source of additional heat when some species are subjected to cold is due to muscular tension and shivering, there is also evidence that, in other species, increased heat production can occur without an involvement of the musculature. In the latter species heat production is increased on continued exposure to cold without shivering and without electrical activity of the skeletal muscle (Sellers & Scott 1954). Additionally, the newborn of a number of those species which shiver when adult, respond to cold by increasing their heat production without overt shivering, examples being human babies, infant rabbits and kittens.

The origin of this non-shivering thermogenesis in animals acclimatised to cold was originally attributed to the increase in heat production that occurs when voluntary food intake increases on continued exposure. The increased heat production due to the heat increment of feeding did not, however, seem sufficient to account for the non-shivering thermogenesis observed. Another suggestion was that the increased heat production arose from a change in thyroidal status, and there is no doubt that increased thyroid hormone output does occur. Cottle & Carlson (1956) showed that in the cold-acclimated rat an increase in heat production took place even when the animal was given curare to prevent muscular contraction. It was further shown that the sympathetic nervous system was involved since the increased heat production of the curarised rat could be blocked by hexamethonium, a ganglionic blocking agent. Hexamethonium also blocked the increase in heat production of the newborn animal in the cold. Additional evidence was that the cold-acclimated animal had a greater response to noradrenaline than had the same animal kept in the warm (see Jansky 1973 and Wallis 1979).

The least equivocal way in which non-shivering thermogenesis can be shown to occur is to estimate whether an increase in heat production due to cold persists in a curarised animal maintained by artificial respiration. As an example, studies with curarised birds have shown that non-shivering thermogenesis is absent (Hart 1962). Less acceptable, since it cannot be fully quantified, is the demonstration of a lack of parallelism between electromyographic recordings and heat production in animals kept in the cold. Again, as an example, it has been found in birds that such a parallelism exists suggesting an absence of non-shivering thermogenesis (Sellers & Scott 1954). A further method is to measure the response of heat production to physiological amounts of noradrenaline. Interpretation of such results is somewhat complicated by a relationship of the response to body size and by the fact that in some species responses, while positive, are small. Another method is to measure the response of animals to additional food or exercise, both in the warm and in the cold. The thermogenic effect of food has been stated to reduce the effect of cold in dogs, sheep, hedgehogs and other species which do not exhibit non-shivering thermogenesis but to have additive effects in species which rely on this method of augmenting heat production in the cold (Jansky 1973). As far as adult man is concerned, there is no evidence that non-shivering thermogenesis occurs.

a. Brown adipose tissue

In the species in which non-shivering thermogenesis has been observed, either in the newborn or in cold-acclimated individuals, a major site of the augmented heat production is the brown adipose tissue (Nicholls & Locke 1983). Earlier work suggested that muscle tissue was in some species the more important site but the methods used to substantiate this suggestion are suspect (Foster & Frydman 1978). Brown adipose tissue is characterised by its colour, multilocular appearance and particularly by its characteristic mitochondria. As discussed in Section 6.1.4, it is equally characterised by a proton conductance pathway readily uncoupled from electron transport and phosphorylation. The tissue tends to be localised in the interscapular and cervical region of the body and in newborn animals accounts for about 5% of body weight (Hull & Segall 1965). It is not found in all natural orders of animals. There is no doubt about its presence in rodents, lagomorphs, carnivores, bats and primates, but in artiodactyles it is only present in the newborn of some species. It is not found in birds or in aquatic mammals and its presence in marsupials is probably limited to a very few species (Rothwell & Stock 1985).

During cold acclimation, the weight of brown adipose tissue increases. In one experiment the increase in its weight in cold-acclimated rats was two-fold and the increase in its mitochondrial content, as measured by its cytochrome oxidase content, was seven-fold (Pospisilova & Jansky 1976). The temperature of the subscapular brown adipose tissue pads increases markedly in the cold suggesting a high metabolic rate of the tissue itself and it has been estimated that 60% of the caloric response to noradrenalin takes place within it. If in cold-acclimated animals kept in the cold the tissue is in part surgically excised, heat production falls markedly in broad proportion to the amount removed. All of these observations show that brown adipose tissue is of major importance in non-shivering thermogenesis in those species which possess it.

The ultimate source of heat, whether it is invoked by shivering or non-shivering processes, appears to be the fat of the body. There does not appear to be a loss of body protein when animals are subjected to cold. Thus the respiratory quotients of sheep kept in the cold for long periods decline without any parallel increase in urinary nitrogen excretion (N.McC. Graham *et al.* 1959). Earlier work with man in which cold was imposed for very short periods suggested that the respiratory quotient increased; this was undoubtedly due to 'auspumpung' consequent upon bodily tension and shivering.

10.4 Responses to hot environments

10.4.1 *Measurement of the temperature of hyperthermal rise*

In hot environments the body temperatures of animals rise, reflecting an imbalance between heat production (plus any heat load acquired through receipt of solar radiation) and heat loss. This increase in body temperature augments metabolism and heat production increases as a result – according to the Arrhenius–van't Hoff equation. The air temperature at which the increase in heat production occurs is often called the *upper critical temperature*. The upper critical temperature cannot be so readily estimated as can the lower one, nor can it be predicted precisely from knowledge of the mechanisms of heat loss. One of the reasons for this is that air temperature (dry bulb temperature) is not the best indicator of the difficulty that the animal has in losing heat in hot conditions. At high environmental temperatures, in the absence of solar radiation, heat losses by convection and infra-red radiation are small because the gradients of temperature from the animal's surface to the air or to the surroundings are small. Most heat is lost by evaporation of water and water vapour pressure gradients are more important than temperature gradients. High air humidity reduces the magnitude of the vapour pressure gradient between skin or respiratory passage surfaces and the air. Thus the upper critical temperature is lower in air which is saturated with moisture than it is in dry air. In some respects it is perhaps better to define the upper critical temperature in terms of a wet bulb temperature. Scales of 'effective' temperature have indeed been devised which take into account these different avenues of heat loss and the effects of humidity (Kerslake 1972). These scales are analogous to the operative temperature discussed in Section 7.3.2, which is most useful in assessing the 'coldness' of an environment, that is, when air humidity is of little consequence.

Air temperature alone is an even more unsatisfactory measure of critical temperature if the animal is exposed to solar radiation. Table 7.6 showed the heat load on a sheep at different solar elevations expressed per m^2 of its skin surface area. At high solar elevations these were in excess of 350 W/m^2, or about seven times the minimal metabolic rate. Surface temperatures of the hair coats of animals can rise to over 70 °C in sunshine. The gradient of temperature from the skin surface to the coat surface is then reversed and heat flows from the coat to the skin resulting in a heat gain. This can occur at air temperatures which are lower than skin temperatures. The upper critical temperature measured in sunshine and defined as an air temperature, will be different to that measured in the absence of solar radiation.

Despite these difficulties, upper critical temperatures can be measured in the laboratory by serially subjecting animals to different temperatures in air of defined humidity and ascertaining when there is an increase in heat production. It is common to find in such experiments that deep body temperature has risen before an increase in heat production can be discerned. This is particularly evident in birds. Weathers (1981) has noted hyperthermia without a concomitant increase in metabolic rate in a number of small birds. This finding suggests that to call the temperature range between the lower and upper critical temperatures a thermoneutral zone is somewhat of a misnomer. The ability to increase body temperature without a concommitant increase in heat production has an advantage since it will marginally increase the temperature gradient for loss of heat by convection and infra-red radiation.

It is also a common finding in many species that when repeated experimental exposures of animals to environments which elicit a response in heat production are made the responses are lower on the repeats. This adaptive change is found in man, ungulates, and rodents and is associated with an increase in cutaneous losses of moisture on repeated exposure to the same environment (Fox 1973). The phenomenon, which is not fully understood, again makes precise definition of the upper critical temperature difficult.

10.4.2 *The problems of hot environments*

Provided that an animal can lose heat by evaporating water at the rate at which it produces heat and gains it from solar radiation, thermal balance can be maintained. Species of animals differ in their ability to evaporate; this was evident in Table 7.12 where maximal rates of heat loss by evaporation were displayed and where actively sweating species, such as man, were seen to be superior evaporators to panting species such as birds. Quite apart from these differences in the ability to vaporise water, a body size dimension is involved. Heat production per kilogram body weight is greater for small animals than it is for large ones and the incident solar radiation on them per unit weight is also greater. Thus for a small animal to maintain thermal equilibrium under hot conditions more water per unit weight per unit time must be vaporised than for a large animal. A small animal is thus at greater risk from dehydration than a large animal in the same environment. Species which live in hot environments reveal a number of adaptations which can be interpreted as devices to economise water by routes other than evaporation. For example, they produce faeces with a low water content and a concentrated urine with a high osmotic pressure. It appears that in hot climates, survival depends more on avoidance of dehydration than on avoidance of hyperthermia. In man, who loses sodium chloride in sweat, survival equally depends on adequate provision of salt.

Body size is of importance in combatting effects of heat in other ways. Most hot environments are not equally hot all the time. In deserts the intense heat load on animals during the day due to solar radiation is followed by cold nights when incoming infra-red radiation from the clear skies is extremely low. Storage of heat in the body during the day and its dissipation at night could mitigate the effects of the heat load. Again, the small animal with its higher heat production per unit mass can store a much smaller proportion of the heat it produces per unit time than can a large one.

As previously mentioned some birds do store heat but the amount relative to metabolic rate is small. Generally, heat storage as a device for dealing with high environmental temperatures is a possible strategy for large animals, but is not one which, quantitatively, is of major importance. The role of body heat storage as a buffer against change in environmental temperature was discussed earlier in relation to the time taken to achieve equilibrium (see Section 10.3.5).

10.4.3 *Behavioural strategies of animals in hot environments*

The most sensible strategy for an animal to adopt in hot environments is avoidance. After all it is only mad dogs and Englishmen who go out in the midday sun! These avoidance strategies include adoption of nocturnal activity patterns, the seeking of shade, sheltering in declivities, immersion in water and wallowing in mud. Some mammals and birds burrow and form underground refuges. The temperature in the burrow of a desert species is considerably below daytime air and ground surface temperatures and does not exhibit the same large diurnal variation. Desert surface temperatures may vary between −10 and 75 °C, whereas burrow temperatures at a depth of 1 m vary from about 20 to 33 °C. At this depth other problems emerge due to lack of ventilation. Carbon dioxide concentrations in the air of the deep part of burrows increases to values in excess of 3% and there is a similar decrease in the oxygen concentration (Kuhnen 1986).

Another behavioural change associated with long-term existence under hot conditions is the reduction in the voluntary intake of food. This reduces the increment of heat which follows its ingestion and hence the total amount of heat to be dissipated. The reduction of milk yield, noted in cows subjected to heat, is due to the reduction in the intake of metabolisable energy rather than to any specific effect of high temperatures on milk secretion. The same is true of the reduction of growth rates of animals in hot environments. The selection of livestock with enhanced ability to lose heat is of considerable importance in tropical and semi-tropical countries. The success of Brahmin cattle compared with European breeds in meat production enterprises in such countries is largely because of their greater ability to lose heat.

An interesting adaptation to heat which is physiological rather than behavioural is the incidence of *torpor*. Torpor can be defined as any state in which body temperature, and consequently heat production, is reduced and from which arousal can occur spontaneously, so that body temperature returns to that normal in the thermoneutral state. *Hibernation* is simply defined as torpor during winter but has come to be regarded as a state in which body temperature is reduced to at least 10 °C below normal and is maintained there for several days. Hibernation can be regarded as a defence mechanism against environmental cold (Hudson 1978). To *aestivate* is to spend the summer in a state of torpor and this is what occurs in a number of desert mammals, in bats, in small marsupials and in some birds. Probably the most studied of the species which adopt aestivation to escape from the acute daytime heat and night cold of summer is the Californian pocket mouse (*Perognathus californicus*) which can become torpid at any temperature below its *lower* critical temperature of 32 °C. Its body temperature then falls to close to environmental temperature and its heat production is commensurate with that temperature. The factors which induce torpor are both a lowered environmental temperature and a reduction in the food supply. The periods of torpor are usually limited to a few hours and obviously the reduction in metabolism effects an economy in the energy requirement. Whether the economy is major is another matter and the studies of Fleming (1985) on the feathertail glider (*Acrobates pygmaeus*), a small Australian marsupial weighing about 14 g, illustrate this point. Fleming kept a colony in his laboratory and found that torpor could occur at any temperature below 30 °C. The duration was usually less than 24 h, being longer at low air temperatures. The onset of torpor was preceded by an increase in oxygen consumption and arousal was also associated with an increase of metabolism over the normal metabolic rate at the temperature of the study. Not all individuals adopt torpor, and this enabled a comparison to be made of the energy economy of torpor. The oxygen consumption during an episode of torpor (including the initial and final periods of increased metabolism), when summated, was greater than that of a normothermic animal in the same environment. This same phenomenon has been noted in other species which indulge in torpor; there is not necessarily a major economy in energy utilisation unless the periods of torpor are long. In an environment which for part of the day is very hot, aestivation has an advantage in that the animal simultaneously economises in water. An interesting discussion of the biological advantages of torpor is given by Schmidt-Nielsen (1985).

10.4.4 *Fever*

A large series of calorimetric experiments was conducted in the United States in the 1920s on patients with febrile diseases of various origin (DuBois 1936). They showed increases in both body temperature and heat production, the relation between the two agreeing with the Arrhenius–van't Hoff equations. On average the studies showed that a rise in body temperature of 1 °C resulted in an increase in heat production of 13%. Similar increases in temperature and heat production have been produced in animals and in man by artificially inducing fevers using bacterial endotoxin or other pyrogens. Heat production is also increased by trauma in both animals and man. Minor surgery increases heat production by less than 10%. Multiple bone fractures increase heat production by 15–30% and the effect lasts for two weeks or more. If sepsis ensues, heat production can be increased by 50%. Extensive burns cause the largest and most sustained increase in heat production, with increases well in excess of 100%.

The pyrexia in infections is due to changes in the hypothalamic control of body temperature. At the onset of an infection, skin temperature is lowered and heat loss consequently reduced. The lowered skin temperature results in shivering which increases heat production; this heat contributes to heat storage and a further rise in body temperature. Later during the infection and at a particular body temperature, the depression of the central temperature regulating centre is overridden, profuse sweating occurs and the typical continuing fever ensues. This is still associated with an elevation of the 'set-point' of the temperature regulating system. There is no evidence that the increase in metabolism precedes the increase in temperature; the initial effect of the infection is on the neural mechanisms which determine the temperature at which heat loss mechanisms are operative.

11

Reproduction and growth

Reproduction can be broadly defined as all those processes which result in the production of young which are themselves capable of breeding. The bioenergetics of reproduction is concerned with the energy exchanges and transfers involved in such processes. Such a broad definition obviously encompasses the growth of the newborn or newly-hatched animal until the time at which it is independent of its parents. Lactation in mammals and the feeding by birds of their chicks are thus included. The definition could also include growth until puberty and the onset of reproduction, that is, the whole generation interval. In some species the length of the juvenile phase between complete dependence on the parents for sustenance and the attainment of breeding age can be considerable. An obvious example is man. Another is the wandering albatross which has a juvenile period of 9–11 y. In the instance of the albatrosses this follows a period of about 11 months between hatching and fledging of the chick. The result is a very long generation interval exacerbated by the fact that these birds, as a result of the long rearing interval, usually breed once every two years. In other species the interval is very short; in the Japanese quail (*Coturnix coturnix*), for example, the generation interval is measured in weeks.

11.1 Strategies of reproduction

The quail and wandering albatross are two examples of different reproductive strategies. Reproductive strategies can be classified in different ways. One way derives from considerations of population growth (McArthur & Wilson 1967). It can be described by the following equation (see May 1981):

$$dN/dt = rN(1 - N)(t - T)/K \quad (11.1)$$

where N is the size of the population; t is time; T is a time delay (due to the generation interval or other factor); K is the limit to the population set by resources; and *r* is the intrinsic rate of population increase. Two types of population can be postulated, the '*K*-selected' and the '*r*-selected'. the *K*-strategists are species which attempt to stabilise their populations to avoid overshooting the carrying capacity of their environment. They are characterised by having low recruitment rates, low fecundity and hence low values of *r*. They invest much care in raising their offspring to minimise mortality and are usually large-bodied species living in relatively stable environments. The *r*-strategists are opportunists, exploiting habitats which are sometimes short-lived. They have very high rates of reproduction and hence high values of *r*, and little investment in their young. Their populations often overshoot the carrying capacity of their environment, resulting in death or necessitating migration. The species concerned are usually small in body size. The classification is obviously not absolute; there is a continuous gradation. As Southwood (1981) has pointed out, some birds such as blue tits (*Parus caerulus*) incline towards the *r*-strategy with very short breeding cycles, while others such as gannets (*Morus bassana*) incline towards the *K*-strategy. In the mammals, whales and elephants are *K*-strategists while lemmings are *r*-strategists. The many ecological and behavioural inter-relationships which can be inferred to occur under these different strategies have been described by Horn & Rubenstein (1984).

Another way to classify the variety of modes of reproduction relates not so much to the whole process as to the stage of development of the young at birth or hatching – a matter which implies differences in the relative investment made in the uterine or egg stages compared with the postnatal stages of reproduction. Several types can be distinguished. The first includes species which produce extremely immature young such as the monotremes and marsupials. In *Echidna*, which lays eggs, the egg is transferred to the pouch and hatches after 10 d to produce an embryo weighing about 300 mg. This obtains milk from specialised areas of the skin of the pouch (Griffiths 1968). In the diprotodont marsupials, the young migrate unaided from the urogenital opening to the pouch where they attach to a teat. In the red kangaroo (*Macropus rufus*), which weighs about 40 kg, the young at this stage weigh about 800 mg (Tyndale-Biscoe 1973).

A second type of young consists of those which are *atricial*, that is, born naked, blind, deaf and immature in other developmental ways. This terms is not only used to describe birds but can equally well be applied to mammals, examples of the latter being the naked young of insectivores and many rodents. This class can be divided to distinguish semi-atricial species which are born with a pelage or down coat but are blind, examples being the young of many of the carnivores (Walser 1977). The final class are those species which produce precocious young, possessing hair, young which are fully sighted, have an ability to thermoregulate and possess considerable neuromuscular coordination. Ungulates generally come within this class. In terms of energy relationships, different species thus partition the energy required for development of their young between prenatal and postnatal stages in different ways.

In mammals there is a major difference between marsupials and placental animals in the relative importance of the uterine and lactational phases of reproduction. In marsupials the gestation period is very short and lactations are long, while in the placental mammals the reverse is true. As discussed in Section 8.2.1, marsupials have a significantly lower minimal metabolic rate per kilogram metabolic body size than placental animals. It has been suggested that there is a broader causal relationship between metabolic rate and reproduction and that mammals with high metabolic rates have greater reproductive potentials and greater rates of natural increase (McNab 1980; Hennemann 1983). Nicoll & Thompson (1987) undertook experiments which apparently showed that species which have high metabolic rates, relative to the Brody–Kleiber norm, did not increase their heat production during either pregnancy or lactation. This seems highly anomalous, indeed the observations can only be explained if there was a massive reduction in physical activity. There is no doubt that marsupials have lower metabolic rates and shorter gestations than placental mammals of the same size. It is doubtful if this association reflects some causal mechanism.

Obviously there is immense interspecific variation in the patterns of reproduction and concomitant growth and development of the young – even within a single phylum. Further discussion of this is given in Austin & Short (1984). The bioenergetics of reproduction and growth are, accordingly equally variable. In the sections that follow, the bioenergetics of reproduction and growth of birds and mammals are considered separately. Major emphasis is given to the enthalpy of combustion of the products of conception and, in mammals, to the enthalpy of combustion of the milk produced. However, there is a number of features of reproduction involving changes in energy metabolism which are common to both birds and mammals and these are dealt with first.

11.2 Pre-mating behaviour and energy metabolism

In species with well defined breeding seasons, the onset of sexual activity clearly involves growth of the gonads together with considerable increases in physical activity and hence in heat production. In male birds growth of the testis takes place and secondary sexual features develop at the onset of the breeding season. In territorial birds, activities which are closely related to reproduction include the delineation and defence of a territory, and in most species, display and advertisement of presence. Walsberg (1983) has estimated that these activities in territorial birds constitute about 11% of total heat production. In seasonally breeding mammals similar changes occur. In red deer stags, for example, the testis increases in size – its diameter almost doubles – the neck develops and the stag collects and defends a group of hinds with which he will eventually mate during the rut. This mating period lasts at its full intensity for about a month and then slowly declines. During the rut the red deer stag does not eat; the source of energy for the whole of what Sir Thoman Browne (1646) called the 'immoderate ſalacity and almoſt unparalleld exceſs of venery which every September may be obſerved in this Animal' is from catabolism of its tissues. There have been no direct measurements of this energy loss; the stags lose about a quarter of their body weight during this time and much of this undoubtedly consists of fat (Kay 1979).

In female animals a number of parallel changes occur. In seasonally breeding birds growth of the ovary, shell gland and oviduct precedes sexual activity. Nest building is commonly, but not invariably, undertaken by the female. The energy costs of these activities are included in the estimates of field metabolic rates discussed in Section 8.3.2. In mammals males usually have greater rates of minimal metabolism than females of the same body weight. In women the metabolic rates, measured during sleep, has been found to increase by 6% during the premenstrual phase (Bisdee & James 1983). A change of somewhat greater magnitude has been noted in female rats during the oestrus cycle, but there are few data for other species.

11.3 Egg production and incubation in birds

11.3.1 *Egg production by different species of bird*

There is an enormous amount of observational information about the numbers and weights of eggs which birds of different species produce, and understandably, there have been many attempts to summarise such data. These have mostly been through allometric analysis, and in view of the wide variety of avian reproductive strategies discussed above, it is equally understandable that many species diverge considerably from allometric generalisations. The basic allometric equation relating egg weight, (W_{EGG}) in grams, to weight of female birds (W_M) in kilograms, is:

$$W_{EGG} = a(W_M)^b \qquad (11.2)$$

where a and b are fitted constants. Brody (1945) found that a had the value 40 g and $b = 0.73$ g. Rahn, Paganelli & Ar (1975) found a and b to be 57 g and 0.77 g respectively, and Blueweiss *et al.* (1978) using far more data found a to be 53 g and b, 0.77 g. The power of body weight with which weight of the egg varies is indistinguishable from the power with which minimal metabolism of different species varies. Walsberg (1983) has made the point that the precision of these intra-specific equations is low; the petrels (Procellariiformes) for example, produce eggs which weigh twice the weight predicted by Rahn *et al.* (1975). A similar point can be inferred from the compilation of Peters (1983) which shows that an inter-specific generalisation can be made that clutch size is independent of body weight at 4.85 eggs/clutch. Immediately one can think of major exceptions. For example, the average domestic fowl, weighing about 2 kg, lays continuously and produces over 220 eggs/y.

Walsberg has also shown that the specific enthalpy of combustion of the eggs of precocial species is 7.76 ± 1.54 kJ/g while that of altricial species is considerably and significantly less at 4.78 ± 0.54 kJ/g. More energy is packaged to provide for the development of the precocial chick.

The product of the weight of the egg and its specific enthalpy of combustion is an estimate of the enthalpy of the product of conception. To convert this to a rate, the time taken to produce the egg must be known. The length of the rapid follicular growth phase of the egg varies with egg mass, from 3–4 d in small songbirds, to 11–18 d in large oceanic species. Walsberg (1983), by combining inter-species relationships for this time interval with those for the relationship between egg weight and bird body weight, estimated that the *rate* of production of egg mass varied with a $W_M^{0.54}$, suggesting that, because minimal metabolic rate varies with $W^{0.75}$, the daily demand on metabolism for the production of eggs is proportionally smaller for larger species of birds. This may be so; the inter-species variation in reproductive behaviour, however, means that the generalisation is not particularly precise or informative.

Reasonably precise information can be obtained for domestic species. The domestic hen, during her extended laying season produces an egg a day. Such eggs weigh about 60 g, weight of egg varying with breed, age and body size. The mean enthalpy of combustion of the egg is about 370 kJ. The minimal metabolism of a hen weighing 2.0 kg can be estimated from the data in Table 8.2 to be 562 kJ/d. The daily storage in the egg is thus 0.66 times the minimal metabolism. Ricklefs (1974) has made similar estimates for a number of wild species when they are at their peak rate of egg-laying, to conclude that the enthalpy of combustion of the egg relative to minimal metabolism is low in raptors, and particularly high in ducks, a conclusion supported by Drobny's (1980) observations.

11.3.2 *Incubation*

Before it is incubated the egg has to be laid. Van Kampen (1976) attempted to estimate the cost of oviposition in the domesticated fowl using calorimetric methods. It was virtually impossible to separate the heat produced by general movement from that unequivocally due to oviposition. There was a sharp increase in locomotor activity – and a consequent rise in body temperature – 15 min before laying; van Kampen attributed this increased activity to the hen trying to make itself comfortable. When the hen settled, heat production dropped. The energy cost of expelling eggs was minimal.

There is, however, a small increase in heat production due to the activities associated with nesting. The incubation time of the egg is greater the larger the species. Rahn *et al.* (1975) obtained data on 475 species and showed that incubation time, t_X (d), was related to parental weight, W_M (kg), by the allometric expression:

$$t_X = 29 W_M^{0.167} \qquad (11.3)$$

The variation about this relationship was considerable. Some reduction of the variation can be obtained by separation of birds which produce precocial chicks from those producing altricial ones.

During incubation the bird provides heat from her body and insulation through her feathers and the nest to maintain the egg at the temperature necessary for its development. In addition she turns the eggs from time-to-time. These activities produce little or no increase in heat production. Croxall (1982) has estimated from his own and other data the heat production associated with incubation in Antarctic species, notably petrels and penguins. These birds fast during lengthy 'shifts' of incubation. Thus the Manx shearwater (*Puffinus puffinus*) fasts for about 6 d, the wandering albatross for almost 3 weeks and Macaroni penguins for even longer. Their rate of weight loss and estimates of their specific enthalpy of the tissues, provide estimates of heat production. Croxall found the ratio of estimated heat production during incubation to minimal metabolism was 1.29 in petrels and 1.37 in penguins. The errors which he presented were such that the estimates do not differ significantly from the ratio 1.0. The true relationship is likely to be above 1.0 simply because muscular activity is likely to be greater than it is when minimal metabolism is measured. Furthermore, it is likely that an elevation of metabolism of birds nesting in the Antarctic would take place in response to environmental cold; such an increase might well be further augmented by the partial removal of down to make the nest.

Table 11.1. *The mass and energy changes which occur during the incubation of a chicken egg*

Mass balance	(g)	Energy balance	(kJ)
Initial weight	60.0	Initial enthalpy	369
Oxygen uptake	6.6		
Total input	66.6	Total input	369
Embryo weight	31.3	Embryo enthalpy	143
Residual weight	18.5	Residual enthalpy	132
Water loss	9.3	Latent heat loss	23
Carbon dioxide loss	7.6	Sensible heat loss	71
Total output	66.6	Total output	369

Note:
The residual weight and enthalpy refer to the weight and enthalpy of combustion of the shell and unabsorbed yolk. The sum of the residual term and the embryo weight represents the final weight (or enthalpy of combustion) of the whole egg. The sum of the latent heat term and the sensible heat loss term is the total heat produced during incubation.
Source: The data are derived from a summary of published sources (Briedis & Seagrave 1984).

What is clear, however, is that there is no major increase in metabolism specifically related to incubation.

11.3.3 *The development of the egg*

During the course of incubation the egg loses water and carbon dioxide and consumes oxygen. Heat is lost from the egg by convection, conduction, radiation and as latent heat from the evaporation of the water. These components have been measured during the incubation of the domestic hen's egg and the observations from different sources have been summarised by Romanoff (1967) and Briedis & Seagrave (1984). In Table 11.1 the overall mass and energy changes on incubation of the hen's egg are presented. The set of data is self-consistent. It has been selected with some rounding from different sources to be so. The perfect balance between ingo and outgo cannot be used as supporting the mass and energy conservation laws discussed in Section 5.4.

The respiratory quotient calculated from the data in Table 11.1 is 0.81, showing that the main source of energy for the development of the embryo is the yolk lipid, which is taken up by the yolk-sac membrane by phagocytosis. The biochemical steps in the subsequent oxidation are the same as those in adult mammalian tissues (Noble 1986). In energy terms, the conversion of egg to embryo can be expressed in different ways. As a percentage of the initial enthalpy of combustion of the egg, that of the embryo at hatching represents 39%; as a percentage of the change in the enthalpy of combustion of the whole egg during incubation it is 52%. There is some evidence that the latter is lower in the earlier phases of embryonic development (J. Needham 1931). The latter efficiency is, of course, the same as the enthalpy of combustion of the embryo expressed as a percentage of the sum of the enthalpy of combustion of the embryo and the heat produced. Heat production represents the enthalpy changes associated with the biosynthesis of the tissues of the embryo and the provision of ATP for their maintenance in a dynamic state.

11.3.4 *The energy cost of feeding the young*

The time period over which the infant bird has to be fed is determined by the time at which it is fully fledged and can fly to obtain its own food. As already mentioned this can vary appreciably. The demand on the energy expenditure of the parent bird is perhaps best estimated using the double-labelled water technique and comparing the field metabolic rates of birds foraging for their nestlings with rates of those without young. Thus Gabrielsen, Mehlum & Nagy (1987) found that in foraging black-legged kittiwakes (*Rissa* sp.) field metabolic rate was 3.2 × minimal metabolism whereas for those not foraging the ratio was 2.4. Furthermore, those with two chicks to feed had a higher metabolism than those with one. Similar observations have been made with other species. Thus, in house martins (*Delichon urbica*), field metabolic rate is highly correlated with the size of the brood, leading to an increase of about 20% in metabolism when the number of chicks increases from 3 to 4 (Hails & Bryant 1979).

11.4 Pregnancy

11.4.1 *The birth weight of placental mammals*

Emphasis has already been given to the great variety of reproductive strategies adopted by different species and the consequent difficulty of resolving the considerable mass of information which has accrued about their reproduction. This difficulty is well illustrated by the relation between maternal body size and weight of offspring at birth in placental mammals. An early analysis of this relationship was undertaken by Leitch and her associates (Leitch *et al.* 1959) which produced the allometric expression:

$$nW_I = 167 W_M^{0.83} \tag{11.4}$$

where n is the mean number in the litter, W_I (g) is the weight of the individual at birth and W_M (kg) is maternal weight. The correlation was respectably high, but there was a number of species (notably the bears) which did not conform. This equation suggests that a 50 kg placental mammal produces a litter weighing 4.3 kg or 8.6% of maternal weight and that a 500 kg mammal produces a litter weighing 29 kg or 5.6% of maternal weight. As pointed out by Robbins & Robbins (1979) within those ungulates weighing about 50 kg, the weight of the *individual* offspring varies from 15% of maternal body weight in the goat to 4% in the bighorn sheep (*Ovis canadensis*). In a woman weighing 55 kg her baby weighs at birth 6.5% of maternal weight. The overall relationship between maternal weight and litter weight thus does not have high predictive power. The reason for this may be due to an association between maternal size and the number in the litter, small species having larger litters. R.D. Martin (1984) analysed data for 92 typical altricial mammals showing that the weight of the individual newborn was:

$$W_I = 29 W_M^{0.78} \tag{11.5}$$

This predicts that the weight of a newborn individual produced by a 50 kg altricial mammal is 613 g. He showed that precocial mammals produced larger offspring than altricial ones of the same weight. Litter size declines with increasing weight and also increases with the latitude at which the species breeds (Lord 1960), temperate mammals having larger litters for their size than tropical ones. Further divisions can be made. Monogamous primates have larger infants for their size than polygamous ones and ungulates have larger young than carnivores of the same weight (Rubenstein 1985).

Some of the difficulties confronting those who attempt to make inter-species generalisations relates to the selection of data on birth weight and litter number. Table 11.2 shows the relation between the weight of the individual neonate in litters of different size. In the species listed, neonatal weights decline as the number in the litter increases but the total weight of the litter increases. The same relationship applies if the enthalpy of combustion of the newborn and of the litter are considered.

11.4.2 *The duration of pregnancy*

There is a considerable variation in the duration of pregnancy between species of the same body weight. As with inter-species analysis of the relation between body weight and litter weight, single all-embracing generalisations about the duration of pregnancy in relation to body size have broken down as more contributing factors have been identified and taken into consideration. Early estimates of pregnancy duration, t_P (d) from maternal body weight, W_M (kg) were:

Table 11.2. *The total weight of the litter in relation to the number born in some multitocous mammal species*

Number of young born	Total weight of litter				
	Pig (kg)	Sheep (kg)	Guinea pig (g)	Large rabbit (g)	Small rabbit (g)
1		3.2	111		53
2		4.5	220		87
3		6.1	272		135
4			342		164
5			349		198
6				360	232
7	10.2			443	280
10	13.2			610	
13	15.7			689	
16	16.5				

$$t_P = 68\, W_M^{0.25} \text{ (Blaxter 1964)}$$

or

$$t_P = 68\, W_M^{0.26} \text{ (Sacher \& Staffeldt 1974)} \qquad (11.6)$$

Kihlstrom (1972) concluded that differences in gestation time varied with type of placentation; species with endotheliochorial placentas having shorter gestations than those of the same weight with villous haemochorial placentas. In addition, May & Rubenstein (1984) showed variation associated with the type of young born – altricial species which build nests having shorter gestations for their size – while Short (1984) has pointed out that whales have relatively short gestations and that hystrichomorph rodents (guinea pigs, coypus, capybaras etc.) have long ones relative to other groups of the same weight. Primates too have long gestations relative to their size; they also produce small young.

The inter-species variation in the most easily measured attributes of gestation, its duration and the weight of the neonate, implies that the bioenergetics of pregnancy must be equally variable. For this reason it is more useful to consider the energy exchanges during pregnancy in a few selected species rather than to attempt broad generalisation. Most emphasis in what follows has therefore been given to observations on farm animals for which there is extensive quantitative information.

11.4.3 *The enthalpy of combustion of the neonate*

The crude chemical composition of the animal body was discussed in Section 5.1.2 and Figure 5.3 showed that in several species the fat-free body at birth is characterised by a higher water content and a lower protein content than is found at maturity. The species differences in protein content of the fat-free weight of the newborn are considerable, ranging from 110–120 g/kg in the mouse, rat and rabbit, through 140–160 g/kg in man and cat, 160–170 g/kg in the guinea pig and pig to 180–200 g/kg in the cow, sheep and other ungulates. An even more variable component is the lipid content of the newborn. In the mouse, rat, rabbit, cat and pig the fat content of the body at birth is 20 g/kg or less; in the sheep and cow it is between 30 and 50 g/kg, in guinea pigs and seals it is about 100 g/kg and in man can be as much as 160 g/kg. The specific enthalpy of combustion of the newborn thus varies appeciably. Values of 3–4 MJ/kg are found in species characterised by low protein and fat contents at birth, such as the rat and rabbit. The enthalpy of combustion of the bodies of species which are more mature at birth, in terms of their protein content, but store little fat – exemplified by sheep and cattle – is about 5–6 MJ/kg. When there is appreciable fat storage, as in the human baby, the enthalpy of combustion of the newborn is 8–9 MJ/kg. Weight of the newborn alone is thus not a reliable index of its energy content.

The compositions of newborn animals with respect to their glycogen and lipid contents is of interest since these compounds dictate the extent to which the animal can survive immediately after birth if they fail to obtain milk. This aspect has been explored by Mellor & Cockburn (1986) and their results are summarised in Table 11.3. The considerable reserve of lipid in the human baby and the preponderance of glycogen in the piglet are very evident. Mellor & Cockburn point out that the high content of lipid in newborn humans in part compensates for the low lipid content of human colostrum.

11.4.4 *Rates of growth of the foetus and other tissues during pregnancy*

The foetus is nourished via the placenta and its major energy supply consists of glucose and lactate, the latter being produced from glucose in the placenta. The development of the placenta, hypertrophy of the uterus and growth of the mammary glands occur concomitantly with foetal development. Table 11.4 summarises a number of analytical studies with ewes which show the weights and enthalpies of combustion of the foetus, the placenta, fluids and hypertrophied uterus, and of the total conceptus. Growth of the placenta is rapid initially and then slows considerably. The foetal growth rate accelerates throughout pregnancy. Pregnancy in the ewe lasts for 147 d; it takes about 125 d for the foetus to reach a weight and an enthalpy of combustion which is half that reached at term.

Table 11.3. *The energy reserves in newborn human infants, lambs and piglets*

Forms of energy reserve	Human infant	Piglet	Lamb
Weight at birth (kg)	3.50	1.25	4.25
Concentrations in whole body (g/kg)			
'Available' glycogen in liver	65	43	26
'Available' glycogen in muscle	103	209	165
'Structural' lipid	10	6.5	13
'Available' lipid	140	4.5	12
'Available' energy in whole body (kJ/kg)			
Glycogen	168	252	191
Lipid	5446	175	467
Composition of first milk (kJ/l)			
Lactose	516	585	602
Lipid	778	2140	4668
Estimated energy intake on first day (kJ/(kg d))			
Lactose	5	136	140
Lipid	7	496	1083

Note:
'Available' glycogen recognises that only 90% of the total glycogen of liver and 60% of glycogen in muscle can be mobilised in the first 24 h of life. 'Available' lipid is the total lipid less that in structural lipids represented by phospholipids.
Source: From data of Mellor & Cockburn (1986).

Figure 11.1, illustrates the comparative rates of growth of the foetus, placenta, fluids and uterus of the ewe bearing triplets.

The growth of the foetus and associated structures can be described algebraically in a variety of ways. Huggett & Widdas (1951) were impressed by the fact that the crown–rump length of the foetus increased linearly with time, a fact that suggests that foetal weight increases with the cube of time, and such a relationship fits data for different species surprisingly well. Others have employed simple exponential relationships or polynomial equations to describe foetal growth. The most successful equation appears to be the Gompertz function used by Robinson *et al.* (1977):

$$\log_e(W) = A - Be^{-Ct} - Dnt \tag{11.7}$$

where $\log_e(W)$ is the natural logarithm of foetal weight; t is time; n is the number of individuals in the litter; and A, B, C and D are constants. This equation states that the relative growth rate, $(1/W)(dW/dt)$ falls exponentially with time and that each additional foetus in the litter reduces the weight of the others by a proportional amount.

Once good descriptive equations for the relation between time and the

Table 11.4. *The weights and enthalpies of combustion of the products of conception in a sheep producing a single lamb at birth*

	Foetal age in days			
	63	91	119	147 (term)
Weight (g)				
Placenta and other structures	548	1045	1412	1760
Foetus	40	408	1664	4000
Total for gravid uterus	588	1423	3076	5760
Enthalpy of combustion (MJ)				
Placenta and other structures	0.9	2.2	3.2	4.0
Foetus	0.1	1.3	6.9	20.2
Total for gravid uterus	1.0	3.5	10.1	24.2
Daily rate of accretion of the products of conception (kJ/d)	49	145	347	699

Note:
The entries for 'placenta and other structures' include fluids and values for the uterus above that in the non-pregnant animal.
Source: The data derive from an analysis of published data (Agricultural Research Council 1980).

Fig. 11.1. The weight increases of the products of conception in ewes bearing triplet lambs, expressed relative to their weights at term. Note that the placenta reaches its maximal weight early in pregnancy and that the foetuses increase in weight throughout. The data are those of Robinson *et al.* (1977).

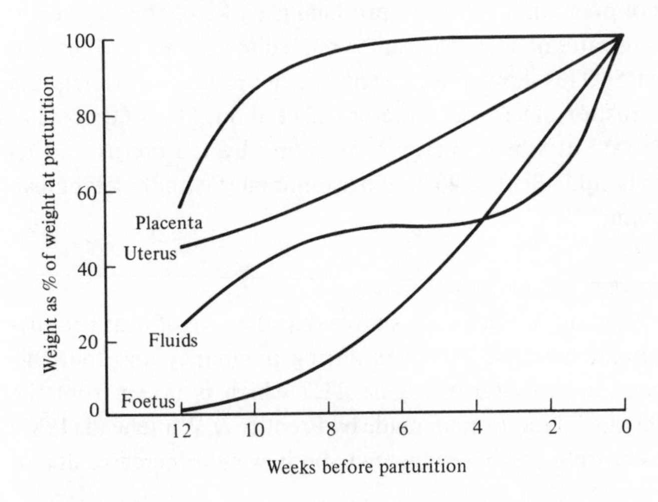

Table 11.5. *The energetics of pregnancy in highly prolific sheep producing two, three or four lambs*

	Number in litter at term		
	2	3	4
Weight of litter (kg)	8.9	12.0	14.2
Enthalpy of combustion at term (MJ)			
Foetuses	39.7	52.0	60.5
Placentas	3.4	4.4	5.2
Foetal fluids	1.1	1.5	1.8
Total conceptus	44.2	57.9	67.5
Uterus	5.4	6.0	6.6
Total gravid uterus	49.6	63.9	74.1
Maternal body	705	599	574
Rate of accretion at term (MJ/d)	1.6	2.1	2.5
Ratios × 100			
Gravid uterus/maternal body	7	11	13
Rate of accretion/minimal metabolism	29	39	46
Weight of foetus/maternal weight	13	17	20

Notes:
The ewes initially weighed about 70 kg. The maternal body enthalpy is the enthalpy of combustion of the whole body less that of the uterus and its contents.
Source: Data from Agricultural Research Council (1980); and Robinson *et al.* (1980).

weight or enthalpy of combustion of the foetus and the adnexa are available, differentiation of them estimates time rates of accretion. These rates increase throughout pregnancy until maxima are reached at term. Rates at different stages of pregnancy for a ewe producing a 4 kg lamb are given in Table 11.4. Means rates of accretion can be related to various attributes of the ewe (Table 11.5). This shows that, at term in a ewe producing triplets, the enthalpy of combustion of her gravid uterus will be about 11% of that of her own body. The rate at which energy is accreting by the uterus and its contents at term is equivalent to 40% of her minimal (fasting) metabolism when non-pregnant.

11.4.5 *Heat production during pregnancy*

Studies in many species have shown that the rate of heat production of the pregnant mammal increases during pregnancy and that this increase is greater towards term. Figure 11.2 which is taken from the summary of published data for man made by Prentice & Whitehead (1987) illustrates this very well. During pregnancy, body weight increases due to the presence of the conceptus within the hypertrophied uterus and the growth of the mammary glands. Brody (1945) measured heat production in a number of species at intervals during pregnancy and calculated what he termed 'the heat increment of gestation'. This was the summation, over the whole of pregnancy, of the difference between the measured heat production and that of a non-pregnant animal of the same weight. Brody concluded that for all the species he examined – from rats and rabbits to cattle and horses – the increase in heat production during pregnancy was probably constant at about 18 MJ/kg body weight of the litter at birth. He further suggested that this value would increase in those species with relatively long gestation periods for their size. Such conclusions are complicated by the fact that in Brody's studies food intake was not controlled; heat could have been affected by the fact that more food or less food was consumed during pregnancy than in the non-pregnant state.

There is, however, no doubt that an increase in the rate of heat production occurs during pregnancy. There may be a small decline in metabolism early in pregnancy; this has been noted in the dog, rabbit and in woman, probably associated with reduced physical activity, but in all species which

Fig. 11.2. The rates of basal and resting heat production of women during pregnancy and lactation. The data refer to studies undertaken during the last 70 years as summarised by Prentice & Whitehead (1987).

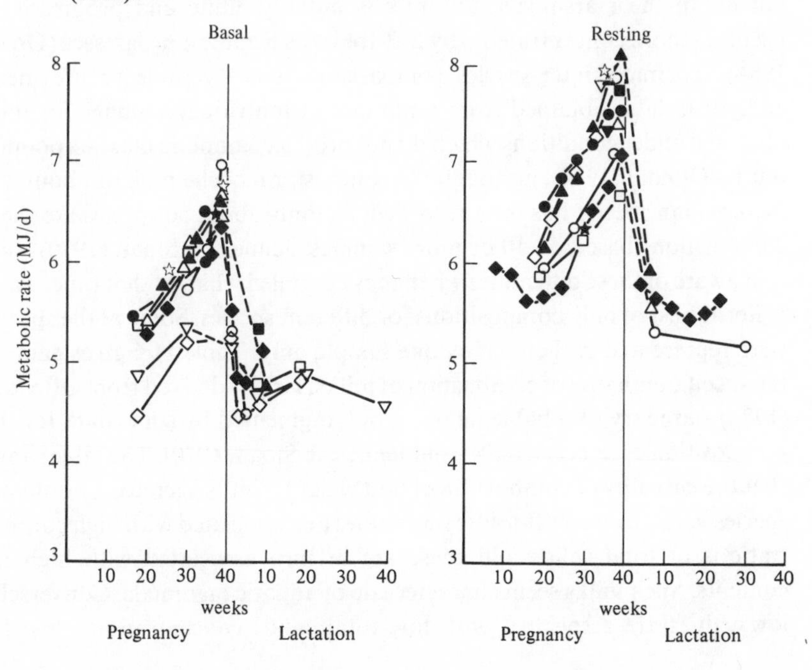

have been studied an increase occurs in the later stages. Part of this increase reflects the continuing metabolism of the foetus and placenta. Direct observation of the metabolism of the foetus, by measurement of arterio-venous differences in the oxygen content of the blood and its rate of flow, show that, in the latter part of pregnancy, the metabolism of the foetus per unit weight is considerably greater than that of maternal tissue. Further consideration of the efficiency of energy utilisation during pregnancy is given in Chapter 12.

11.5 Lactation

11.5.1 *The composition of milk*

There are difficulties in quantifying the composition of the milk produced by a mammal. First there are sampling problems; the milk to be analysed should be representative of what the suckling obtains. In some species there is a marked increase in fat content in the successive portions of milk removed from the mammary gland during a nursing period. Secondly, during the course of lactation the composition of the milk may change. The first milk or colostrum usually contains more fat and protein than that produced later. In late lactation fat content rises in some species such as rodents, carnivores and ruminants, but falls in others – notably in pigs and horses. In the marsupials, the milk is initially dilute and progressively becomes more concentrated – by 2–3-fold – as lactation progresses (Green 1984). To make inter-species comparisons it is desirable to use mean analytical data, obtained from a number of individual animals, for milk obtained under conditions which do not produce sampling bias. As pointed out by Oftedal (1984), although the composition of the milk of about 200 mammalian species has been recorded, for only about 50 species are these compositions based on 10 or more samples. Jenness & Sloan (1970) were well aware of these difficulties when they compiled what, at that time, was a definitive list of milk compositions for different species. Some of the species were represented in their list by one sample only. Table 11.6 gives data on the specific enthalpy of combustion of milks, largely derived from Oftedal's (1984) summary of reliable analyses but augmented by some data for the composition of cetacean milk from Jenness & Sloan (1970). The table shows that the enthalpy of combustion of unit weight of milk secreted by different species varies by over 10-fold. High values are associated with high concentrations of total solids and these are in turn associated with high fat contents. Such milks seem characteristic of aquatic mammals. Conversely, low values are associated with low total solids concentrations, low fat

Table 11.6. *The enthalpy of combustion and its distribution between fat, protein and carbohydrate of the milks of different species of mammal during established lactation*

Species	Dry matter (%)	Enthalpy of combustion (MJ/kg)	Distribution of enthalpy (%)		
			Fat	Protein	Carbohydrate
Marsupials					
Red kangaroo (*Macropus rufa*)	22.8	5.06	36	31	32
Primates					
Baboon (*Papio hamydryas*)	14.0	3.35	52	9	38
Human	12.4	2.89	54	7	39
Lagomorphs					
Rabbit (*Oryctolagus cuniculus*)	31.2	8.53	68	29	3
Rodents					
Brown rat (*Rattus norvegicus*)	22.1	5.98	56	33	10
House mouse (*Mus musculus*)	29.3	7.74	65	29	6
Guinea pig (*Cavia porcellus*)	17.5	4.52	48	34	18
Carnivores (terrestial)					
Arctic fox (*Alopex lagopus*)	28.6	8.28	62	32	6
Dog (*Canis*)	22.7	6.10	59	31	11
Carnivores (aquatic)					
Northern fur seal (*Callorhinus ursinus*)	61.0	21.29	88	11	0.1
Hooded seal (*Cystophora cristata*)	51.7	18.16	89	11	0.1
Proboscids					
Elephant (*Loxodonta africana*)	17.3	3.68	52	24	24
Perissodactyls					
Ass (*Equus asinus*)	8.5	1.59	14	22	63
Horse (*Equus caballus*)	10.5	2.13	23	22	54
Artiodactyls					
Pig (*Sus scrofa*)	20.1	5.19	61	22	16
Red deer (*Cervus elaphus*)	21.1	5.73	56	30	13
Cow (*Bos taurus*)	12.4	2.97	48	26	26
Goat (*Capra hircus*)	12.0	2.89	50	22	27
Sheep (*Ovis aries*)	18.2	4.64	59	22	18
Cetaceans					
Blue whale (*Balaenoptera musculus*)	57.1	19.0	85	14	1

contents and milks in which a considerable proportion of the non-fat energy is derived from lactose rather than protein.

Many attempts have been made to make inter-species generalisations concerning milk composition. On the assumption that gastro-intestinal capacity of the neonate would vary directly with body weight but energy requirements would vary with weight raised to the power with which minimal metabolism varies (0.73–0.75), Blaxter (1961a) suggested that smaller species should produce more concentrated milks and that the concentrations of nutrients in milk should vary with the weight of the newborn raised to the power (−0.25 to 0.27). This suggestion was examined by Payne & Wheeler (1968) who found that the specific enthalpy of combustion of milk was related to birth weight raised to the power −0.28, a value in broad agreement with the hypothesis. The blue whale (*Balaenoptera musculus*), grey seal (*Halichaerus grypus*), porpoise (*Phocaena, phocacha*), and tree shrew (*Tupaia* sp.) were, however, exceptions. Payne and Wheeler drew attention to the fact that the tree shrew suckles only once every two days and at that time the suckling, weighing 9 g, can consume 6 g milk. In this respect, Ben Shaul (1962) distinguished groups of mammals by their nursing behaviour and their maturity at birth, and showed that species which nursed their young on demand produced dilute milks. It has been suggested that the high concentration of fat in milks of marine mammals relates to a need to conserve water in a salt water environment and/or a need to provide energy to combat heat loss in the sea. None of these explanations appear to be entirely sufficient in accounting for the considerable inter-species variation in the specific enthalpy of milk. This situation is analogous to explanations of inter-species variation in other attributes of reproduction in animals.

11.5.2 *The rate of secretion of the energy-yielding constituents of milk*

The rate of secretion of energy in milk is simply the product of milk yield and the specific enthalpy of combustion of the milk. Milk yield is difficult to measure in animals other than those domesticated for milk production. The methods for measuring yield have been critically appraised by Linzell (1972) and by Oftedal (1984). Most are prone to systematic error. More important, however, is the fact that the rate of secretion varies with the stage of lactation. In the normal pattern of secretion, yield rises to a peak and then declines to zero when the offspring are weaned. In making comparative studies it has been the practice to measure milk yield as the rate of secretion at the peak of lactation. A further factor is that the rate of secretion is dependent on the number of young that suckle the mother. This is illustrated for the mouse in Table 11.7. During the 10-d study the infant

Table 11.7. *The adjustment of lactation in the mouse to the number to be suckled: total weight of the litter and mean weight of individuals 10 days after birth*

Number in the litter	Mean weight of individual (g)	Total weight of litter (g)
4	5.9	23.7
5	5.7	28.5
6	6.1	36.4
7	5.5	38.7
8	5.6	45.0
9	5.8	51.8
10	5.3	52.9
11	5.8	63.9
12	5.1	61.1

Source: From Blaxter (1961b).

mice did not receive any food other than milk. When the number to be fed was tripled there was little change in the weight of the individual young, showing that the rate of milk secretion had accommodated almost perfectly to the demand of the litter.

Commencing with the studies by Brody & Nisbet (1938) when they compared the daily yield of energy in milk of the rat, goat and dairy cow, there have been several investigations of the effect of maternal weight on the enthalpy of combustion of the milk secreted each day at the peak of lactation. All of these studies, which are admittedly with small numbers of species, show that milk secretion varies with a power of weight close to 0.75. They have been summarised by R.D. Martin (1984) who accepts that $W^{0.75}$ can be used as a scaling factor when comparing milk secretion in different species. This approach is similar to that adopted in Chapter 8, and in particular in Table 8.2, when $W^{0.75}$, termed metabolic body weight, was used as a common scaling factor in the presentation of measurements of minimal metabolism. In Table 11.8 the scaled rates of secretion of energy in milk by different species are compared. As pointed out by Oftendal (1984), groups of species can be discerned in such a tabulation. Species producing litters of many young produce milk at the rate of 841 kJ/d, ungulates producing single young produce at the rate of 344 kJ/d and primates producing single young produce at the rate of 154 kJ/d. There is no doubt that the differences between these groups is significant; whether extension of the species coverage will confirm them remains to be seen.

Rates of milk secretion can be compared with rates of minimal metabolism. In species producing many young the rate of secretion of energy in

Table 11.8. *The daily secretion of energy in milk by different species at the peaks of their lactations expressed relative to the metabolic body weight ($W_M^{0.75}$) of the mother and also to the metabolic weight of the litter ($nW_I^{0.83}$)*

	Milk energy/ maternal metabolic weight (kJ/kg $W_M^{0.75}$)	Mean litter size	Milk energy/ metabolic weight of litter (kJ/$nW_I^{0.83}$)
Species producing many young (n=5 or greater)			
Pig	1029	9	953
Dog	958	5	1109
Rabbit	766	7	1013
Striped skunk	686	5	1046
Mink	602	5	887
Rat	1008	8	892
Mean	841	—	983
Ungulates producing single young			
Horse	344	1	879
Beef cow	278	1	816
North American elk	301	1	925
Reindeer	332	1	1188
Red deer	320	1	945
Ibex	336	1	891
Bedouin goat	473	1	853
Dorcas gazelle	371	1	883
Mean	344		935
Ungulates producing two young			
Black-tailed deer	494	2	979
Sheep	586	2	1054
Mean	540		1016
Primates producing single young			
Human	146	1	715
Baboon	162	1	845
Mean	154		780
Other species producing less than five young			
Guinea pig	354	3	640
Brown hare	376	3	862

Source: Data recalculated from Oftedal (1984).

milk is almost three times the minimal metabolism. In ungulates producing single young it is about 1.2 times minimal metabolism and in primates it is only 50% of the heat produced by minimal metabolism. A superlative cow of a highly selected dairy breed can secrete at the peak of her lactation about five times as much milk energy as the beef cow depicted in Table 11.8. This suggests that her rate of secretion of energy in milk is about five times her minimal metabolic rate.

The final column in Table 11.8 expresses the secretion of energy in milk relative to the metabolic demand made by the litter. This approach, which is due to Oftedal (1984), involves estimating the metabolic weight of the individual neonate and multiplying this weight by the number in the litter. It is immediately evident that there is less spread of values for the ratio milk energy:litter metabolic weight than there is for the ratio milk energy:maternal body weight. This suggests in turn that, as between species, an accommodation of milk secretion takes place based on the demand made by the litter. This seems analogous to the accommodation of yield to the demand exerted by the litter in the individual animal as depicted in Table 11.7.

11.5.3 *Maternal heat production during lactation*

In most mammals, and particularly in those in which lactation results in the secretion of a considerable amount of energy as milk per kilogram of metabolic body weight, lactation is associated with an increase in food intake. In the rat, for example, the voluntary consumption of food is 3–4 times that observed in a non-lactating and non-pregnant animal of the same weight. Heat production concomitantly increases, and, as is discussed in Section 12.1.1, the increase is proportional to the amount of metabolisable energy consumed. The question arises whether this is a change in metabolism specifically associated with the lactating state as distinct from the rise associated with the increase in the amount of food consumed. This question is difficult to answer. In species which make nests and suckle their young frequently, muscular activity is reduced and this should depress metabolism. Lactation continues when food intake is reduced. The animal increases the catabolism of its body tissues and this catabolism is again associated with an increase in heat production. Fasting the animal, with a view to measuring whether the minimal metabolism is increased by lactation, results, after a short delay, in the cessation of lactation and is thus not a suitable approach. Extrapolation of the relationship between heat production and milk energy secretion to zero energy secretion does not suggest that there is a change in metabolism that cannot be accounted for by the heat increment associated with the increase in food intake or the catabolism of body tissue.

11.5.4 *The energetics of human pregnancy and lactation*

It is of interest to compare the energetics of reproduction in humans with that in other species of about the same size. Since the sheep has been used previously as an illustrative example, Table 11.9 makes such a comparison. The total energy accretion during pregnancy is not very

Table 11.9. *A comparison of the energetics of reproduction in woman with that in the sheep – two species of similar body weight*

	Woman	Sheep
Weight of newborn (kg)	3.5	8.9 (twins)
Length of pregnancy (d)	266	147
Length of lactation (d)	180	112
Total accretion in pregnancy (MJ)	43	50
Additional heat produced[a] *(MJ)*	83	260
Cumulative yield of milk (MJ)	320	910
Total cost of reproduction (MJ)		
Pregnancy	126	310
Lactation	320	910
Total	446	1220
Crude rates (MJ/d)		
Pregnancy	0.47	2.10
Lactation	1.78	8.13
Reproduction	1.00	4.71

Note:

[a] The additional heat for woman was calculated according to Brody (1945) rather than accepting the estimate of Hytten & Leitch (1971). The value for the ewe was the experimentally observed value from experiments (Robinson *et al.* 1980); it agrees with the Brody estimate (see Section 11.5.4).

different between a woman producing a 3.5 kg baby in 266 d and a ewe producing twin lambs weighing 8.9 kg, the reason being the very high fat content of the human baby at birth. The additional and obligatory heat production associated with human pregnancy is estimated to be less in a woman than in the ewe. The human value is estimated from Brody's relationship (Brody & Nisbet 1938) which states that the heat increment is 18.4 MJ/kg of weight of the neonate raised to the power of 1.2. Hytten & Leitch (1971) give a value of 149 MJ. The value for the ewe (260 MJ) was computed from the results of Robinson *et al.* (1980); Brody's equation would predict a value of 254 MJ. The overall energy required throughout pregnancy is thus 2.5 times as great in the ewe as in the woman. The total energy secreted as milk during normal lactations is probably about three times as great in the ewe as in woman, but this is not a very precise figure since human lactation can be protracted. Generally, the overall crude rate of synthesis and obligatory expenditure of energy is over four times as great in the ewe as it is in woman.

This example reflects the considerable inter-species variation in the energetics of reproduction. Perhaps one of the more extreme examples of reproductive behaviour in placental mammals is to be found in the lactational behaviour of the phocid or true seals (Bonner 1984; Oftedal, Boness & Tedman 1987). The harp and hooded seals (*Phoca groenlandica, Cystophora cristata*) are phocid seals which whelp on the unstable pack ice rather than on fast ice or land. The harp seal weighs about 130 kg and the hooded seal, 180 kg. Lactation is very short in both species – 4 d in the hooded seal and 12 d in the harp seal. The rate of milk energy secretion is 4150 kJ/(kg metabolic weight d) in the hooded seal and 1700 kJ/(kg metabolic weight d) in the harp seal, values which are massively above those given in Table 11.8 for terrestial species. The ratio of rate of milk energy secretion:minimal metabolism for the hooded seal approaches 15, three times that of a superlative dairy cow! The lactating females enter the water during the period in which they suckle but their food intake is considerably less than that of non-lactating females and quite insufficient to sustain this output of energy in the milk. The sources of energy for the very high rate of milk secretion are the reserves of fat and protein laid down before parturition, when food is abundant. Weaning is abrupt; the phocid females simply depart from the rookery at the end of lactation and do not return until the following spring. The young harp seal increases in weight during lactation from a birth weight of 10.8 kg at a rate of 2.5 kg/d, storing considerable amounts of fat – mostly in blubber – until it weighs about 35 kg at weaning. After its abandonment, the harp seal pup consumes no food, or at the most obtains a minute amount, for a period of about six weeks. It loses about 10 kg during this period, about half of which is from the blubber and skin. Metabolic rate per kilogram of body weight does not decline appreciably during the post-weaning fast and when it ends the pup gains weight at a rapid rate (Worthy & Lavigne 1987).

11.5.5 *Food energy supply and reproduction*

In the account of reproduction in birds and mammals given above, little mention was made of the effect of the diet on reproduction. It might almost be inferred that reproduction occurs irrespective of the nutritional state of the animal, but this is not so. A very brief account of the effects of the dietary energy supply on some aspects of reproduction is therefore desirable.

The attainment of puberty in mammals is delayed by under-nutrition, though in some species it can occur at a reduced body weight. Conversely, it seems probable that the secular decline in the mean age of menarche which is seen in many human populations is largely due to better nutrition during early life. Under-nutrition also interferes with the oestrus cycle, examples being the amenorrhoea which occurs in young women with primary anorexia nervosa and a similar condition in over-trained women athletes

(Kirkwood, Cumming & Aherne 1987). In all species that have been studied, under-nutrition in pregnancy affects the birth weight of the young, the effect being most pronounced in the later stages of pregnancy when foetal growth is maximal. In sheep, for example, under-nutrition in the last six weeks of pregnancy reduces birth weight by 40%, and, during the siege of Leningrad in World War II, when women were severely under-nourished, the birth weight of their babies fell by about 15%. The effect of under-nutrition in late pregnancy is greater on the placenta than on the foetus. There is also evidence that under-nutrition in the early stages of pregnancy affects placental development (Robinson 1977). Growth of the mammary gland before parturition is also affected by the maternal energy supply and so too is milk production in the subsequent lactation. The effects of nutrition on lactation have, for economic reasons, been most studied in the dairy cow. The findings, summarised by Broster & Broster (1984) indicate the complex inter-relations involved. Lactation can and does proceed at the expense of the cow's body fat and protein when the food supply is insufficient. This also occurs in other mammals including man. Conversely, an increase in milk energy output occurs in response to an increase in the food supply. Similar sensitivities of reproductive phenomena to dietary energy supply are seen in birds, and again these have been most studied in the domestic fowl (Hocking, 1987).

11.6 Post-natal growth

11.6.1 *The description of growth in weight*

In Section 5.3 the limitations of measurements of gain in body weight as an index of the gain in the enthalpy of combustion of the body were fully discussed; weight increases, particularly those measured over short periods of time, can vary considerably in composition and hence in enthalpy. Measurements of weight increases, however, are the most common ones employed in biological studies, and some discussion of them is warranted.

In growth studies it is usual for the animal to be weighed at frequent intervals thus accumulating a considerable amount of information. The data are then fitted to a mathematical function which summarises them. The most common mathematical expression is a linear function of time when the coefficient is the mean rate of gain in weight per unit time. This may be satisfactory for segments of the overall course of growth but is hardly a device for summarising growth from birth to maturity. Growth in weight is characteristically sigmoid; it accelerates during a short initial period and then declines until, as maturity approaches, it approaches zero.

A large number of different functions have been used to describe this relationship between weight and time. Virtually all of them state that dW/dt, the rate of change in weight with time (or *rate of growth*) is a function of weight at the time that dW/dt is measured. Such functions include the logistic equation, the Gompertz function and the Bertalanffy function. These are derived algebraically in most textbooks of biomathematics (see Causton 1977). They were generalised by F.J. Richards (1959) and have been reviewed and critically analysed by Parks (1982). One simple growth function which applies specifically to the phase of growth when the rate is diminishing is that of Brody (1945). It states that growth rate is proportional to the difference between body and weight at maturity, A, and the weight achieved at the time:

$$dW/dt = k(A - W)$$

or

$$dW/dt = (1/\tau)(A - W)$$

which on integration gives:

$$W(t) = A - B\exp(-kt)$$

which can be written: (11.8b)

$$W(t) = A[1 - \exp(-k(t - t^*))]$$

where $W(t)$ is weight at time, t, usually measured from conception; A is mature weight; B is a constant of integration; k is a constant which can be expressed as $1/\tau$ in which τ is termed the time constant and has the dimensions of time; and t^* is the intercept of the relationship when $W = 0$.

As Parks (1982) has pointed out these various functions, even when they were derived from assumptions about the determinants of growth, do not include any explicit expression of the fact that the supply of energy and nutrients as food are determinants of growth. The involvement of food in growth can be stated generally as:

$$dW/dt = (dW/dF)(dF/dt) \quad (11.9)$$

where W is body weight; and F is cumulative food intake – dF/dt being the rate of intake of food per unit time.

a. Parks' theory of growth

Parks (1982) in deriving his theory of growth considered first the term dW/dF in equation (11.9). He adopted the approach used by Titus & Yull (1928) who showed that the law of diminishing returns applied to the relation between gain in weight and food consumed. In effect they stated that the relation of weight to cumulative food intake could be expressed as:

$$W(F) = (A - W_0)[1 - \exp(-bF)] - W_0 \quad (11.10)$$

where $W(F)$ is body weight at any particular value of cumulated feed intake, F; A is mature body weight; W_0 is initial body weight when $F=0$; and b is a constant. Parks' approach was to devise a descriptive exponential equation relating the rate of food intake, dF/dt in equation (11.9), to time. This equation assumed that the rate of food intake increased with time toward an asymptote. By substituting the integral of this equation into the equation of Titus & Yull (equation 11.10), Parks arrived at a Gompertz-type equation relating body weight to time and the constants of the equation explicitly related to food intake. Parks showed that if rate of food intake is constant the equation reduced to Brody's equation (equation (11.8)).

b. A general differential equation describing growth

The problem of relating growth to both time and food intake can be approached in another way and specifically in terms of energy (Blaxter 1968). The rate at which metabolisable energy is ingested by an animal has hitherto been depicted as $\dot{M}_E$. For algebraic convenience it is now expressed as dM_E/dt. The rate at which metabolisable energy has to be supplied solely to meet maintenance demands (that is the rate when energy retention is zero) can be stated to be some function of the animal's body weight, $f(W)$. The retention of energy in the body of a growing animal is simply the rate of gain in weight multiplied by the specific enthalpy of combustion of the gain, c. The metabolisable energy required to promote unit gain of energy is the energy gain divided by the efficiency with which metabolisable energy is used to promote gain, k_{f+p}, where the subscript, f+p, indicates that the gain conventionally consists of fat and protein. The factors which determine k_{f+p} are discussed in Section 12.1. A relationship can thus be devised:

$$\frac{c}{k_{f+p}}\frac{dW}{dt}+f(W)=\frac{dM_E}{dt} \tag{11.11}$$

This relationship is perfectly general; it does not preclude the enthalpy of combustion of unit gain varying with body weight or predicate any particular algebraic form for the maintenance function or for the rate of supply of metabolisable energy. Several solutions to the equation have been given (Blaxter, Fowler & Gill 1982). One of interest is when rate of metabolisable energy, dM_E/dt, is constant at Z; the rate of maintenance metabolism is aW^n; and the term (c/k_{f+p}) is constant at b. Then the solution is:

$$W-W_0=\frac{Z}{anW^{n-1}}\left[1-\exp\left(\frac{-anW^{n-1}}{b}\right)t\right] \tag{11.12}$$

Other solutions making different assumptions about the algebraic form of the maintenance function and of the rate of input of metabolisable energy all show that the asymptotal weight is the rate of consumption of metabolisable energy divided by the rate of maintenance metabolism per unit weight. In the above example it is Z/anW^{n-1}. The rate at which this asymptote is attained is the rate of maintenance metabolism per unit weight divided by the metabolisable energy required to promote unit gain in weight, c/k_{f+p}. The implications of this finding will be dealt with later.

c. Growth in different species and Taylor's generalisation

Many studies of growth in weight in different animals have been made and summarised algebraically. The most complete compilation is still that of Brody (1945) who, using equation (11.86), calculated the constants A and k (or $1/\tau$) from 93 sets of data, ranging from those for mice to those for horses. The growth data for all the species examined were adequately described by equation (11.8b) – with the exception of the data for man, who has an excessively long juvenile period of growth. Apart from man, Brody showed that the growth curves for all the species he examined could be superimposed by expressing weight as a proportion of the mature weight of the species and time on a scale of $k(t-t^*)$. Other primates also have protracted juvenile phases of growth. Figure 11.3 (from Evans & Miller 1968) illustrates the divergence of five primate species from equation (11.8c) and also data, not available to Brody, for African elephants and polar bears. The latter two species agree with Brody's contention.

It was evident to Brody that, between species, the constants A and k were related and that large species (with large mature weights) approached mature weights more slowly than small species. St.C.S. Taylor (1965) examined this relationship and showed that:

$$\tau = 1/k = 100A^{0.27} \tag{11.13}$$

where A is mature body weight and the error attached to its exponent, 0.27, is ± 0.04. The mean coefficient ranged from 93 to 115. This finding was a statistical one, and, as in the original data, the relationship did not involve any consideration of energy transformations. It was shown by Blaxter (1968) that the solution to the differential equation relating energy intake to growth in weight (equation (11.11)) was of the same form as Brody's equation and that the time constant of this equation was:

$$\text{time constant} = c/anW^{(n-1)}k_{f+p} \tag{11.14}$$

The question arises whether, inserting reasonable values for the parameters into equation (11.14), the resulting time constant agrees with the statistically determined value of τ found by St.C.S. Taylor (1965). The metabolisable energy required for maintenance in the laboratory species and domesticated animals which Brody used is probably about twice the minimal rate of metabolism, rather than 2.5–3.0 times the minimal metabolism as the field metabolic rates of wild species shown in Table 9.6 might suggest. Maintenance requirements of adult animals can therefore be estimated as $2 \times 300W^{0.75}$ kJ/d. The constant a thus has a value of about 600 kJ/d and n is 0.75. The efficiency of utilisation of metabolisable energy for promoting gain varies from species to species as shown in Table 12.1; k_{f+p} assumes a mean value of 0.70. The enthalpy of combustion of gains in weight also varies appreciably with species, domesticated animals having gains which contain appreciably more fat than laboratory animals; an average value for all species of c, the enthalpy of combustion of a gain in weight, can be taken to be 25 000 kJ/kg. Insertion of these values into equation (11.11) shows that:

$$\text{time constant} = 80W^{0.25} \tag{11.15}$$

The estimate from energy considerations is thus not very different from that found statistically by Taylor. In Figure 11.4 the Taylor constant is related to

Fig. 11.3. A comparison of the form of the growth rates of primates and other animals. The ordinate represents weight as a percentage of that at maturity and the abcissa is scaled time from conception. Most animals conform to the general relationship for non-primates. Primates, and particularly man, have long juvenile phases of growth. From Payne & Wheeler (1968).

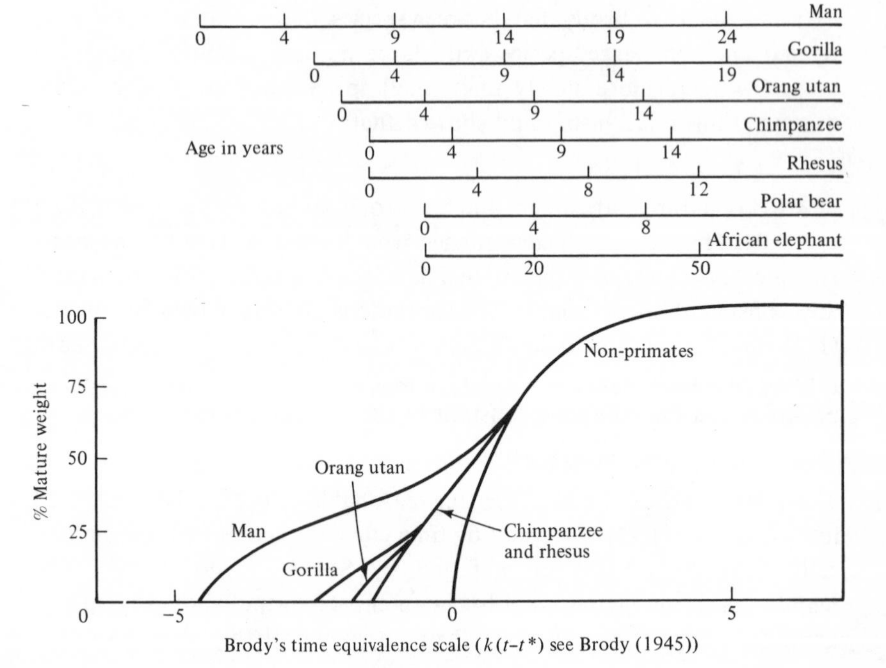

that computed from energy relations together with some of the original observations by Brody.

11.6.2 *The composition and enthalpy of combustion of gains in weight*

No doubt influenced by Huxley's (1932) studies in which he used an allometric equation to relate the weights of organs of animals to the weights of their whole bodies, many of the studies which have been made relating chemically-defined attributes of the body to body weight have also employed allometry. The allometric equation is simply:

$$\text{weight of part} = a(\text{weight of whole})^b \qquad (11.16)$$

If the exponent, b, is greater than 1.0 then the part increases in weight at a rate which is greater than that of the body as a whole; if it is less than 1.0 it increases in weight at a lesser rate than that of the whole body.

Studies, mostly with domesticated animals and laboratory ones, show that during growth, $b > 1.0$ for fat and $b < 1.0$ – usually between 0.7 and 0.9 – for protein. The exponent for fat is about 1.8 in cattle, 1.9–2.1 in sheep and

Fig. 11.4. A comparison of (solid curve) time constants (τ) of growth curves of animals of different mature size as estimated statistically by Taylor (1965) and those derived from energy relationships according to equation (11.14). The points refer to the determinations by Brody (1945) of the time constants of growth for the species concerned.

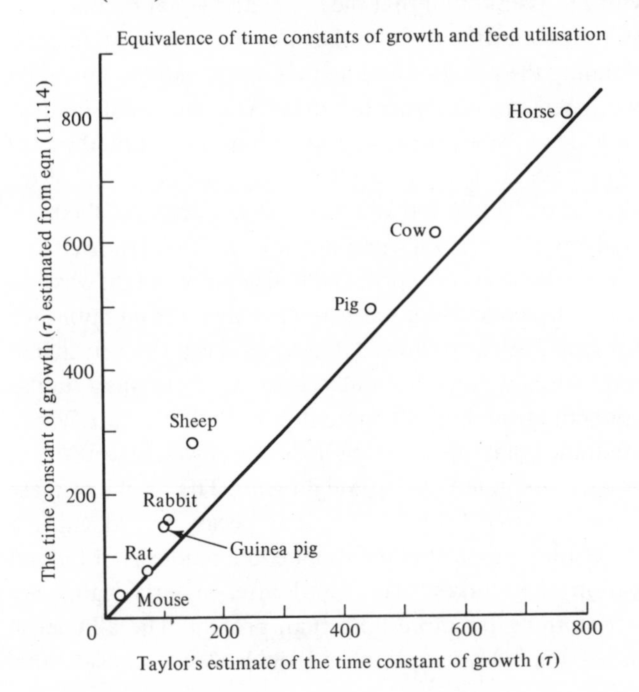

1.5–1.7 in pigs. Differentiation of these allometric relationships gives the fat and protein contents of the unit gain; fat content of unit gain thus increases and protein content decreases with increase in body weight. Obviously, according to this descriptive model of fat and protein growth, the enthalpy of gains in weight increases continuously with body weight. An example clarifies matters. The allometric equations for the fat and protein contents of sheep and cattle were averaged by the Agricultural Research Council (1980). They show that in a lamb weighing 10 kg (digesta-free weight) the enthalpy of combustion of the gain would be 8 MJ/kg but at an ingesta-free weight of 40 kg it would be 20.5 MJ/kg.

A considerable amount of dissection and analytical work on a large number of animals is needed to obtain sufficient information to enable an allometric relationship to be fitted. Domestic animals destined for meat production are usually killed when far from their ultimate size and much of the analytical data for them relates to animals which have been killed at these 'economic' weights. A notable advance was made when Australian workers used indirect methods to determine the fat and protein contents of the bodies of individual animals at intervals during their growth (Searle, Graham & O'Callaghan 1972). They gave sheep unlimited access to good quality food and showed that after puberty, at 25–30 kg, the composition of the gain in weight was constant irrespective of body weight. This conclusion does not agree with the assumption that the allometric equation describes growth. Although the proportion of fat in the body increases with increasing weight this is simply the result of adding to a body, initially low in fat content, successive increments which are rich in fat. The Australian findings – which might be criticised since they were based on indirect methods of determining body composition – were confirmed by direct analysis of the bodies of sheep also fed *ad libitum* and killed at different weights (Blaxter *et al.* 1982). From analyses of published work, Searle *et al.* (1972) showed that the same constancy of composition of post-puberal gains of weight was true of cattle and pigs and this conclusion was extended, again from published work, to mice and dogs. They also showed that when sheep were given half the amount of food that they would normally consume, there was a similar linearity of post-puberal growth of fat content but at a lower rate. When sheep were underfed, such that they lost weight, losses of weight were of the same composition as those noted during weight gain. The results of these important studies are shown in Table 11.10. In the species studied, gains in weight during the suckling phase contain about equal amounts of fat and protein. A transition then takes place until after-puberty gains are characterised by containing far more lipid than protein. The allometric model provides a reasonable description of growth over a considerable

Table 11.10. *The composition of gains in weight in cattle and sheep computed from linear regressions of body components on body weight in the post-puberal period of growth*

Species	Composition of gain (g/kg)			Enthalpy of combustion of (MJ/kg) gain	Data source
	Fat	Protein	Water		
Sheep					
fully fed	650	87	242	27.6	Searle *et al.* (1972)
limited	551	97	322	23.9	Searle *et al.* (1972)
fully fed	681	77	241	28.5	Blaxter *et al.* (1982)
Cattle					
fully fed	559	122	305	24.8	Haecker (1920)
fully fed	661	67	237	27.6	Moulton, Trowbridge & Haigh (1922)
limited	404	182	343	20.2	Moulton *et al.* (1922)

Note:
Based on linear regressions of body composition on weight and calculation of the enthalpy of the gain using the factors: fat = 39.3 MJ/kg and protein = 23.6 MJ/kg.

range of body weight from birth; after puberty, however, it does seem that the increments made are constant in composition rather than progressively increasing in fat content as an allometric model supposes. More analytical information on a larger range of species is required to assess the best ways of describing growth.

a. The composition of gains in weight in adult man

Table 11.10 shows that the composition of gains made by sexually mature cattle and sheep given adequate food *ad libitum* contain 60–70% lipid, and less if the amount of dietary energy supplied is less. The same appears to be true of the gains made by adult man. In obese subjects the excess weight appears to contain 30% non-fat tissue and 70% fat, while, in short-term, experimental studies in which subjects have been overfed, gains contain 62% fat and 38% non-fat tissue (Forbes *et al.* 1986). Experimental studies with monkeys show that their gains contain 66% fat. Assuming that the enthalpy of combustion of the fat-free component of the gain is about 23 kJ/g, the enthalpy of combustion of the gains in weight made by adult man range from 25 to 30 MJ/kg.

Realising that weight gains made in adult life contain considerable fat, it is pertinent to ask what gains are reasonable, or, alternatively, what is the ideal composition of the human body, or what constitutes such a departure from normality that a subject should be classed as obese? The criterion which is probably most acceptable relates to the association of mortality with body fat content. There is considerable evidence that mortality is increased as the ratio of weight to the square of height (the *Quetelet* or *ponderal index*) increases. The fat content of the body can be estimated from the Quetelet index by the equations:

$$\left.\begin{array}{l}\text{Men}\\ \text{fat (\% of body weight)} = 1.28(W/H^2) - 10.1\\ \text{Women}\\ \text{fat (\% of body weight)} = 1.48(W/H^2) - 7.0\end{array}\right\} \quad (11.17)$$

where W is body weight (kg); and H is height (m) It is thought that 'normal' or 'ideal' body fat contents are in the region of 14–20% for males and 21–27% for females. Grade I obesity is defined as a Quetelet index of 25–30, corresponding to a body fat content in men of 22–28%. Grade II represents a Quetelet index of 30–40, corresponding to a fat content in men of 28–41%. Grade III individuals are those with a Quetelet index >40 (Garrow 1981). The prevalence of obesity as so defined can be estimated from height and weight statistics for the UK. At present, among men aged 16–64, 34% are in Grade I, 6% in Grade II and 0.2% in Grade III.

The 'ideal' fat content of man can be compared with that for animals. Pitts & Bullard (1968), from analyses of the bodies of 49 species, derived the allometric equation

$$\% \text{ fat in the body} = 6.01\,W^{0.20} \quad (11.18)$$

where W is body weight (kg). This equation suggests that, from the concordance of other species, the percentage of fat in the body of man should be about 14%, a value which is at the lower end of the range of accepted normality.

b. Dietary and other factors affecting the composition of weight gain

The variation in the enthalpy of combustion of a gain in body weight reflects the fat content of the gain, and as was discussed in Section 5.3, this can vary appreciably. In young animals given nutritionally adequate diets, the proportion of fat in the gain increases with increasing amount of the diet given. This is because in young animals an energy retention of zero is associated with a gain of protein, water and minerals and a loss of fat. If the protein concentration in the diet is low, protein retention at the energy maintenance level is depressed, and as the amount of feed given is increased, the ratio of the energy gained as fat to the gain of protein is increased. These effects are illustrated very well in Figure 11.5 which refers to experiments with baby pigs (Campbell & Dunkin 1983). When the intake of the liquid

Fig. 11.5. The effect of protein content of the diet on the retention of fat and protein by baby pigs; (*a*) shows the relation between retention and energy intake from a diet adequate in protein, and (*b*) the relation for a protein-deficient diet.

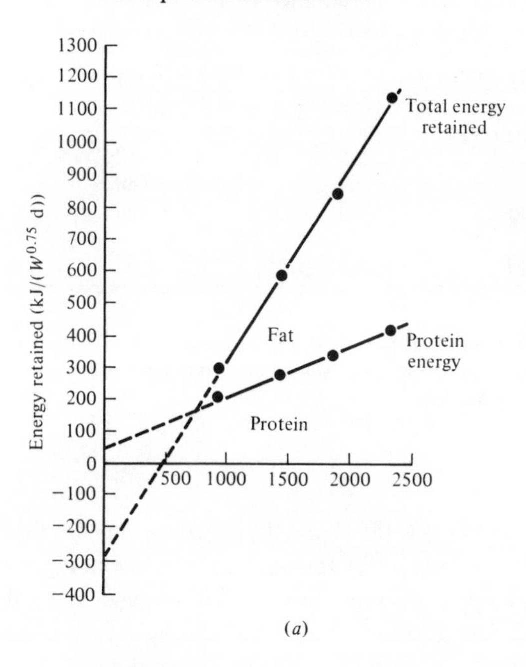

(*a*)

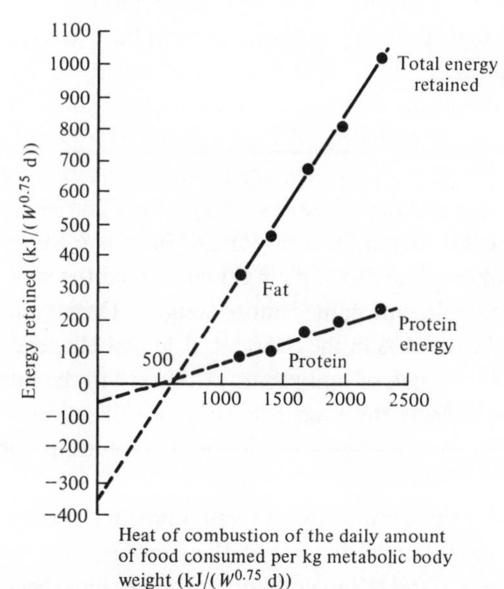

(*b*)

Table 11.11. *The effect of a high-fat diet on the composition of body gain in rats*

	Laboratory diet	High (42–60%) fat diet
Total intake in 60 weeks (MJ)	1510	1511
Final body weight (g)	714	916
Digesta and hair-free weight (g)	666	880
Total body fat (g)	205	452
Fat content of gain (g/kg)	560	880
Non-fat content of gain (g/kg)	440	120
Calculated enthalpy of combustion of gain (MJ/kg)	25.3	35.1

Note:
It was assumed that the fat-free material has a protein content of 20.6%, that protein has an enthalpy of combustion of 23.6 MJ/kg and that fat has a enthalpy of combustion of 39.3 MJ/kg.
Source: The results were recomputed from the experiments of Oscai *et al.* (1984).

diet simulating milk was 2.3 MJ/(kg $W^{0.75}$ d), the piglets given the diet adequate in protein retained 53.2 g/d of fat and 48.9 g/d of protein. Piglets given the same amount of energy from a low-protein diet retained 56.2 g/d of fat and 26.9 g/d of protein. At zero energy retention the piglets given the normal diet retained protein; those given the inadequate diet did not. This type of experiment in which diets differing in protein content are given is discussed further in Section 12.3.2.

Increases in the lipid content of the diet also increase the fat content of gains in body weight and hence the enthalpy of combustion of the gain. In Table 11.11 the results are presented of an experiment in which rats were given unlimited access to normal and high-fat diets for a period of 60 weeks – an extremely long experimental period for a rat (Oscai, Brown & Miller 1984). The rats which were offered the diet high in fat consumed the same amount of energy as did controls, but gained more weight. Their gains contained 88% fat compared with 56% in the controls. A higher efficiency of conversion of food energy into gain of body energy occurred in the rats given the high-fat diet. This reflects the high efficiency of utilisation of metabolisable energy derived from fat in promoting energy retention as is evident from Table 12.1.

Endocrine factors also affect the composition of gain made by animals receiving the same amount of the same diet. Anabolic steroids, notably androgens, oestrogens, progestins and their synthetic analogues have been much investigated in relation to the production of meat low in fat content.

Either singly or in combination they increase the rate of protein and water deposition and depress the rate of fat deposition (Heitzman 1978).

Again with a view to reducing the fat content of meat produced from domesticated animals, there has been an interest in the use of sympathomimetic drugs and particularly of the synthetic adrenergic agonist, clenbuterol. This compound is active by mouth and when given to calves results in a depression of fat retention and an increase in both protein retention and in heat production (Williams *et al.* 1987).

11.7.3 *Compensatory growth*

If growing animals are underfed to the extent that their weight is stabilised or they lose weight, and if, after a period of such underfeeding, they are then given food in adequate amounts, their rate of gain in weight is considerably accelerated above that in normal animals given the same amount of food. This acceleration of gain in weight after under-nutrition is termed *compensatory growth.* The same phenomenon is seen in human babies and children. If after failure to grow due to malnutrition or infection, adequate food is given, growth rate is enhanced until the child reaches the weight expected for its age. Subsequently growth is at the expected rate. In man the term used for compensatory growth is 'catch-up growth'.

Compensatory growth and catch-up growth are terms specifically applied to increases in weight. The evidence from studies in a variety of species is that this growth in weight has a lower than normal enthalpy of combustion. Water and protein contents are higher and fat content is lower than in the normal growth of an animal of the same weight. In addition, there may be a depression of minimal metabolism, muscular activity and maintenance heat production during under-nutrition and these effects are not immediately reversed when food becomes available.

12

The utilisation of the energy of food

12.1 Efficiencies of utilisation of metabolisable energy

As was briefly discussed in Section 2.4 ingestion of food increases both heat production and the retention of energy in the body. Figure 12.1 shows the relation between energy retention and the enthalpy of combustion of the food ingested by adult sheep and adult cattle. Each datum point was obtained by giving the animal a precise amount of food for three weeks and determining the energy retention during the last four days of the period. The results are expressed as daily rates of gain or loss of energy and of total (not metabolisable) energy intake ($\dot{I}_{\mathrm{E}}$). The relationship between daily energy retention and the enthalpy of combustion of the ingested food is distinctly curvilinear; the increase in retention per unit food ingested falls as intake increases. Part of this is due to the decline in the metabolisable energy of unit food as intake increases. This was discussed in Section 3.6 and shown to be particularly evident in ruminants. However, even when changes in metabolisability with increased food intake are discounted by plotting retention against metabolisable energy intake, the relationship is still not rectilinear. In all species for which such measurements have been made, the slope is distinctly higher below maintenance (zero energy retention) than it is above, and there is evidence that a further reduction in slope occurs at very high feeding levels, although this is not invariably observed in all species and with all diets.

The slope of the relationship between rate of energy retention, $\dot{R}$, and metabolisable energy intake, $\dot{M}_{\mathrm{E}}$, is termed the efficiency of utilisation of metabolisable energy and is $\mathrm{d}\dot{R}/\mathrm{d}\dot{M}_{\mathrm{E}} = k$. Since the relationship is not rectilinear, k varies according to the point at which it is measured. It has become conventional to express the slope not as a differential but as an average one. The *efficiency of utilisation of metabolisable energy for maintenance*, denoted k_{m}, is thus the average slope measured within the bounds of zero food intake and a food intake that maintains zero energy balance. The *efficiency of utilisation of metabolisable energy for the net deposition of fat and protein in the body*, denoted k_{f+p}, is the average slope measured with the lower bound at maintenance and the upper bound at some higher level. For ruminant animals this upper bound is usually taken to be an amount of feed which is twice the maintenance level. This 'model' is of course an approximation for it substitutes a continous relationship by two intersecting straight lines. It seems improbable, in view of the dynamic nature of biochemical reactions in the body, that suddenly, at maintenance, the utilisation of the energy of food changes as exemplified by a change from an efficiency of utilisation, k_m, to one of k_{f+p}. Indeed the model of ruminant energy metabolism of Blaxter & Boyne (1978) describes the relationship between $\dot{R}$ and $\dot{I}_E$ by a continuous curve and derives the values of k_m and k_{f+p} from it algebraically.

In passing it should be noted that efficiency of utilisation of dietary

Fig. 12.1. The relation between food intake and energy retention in adult sheep and cattle. Food energy intake is expressed as the enthalpy of combustion of the diet.

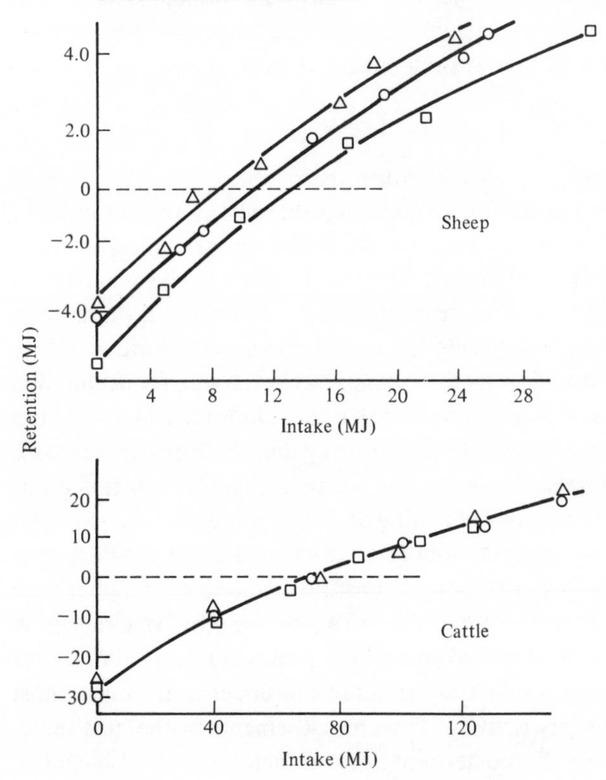

energy is sometimes expressed in other ways. The most common is the ratio of *gain* in energy by the body divided by the metabolisable energy or the heat of combustion of the food energy ingested. The value of this ratio is zero when there is no gain and increases with increasing energy retention.

12.1.1 *Increments of heat associated with food ingestion*

The retention of energy in the body is the difference between the intake of metabolisable energy, $\dot{M}_E$, and the heat produced, $\dot{H}_P$. It follows that the increment of heat per unit metabolisable energy, h_I, is related to k in the following way:

$$k = \frac{\Delta \dot{R}}{\Delta \dot{M}_E} = \frac{\Delta(\dot{M}_E - \dot{H}_P)}{\Delta \dot{M}_E} = \left(1 - \frac{\Delta \dot{H}_P}{\Delta \dot{M}_E}\right) = 1 - h_I \tag{12.1}$$

The value of estimates of efficiencies, k, and of heat increments, $(1-k)$, is that they facilitate calculation of the amounts of heat produced and of metabolisable energy required in a variety of circumstances. Thus the amount of metabolisable energy to be consumed to meet maintenance needs of an animal completely at rest is simply $\dot{H}_B/k_m$, where $\dot{H}_B$ is basal metabolism, and this is equal to the amount of heat the animal produces at maintenance. If the animal consumes 25% more food than maintenance (and still continues in the resting state) it will retain in its body $k_{f+p}[0.25(\dot{H}_B/k_m)]$.

12.1.2 *Measurement of efficiency or heat increment*

Efficiencies of utilisation of metabolisable energy are measured by determining energy retention when two different amounts of metabolisable energy are given. Division of the increment of retention by the increment of metabolisable energy is a measure of efficiency. With large farm animals and laboratory species, the two determinations of metabolism are usually of 24-h duration and are made after the animals have been given constant food for several days. It is implicitly assumed that the minor muscular activities due to body movement which take place during the whole day are the same at both nutritional levels, an assumption which is difficult to sustain if one of the nutritional levels is that of fasting and even more so if the value for energy retention at zero food intake is calculated from a short-term determination of minimal metabolism. Fasting metabolism is higher than minimal metabolism. In the pig, direct measurements by McCracken & Caldwell (1980) showed that its fasting heat production was about 15% above its minimal heat production. In addition, under-nutrition in most species reduces voluntary activity. Thus measurements of the heat increment of food based on 24-h determinations of metabolism can include a component which is the result of increased body movement and activity at higher nutritional levels. Recognising this, in earlier calorimetric work attempts were made to correct all measurements of heat production to a standard day of 12 h standing and 12 h lying. Such corrections are not precise and they are now seldom used. It can be argued that a change in minor activity is a consequence of the ingestion of food and, in an accounting sense, reasonable to allocate its cost to the food.

The above procedure for measuring the increase in retention of energy or of heat production due to an increase in food intake is time consuming because it involves giving the diets for long periods and making measurements of heat production for several days. It is understandable that only a limited number of such experiments have been made with man. With man the usual procedure is to measure the response of heat production to a single meal. Heat production is first measured under standard conditions – sitting comfortably or lying – the meal is then consumed and thereafter heat production is measured at intervals under the same conditions for several hours. The summated increase in heat production over the initial measurement is termed the *specific dynamic effect of food* (SDA) and is related to the enthalpy of combustion of the meal, to the metabolisable energy of the meal, or, more rarely, is expressed as a percentage increase in the rate of metabolism. The term *specific dynamic effect* was first used by Rubner (1902) and the approach was used by Lusk (1928) in his classic studies of metabolism. The method has the advantage over the 24-h measurement that the amount of incidental muscular activity is controlled. Its disadvantage is that to obtain the overall thermic effect of the meal, observations must be continued until heat production has returned to a baseline value. What this baseline value should be may be conjectural in that there are rhythms in metabolism which affect heat production in the absence of a meal. Control experiments in which no food is given could provide a valid baseline, but are rarely conducted.

The time course of heat production in man after a meal is shown in Figure 12.2. Characteristically, metabolism rises to a peak and then declines. The rate of fall after the peak is slow and protracted. In a meal-eating subject it can be envisaged that the effects of meals are superimposed one on another and if the length of time for which the SDA persists is longer than the interval between meals, then the baseline will decline consequent upon the persisting metabolic effects of previous meals. Effects in man certainly persist for 15 h or more and, as Flatt (1985) implied, to observe the total effect may involve more than a day's observation of the heat production of a subject both when fed a meal and when fed none, if the real thermic effect of the meal is to be ascertained. Many of the determinations of the SDA are made, however, for periods of only 4–6 h, some for less, and only rarely are they continued or extrapolated to times after eating which would include the whole of the heat arising from the meal.

An alternative approach has been used with patients who receive nutrients by continuous parenteral administration. Here a steady state of metabolism of food can be assumed and measurements of heat production for short periods at the same time of day when different amounts of the infusion mixture are given overcomes the difficulty encountered when the course of a single meal is followed. The assumption is made that minor muscular activity is the same at each nutritional level. Studies using this approach have usually been made with ill patients rather than with healthy people; whether the results are a true indication of normal metabolism can be questioned.

Both the measurement of the increment of heat following long-term subsistence on two different amounts of food and the measurement of the time course of heat production after a single meal are thus open to criticism. Table 12.1 gives some values for the efficiency of utilisation of metabolisable energy for both maintenance and for protein and fat deposition. These relate to different species of animal when carbohydrate, fat or protein are the energy sources. The values were mostly obtained using the 24-h

Fig. 12.2. Change in the rate of heat production by human subjects above initial values following the ingestion of high fat (●), high carbohydrate (■) or mixed diets (▲). The specific dynamic effect of food or its heat increment is calculated as the heat produced before metabolism returns to this initial, pre-feeding value – in this instance about 15 h after consumption of the meal. Note that control measurements in which no food was given were not made. From Acheson *et al.* (1984).

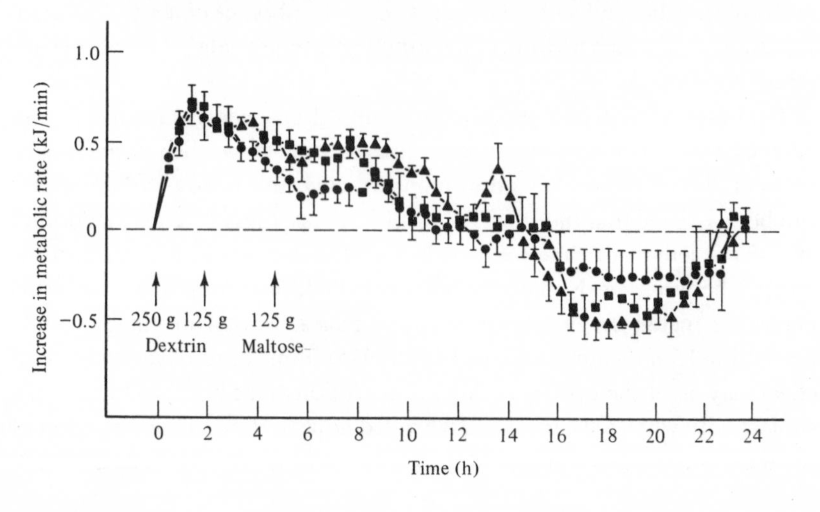

Table 12.1. *Average efficiencies of utilisation of metabolisable energy below maintenance and for deposition of fat and protein above maintenance together with the corresponding heat increments*

		Efficiency		Heat increment	
Nutrient	Species or group	below maintenance	above maintenance	below maintenance	above maintenance
Carbohydrate	Simple-stomached	0.94	0.78	0.06	0.22
	Ruminants	0.80	0.54	0.20	0.46
	Other herbivores	0.90	0.64	0.10	0.36
	Birds	0.95	0.77	0.05	0.23
Fat	Simple-stomached	0.98	0.85	0.02	0.15
	Ruminants	?	0.79	?	0.21
	Other herbivores	?	0.79	?.	0.21
	Birds	0.95	0.78	0.05	0.22
Protein	Simple-stomached	0.77	0.64	0.23	0.36
	Ruminants	0.70	0.45	0.30	0.55
	Other herbivores	0.76	0.50	0.24	0.50
	Birds	0.80	0.55	0.20	0.45
Approximate values for species consuming average diets					
	Man	0.90	0.75	0.10	0.25
	Rat	0.90	0.75	0.10	0.25
	Dog	0.85	0.70	0.15	0.30
	Pig	0.85	0.70	0.15	0.30
	Rabbit	0.80	0.65	0.20	0.35
	Horse	0.75	0.60	0.25	0.40
	Ox	0.70	0.50	0.30	0.50
	Sheep	0.70	0.50	0.30	0.50
	Chicken	0.90	0.75	0.10	0.25

Note:
Values of both efficiencies and heat increments for different species are expressed to the nearest 0.05.

balance technique. The table shows that efficiency of utilisation for maintenance is invariably higher than is efficiency for fat and protein deposition and that the metabolisable energy of protein is used less efficiently than are that of carbohydrate or fat. Table 12.1 also shows that ruminants in particular, and herbivores to a lesser extent, utilise the metabolisable energy of nutrients with a lower efficiency than do species with simple digestive tracts. The final part of Table 12.1 gives some average values for the efficiency of utilisation of metabolisable energy when different species consume what may be regarded as their habitual diets. These data again show the differences in utilisation below and above maintenance and the differences between ruminants, non-ruminant herbivores and omnivores.

12.2 The determinants of the heat increment

12.2.1 *The origin of the heat increment*

Why heat production should increase after food is a question which has intrigued investigators for many years. Several general theories to account for the SDA have been propounded, beginning with those of Voit, Zuntz and others in the nineteenth century. One of these older theories which is still valid is that of Rubner (1902). He ascribed the SDA to the waste heat produced by reactions necessary to support the physiological processes of the body. This is such a general statement that it can accommodate a wide variety of modifications. Mitchell (1962), considering various theories and their subsequent modifications, wrote

> Attempts to determine the cause of the specific dynamic action of food have been largely unsuccessful and ephemeral because they have assumed or implied that there is one cause only, or one factor that dominates all others. As a matter of fact the causes are many, some of undoubted reality while others are highly speculative in nature though often highly probable . . .

An immense amount of information has now accrued about the factors which influence the increment of heat and the efficiency with which the energy of food is used by different animal species. It is thus relevant to consider this information even if a complete and sufficient theory eludes us. The various components of the increment of heat are first discussed separately.

a. Heat related to the act of eating

The act of prehending, masticating, and swallowing food involves the musculature. In addition salivary secretion takes place. In sheep, eating is associated with a reduction in the ventilation of the lung and a very large increase – up to 50% – in oxygen consumption (Blaxter & Joyce 1963b). A similar rapid increase in oxygen consumption takes place in the pig; a doubling of heat production occurs during the combined activity of eating, standing and movement and excitement associated with the anticipation of the meal. Comparison of the heat production of chickens, given the same amount of food by intubation of the crop as they eat normally, suggests a high energy cost of eating. The experiment does not, however, enable the act of eating to be isolated precisely because muscular activity (measured by Doppler radar) was massively increased in the self-fed fowl (MacLeod, Jewitt & Andersen 1987). In man it is also difficult to separate an increase in heat production during eating from the effects of movement. In the sheep, when eating stops, heat production falls rapidly – within a few minutes – to values which are only slightly above the pre-prandial level. If the animal is sham-fed, that is, the food is chewed and swallowed but escapes from an oesophageal fistula, the increase in heat production is the same as that in the normal animal (B.A. Young 1966). If the effect of eating is isolated as the difference in metabolism between the animal eating normally and that found when food is introduced into the gut an elevation in metabolism is still every marked (Osuji, Gordon & Webster 1975). The increase in heat production observed during eating is thus not complicated by events taking place later in the digestive process.

Webster (1972) showed that the effect of eating on the metabolism of sheep varies relatively little per minute but an elevation of about 40 J/kg body weight persists throughout the meal, that is, the magnitude of the increment of heat due to eating depends on the time spent eating. This has been confirmed by Adam *et al.* (1984). It is illustrated in Table 12.2 which summarises observations on both sheep and cattle. In the final column of the table an estimate has been made of the increment of heat as a percentage of the metabolisable energy of the diet. This suggests that the act of eating accounts for an increment of heat of about 3% of the metabolisable energy ingested, varying from feed to feed and being very low in the instance of pelleted feed or grain which is consumed very quickly.

Ruminants ruminate. During rumination heat production also increases but the measured increase is very small compared with the increase due to eating (Osuji *et al.* 1975). It represents 0.3% or less of the metabolisable energy consumed – about a tenth of the cost of eating. The amount of chewing and the rate of salivary secretion when eating or when ruminating are not too dissimilar, suggesting that the increase in metabolism during eating cannot be explained by an increase in muscular work or work of salivary secretion. Webster (1972) has shown that complex neural and hormonal effects are involved, but the precise mechanism remains un-

Table 12.2. *Increments of heat associated with the act of eating in ruminants*[a]

		Increment of heat		
Food or diet	Rate of ingestion of dry matter (g/min)	per unit body weight and time (J/(kg/min))	per unit body weight and dry mass (J/(kg/g))	as a % of the $\dot{M}_E$
Cattle				
Fresh cut forage	20–25	40.2	1.95	5.9
Long, dried forage	17–37	32.5	1.48	4.9
Chopped, dried forage	39	27.6	0.79	2.5
Pelleted food	130–138	23.8	0.23	1.0
Chopped turnip	30	34.7	1.43	3.1
Simulated grazing	13–17	39.0	3.42	2.5
Mean		33.0	1.55	
Standard deviation		6.4	1.09	
Sheep				
Fresh cut forage	4–7	36.2	5.12	2.6
Long, dried forage	8–9	54.9	6.95	3.9
Chopped, dried forage	4–14	38.1	4.01	2.5
Pelleted food	8–58	36.6	1.07	0.4
Simulated grazing	1–2	37.7	25.1	11.4
Mean		40.7	8.45	
Standard deviation		8.0	9.55	

Note:
[a] Adapted from a summary of published work by Adam *et al.* (1984). Body weights and the metabolizable energies of food were assumed.

known. There are no comparable data on the quantitative effects of the act of eating in other animals.

b. Heat arising from events in the lumen of the gut

The enthalpy of the enzymic hydrolyses of lipid, polysaccharides and proteins in the lumen of the gut has already been estimated to be about 0.1–0.2% of the energy of the substrates hydrolysed (see Section 3.6.3). The increment of heat as a percentage of the metabolisable energy can be taken to be the same. This increment of heat does not include any heat produced consequent upon the secretion of enzymes into the gut; it is simply the enthalpy of the hydrolyses. Heat arising from the anaerobic fermentations in the gut also contributes to the increment of heat associated with food ingestion. This was also discussed in Section 3.6.3. It has been investigated mostly in ruminants and particularly by Webster and his colleagues (Webster *et al.* 1975; Weekes & Webster 1975; Webster 1978) as part of their study of the determinants of the heat increment. Webster measured simultaneously the heat produced and the oxygen consumed by the viscera drained by the portal vein as well as the metabolism of the sheep as a whole. The total heat production by the tissues of the digestive tract was obtained using the Fick principle, namely as the product of rate of blood flow, the arterio-venous difference in temperature of the blood and its specific heat. The corresponding arterio-venous difference in oxygen consumption multiplied by rate of blood flow and the heat equivalent of the oxygen, represents the aerobic metabolism of the gut. The difference between the two measurements represents the heat due to anaerobic fermentation. When the sheep were starved for two days total heat and aerobic heat were virtually identical, that is, there was negligible fermentation heat. When food intake was increased, blood flow increased exponentially with metabolisable energy intake and differences between total heat and aerobic heat emerged. The increment of anaerobic heat or fermentation heat as a percentage of the enthalpy of the apparently digested diet (enthalpy of feed less that of faeces) averaged 5.9%. This agrees well with the stoichiometric calculations made by Hungate as given in Section 3.6.3.

These experimental results also suggest that the ratio of fermentation heat to the enthalpy of combustion of the methane produced is about 0.5–0.7, values which agree with calculations derived from equations (3.7), (3.8), (3.9) and (3.10). This implies that both in ruminants and in species which ferment food in the hind gut, the heat of fermentation can be calculated from methane production. Applying a factor of 0.6 kJ fermentation heat/kJ methane produced to the data in Table 3.5 suggests for example, that in sheep given straw the increment of fermentation heat as a percentage of the metabolisable energy supplied is about 13% falling to about 9% when the sheep is given grain. In the horse the comparable values are 3% and 2% for straw and hay, respectively. In man, taking into account the hydrogen produced, calculations suggest that the fermentation heat is considerably less than 1% of the metabolisable energy ingested.

In simple-stomached species attempts have been made to estimate the effect of microbial fermentation on the heat increment of food by using germ-free animals or by reduction of the gut flora by the administration of antibiotics. Such studies have shown that the efficiency of the utilisation of metabolisable energy is increased (and the heat increment depressed) in germ-free birds compared with normal birds (Charlet-Lery & Szylit 1980; Furuse & Yokata 1984). In rats given antibiotics to reduce their gut flora a similar effect has been found (Eggum & Chwalibog 1982). In the latter experiments a high-fibre diet was given. The increase in the efficiency of utilisation of metabolisable energy for fat and protein deposition (k_{f+p}) due to a reduction in the flora was 8 kJ/100 kJ metabolisable energy. Not all the increase can be attributed to the heat associated with the fermentation; part arises from the low efficiency with which the energy of the products of the fermentation is used.

c. Heat arising from the absorption of nutrients

The amino acids and small peptides, which are the products of digestion of protein, are absorbed by active transport processes involving carrier proteins. Four such carrier mechanisms have been identified, catering for the neutral amino acids, the basic amino acids, for dicarboxylic acids and for proline, hydroxyproline and glycine. There is simultaneous transfer of a sodium ion; the transport mechanisms are energy demanding and are called *symports*. On absorption polypeptides are hydrolysed by dipeptidases which are present within the epithelial cells of the mucosa and these increase markedly in concentration as the amount of protein supplied by the diet is increased. Similar active transport processes coupled to sodium ion movement apply to carbohydrate absorption. Glucose and galactose are absorbed by both active transport and diffusion processes while fructose absorption is passive, with the fructose being converted to glucose or lactate within the mucosal cell so as to maintain the concentration gradient for its diffusion. The heat production which accompanies active absorption arises from the ATP-dependent sodium transport component; it is the enthalpy change associated with oxidations which leads to ATP formation – the ultimate source of heat. There is a direct proportionality between sodium movement in one direction and amino acid or sugar movement in the other.

Lipid digestion involves partial hydrolysis to water-soluble micelles. In simple-stomached species these micelles consist of fatty acids, monoglycerides and bile salts with the hydrophobic, fat-soluble components in the interior. They are absorbed passively, and in the cell the triglycerides are resynthesised and covered with a protein coat to form a chylomicron. The protein is specifically synthesised for the purpose of transport of the lipid from the mucosa; it is not simply a non-specifically adsorbed protein. Again an energy cost is involved in the synthesis of the triglyceride from the monoglyceride and longer chain fatty acids. Fatty acids with a carbon chain-length of less than 14 are absorbed actively by a sodium-dependent process.

Thus quite apart from the energy required to synthesise the digestive enzymes, to enable the entero-hepatic circulation of the bile, and to synthesise and secrete the gastrointestinal hormones, the processes of digestion and absorption impose additional costs. These are broadly proportional to the amount of food ingested and thus contribute to the heat increment of unit food. In addition, oxygen must be supplied for the oxidation of the absorbed nutrients. As the amount of food given is increased, the work of the heart also increases and more work is done by thoracic muscles and the diaphragm in respiration. A component of the heat increment obviously involved is one which relates to the provision of oxygen and the excretion of carbon dioxide and other end-products of metabolism.

d. Heat associated with the effects of food on cellular activity

It was pointed out in Section 6.4 that cellular constituents are in a dynamic state as exemplified by the turnover of body proteins. The rate of retention of protein in the body is a net quantity. It represents the difference between the rate of protein synthesis and that of protein breakdown. When a nutritionally adequate diet is given to a fasting animal, net retention of protein increases; the loss of protein from the body which occurs in starvation diminishes and retention becomes positive if sufficient is given. Measurements show that both the rate of protein synthesis and the rate of protein breakdown increase with greater food intake. This is illustrated in Figure 12.3 which shows the results obtained by Reeds & Fuller (1983) and

Fig. 12.3. The increases in the rates of protein synthesis (●) and breakdown (○) which occur as the food energy intake increases. Energy intake is expressed as multiples of the maintenance food intake and so too are those for protein synthesis and breakdown. Below maintenance, breakdown exceeds synthesis; above maintenance, the reverse is true. From Reeds & Fuller (1983).

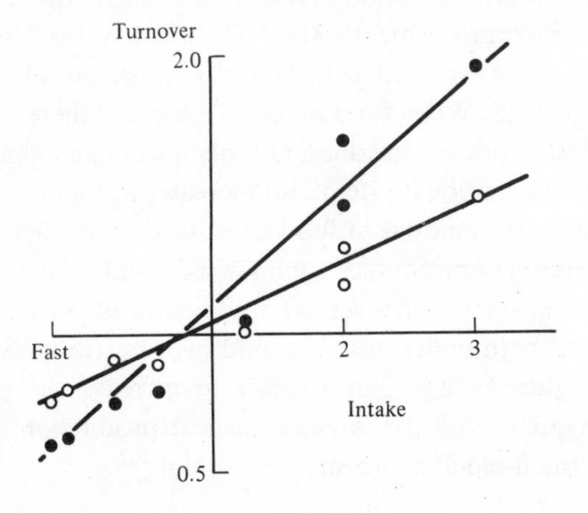

by others. The rate of protein synthesis at the maintenance level of nutrition is about double that in fasting. The analysis of the effect of food made by Reeds & Fuller further suggests that dietary protein and non-protein sources of energy have different but additive effects on the two rates. The augmentation of rates of both protein synthesis and breakdown has been demonstrated in experiments with malnourished children, adult men, pigs, rats and cattle and it seems to be a general phenomenon. Clearly, if the rate of protein synthesis is increased by food, then an additional energy cost is incurred and this must contribute to the increment of heat associated with food ingestion.

A similar increase in cellular metabolism with increase in food intake is evident in the activity of ion pumps. The maintenance of the high concentration of potassium ions and the low concentration of sodium ions within cells relative to the extracellular fluid is maintained by the activity of a Na^+, K^+-ATPase. Extrusion of three Na^+ and the simultaneous uptake of two K^+ against their concentration gradients is accomplished at the cost of hydrolysis of one ATP to ADP and inorganic phosphate. This membrane ATPase is specifically inhibited by the digitalis glycoside, ouabain. Measurement of the inhibition of tissue respiration by oubain provides an estimate of the Na^+, K^+-ATPase-dependent respiration, that is, the energy cost of maintaining the sodium and potassium gradients. Milligan & McBride's (1985) review shows that the activity of Na^+, K^+-ATPase varies with food intake, being lower in starvation than in the fed state. Again there is a component of the heat increment which can be ascribed to changes in ion fluxes consequent on food ingestion.

There is a number of other correlates with the increase in heat production associated with the ingestion of food. There is considerable evidence that thyroid status changes with nutritional level. The monodeiodination of thyroxine (T_4) to form triiodothyronine (T_3) is proportional to the intake of food (Danforth 1983). Increases in membrane ATPase activity and in RNA and protein synthesis are associated with this change in thyroid status (Guernsey & Edelman 1983). When food intake is increased there are also changes in the secretion of other hormones, notably insulin and glucagon. Activity of the autonomic nervous system also increases as exemplified by changes, proportional to the amount of food consumed, in the activity of the sympathetic nervous system. Studies summarised by Danforth (1985) show considerable changes in the turnover of noradrenalin and sympathetic activity associated with both under-nutrition and over-nutrition. Specific effects are involved; glucose ingestion appears to increase the rate of secretion of catecholamines and the increase in heat production can be partially blocked by the β-blocking agent, propanolol.

e. Heat arising from the substitution of endogenous energy sources by nutrients

From Table 12.1 it is evident that efficiency of utilisation of metabolisable energy varies with the chemical nature of the energy-yielding nutrients absorbed. Protein in particular is associated with a low efficiency and consequently a high increment of heat. Considering events below maintenance, when food is given to a fasting animal, nutrients clearly replace body constituents as the source of energy, that is, they furnish the ATP for the work which the body performs. The efficiency of utilisation of metabolisable energy for maintenance can thus be calculated as the ratio of the enthalpy of combustion/mole of ATP of the tissues spared from oxidation to the enthalpy of combustion per mole of ATP of the nutrient which is supplied from the diet.

$$k_m = \frac{\text{enthalpy of tissues spared/mol ATP}}{\text{enthalpy of nutrient supplied/mol ATP}} \qquad (12.2)$$

This approach was used by Krebs (1960, 1964) Blaxter (1961a, 1962b) Schiemann, Hoffman & Nehring (1961), Schiemann *et al.* (1971), Armstrong (1969) and by others to calculate the theoretical efficiency with which nutrients supply energy to meet maintenance needs. More recently, and taking into account newer knowledge about the stoichiometry of oxidations, the same approach has been used by McGilvery & Goldstein (1983), Flatt (1978) and Livesey (1984). The question arises whether these theoretical estimates of efficiency or of heat increment agree with observation. This was systematically examined in ruminants by Blaxter and his colleagues in calorimetric studies in which nutrients were infused into the gut of fasting sheep over periods of days. The increment of heat was then measured and compared with that estimated theoretically. The ruminant is of interest because of its extensive ruminal fermentations and the fact that about 60% of the energy absorbed from the digestive tract is as acetic, propionic, n-butryric and higher steam volatile fatty acids. By infusing nutrients into the true stomach (the abomasum) ruminal fermentation can be avoided. Figure 12.4 shows the results obtained. For two proteins, for sucrose, starch, glucose, propionic acid and mixtures of fatty acids, agreement was excellent. When acetic acid or n-butyric acids were given alone or mixed, agreement was poor. Ancilliary evidence showed that the animals became acidotic, developed ketosis and hypoglycaemia and that urinary nitrogen excretion was enhanced when acetic acid or n-butyric acids were given. Clearly a considerable augmentation of gluconeogenesis from protein occurred. It was significant that only a small amount of propionic acid (which is glucogenic) prevented these changes. These results suggest that the primary hypothesis, namely that nutrients replace one another as energy sources in proportion to the extent that their oxidation yields ATP to meet the energy demanding processes of the body, is a tenable one.

Further verification was obtained by Hoffmann, Klein & Schiemann (1986) in experiments with rats. These workers, using the carbon and nitrogen balance technique (see Section 5.2) determined the relative amounts of glucose, sunflower oil and casein necessary to achieve zero energy retention. Their results are presented in Table 12.3 where they are expressed relative to glucose. Following the stoichiometry adopted by Blaxter (1961b) and Schiemann *et al.* (1971) the yields of ATP from glucose, fat and casein were calculated, to give theoretical efficiencies. The bottom line of Table 12.3 gives these expected efficiencies deduced from biochemical calculation. Agreement is again good. The table also includes similar results which Hoffmann and his associates calculated from Rubner's (1902) experiments with dogs, Schiemann's studies with rats (Schiemann *et al.* 1971) and Steiniger's (1983) experiments with man. Once more there is good

Fig. 12.4. The agreement between efficiencies of utilisation of the metabolisable energy of single nutrients determined calorimetrically and those calculated from knowledge of the stoichiometry of their oxidation. The line represents complete agreement. It will be noted that when acetic and butyric acids were given alone or in combination efficiencies were low. Points labelled 75 C2, 15 C3, 10 C4 etc. refers to mixtures of the acetic, propionic and n-butyric acids.

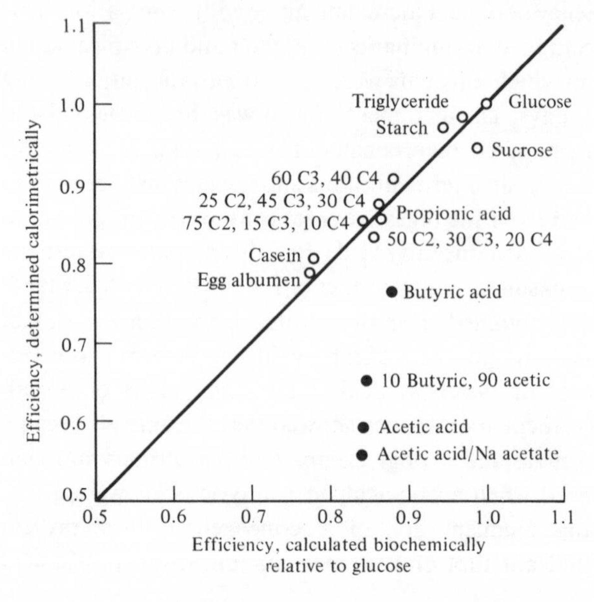

Table 12.3. *Metabolisable energy requirements for maintenance when carbohydrate, fat or protein are the sole sources of energy: Values are expressed relative to carbohydrate*

	Source of energy		
Species	Carbohydrate	Fat	Protein
Rat	Glucose 100	Rat lipid 109	Casein 133
	Glucose 100	Sunflower 105	Casein 135
	Glucose 100	Sunflower 102	Casein 128
Dog	Sucrose 100	Pig fat 108	Meat 136
Man	Starch 100	—	Casein 128
Theoretical	Glucose 100	Stearic acid 105	Casein 128

Note:
The experiments were by Hoffmann *et al.* (1986), Schiemann *et al.* (1971), Rubner (1902) and Steineiger (1983) as summarised by Hoffmann *et al.* (1986). The theoretical calculation was that of Schiemann *et al.* (1971).

agreement between the observed results and theoretical expectation; It thus appears that nutrients replace one another for maintenance in proportion to the amount of free energy which they yield as ATP when they are oxidised in the body. These results, however, are not estimates of heat increments; they are of the replacement rates of dietary nutrients. Table 12.4 summarises values for the enthalpy of combustion of nutrients required to furnish one mole of ATP, this table being an extension of the calculations made in Section 6.2.

f. Intermediary storage and the increment of heat at maintenance

The first law of thermodynamics, as exemplified by the law of Hess, states that the enthalpy change of a chemical reaction is independent of the path of that reaction and is determined simply by the enthalpies of formation of the reactants and final products (see Section 2.1.3). The enthalpy change on the oxidation of glucose to carbon dioxide and water is the same, whether the glucose is oxidised through glycolysis and the tricarboxylic acid cycle, whether it is first incorporated into glycogen and then oxidised, or whether it is used to synthesise fat and the resultant triacyglycerol is then completely oxidised. While the enthalpy change in these oxidations is invariant, the net amount of ATP synthesised from ADP and inorganic phosphate is not. Energy as ATP or as reduced coenzymes has to be furnished to effect syntheses, thus reducing the net yield of ATP.

It was shown in Section 6.3.1 that the incorporation of glucose into glycogen incurs an ATP 'cost' of 2.1 mol/mol glucose. It was also shown

Table 12.4. *The enthalpy change associated with the formation of one mole of ATP from ADP and inorganic phosphate when nutrients are completely oxidised*

Nutrient	$-H_C$/ATP (kJ/mol)
Glucose	78.9
Tri-palmitin	78.0
Protein	89.5
Steam-volatile acids (1)[a]	85.9
Steam-volatile acids (2)[a]	84.9

Note:
[a] Mixture (1) of steam-volatile acids is one which reflects a fermentation which results in a high molar proportion of acetic acid in the total volatile acids produced and mixture (2) produces a low proportion of acetic acid.

that the complete oxidation of glucose yields 35.5 moles of ATP. The net ATP yield for the oxidation of glucose when an intermediary step is the formation of glycogen is thus 35.5–2.1 = 33.4 and the enthalpy change associated with this oxidation, expressed per mol of ATP formed, is 2803/33.4 = 84, rather than 79 kJ/mol for the direct oxidation. If glucose is used as the substrate for synthesis of palmitic acid which is then completely oxidised, the enthalpy change per mole of ATP formed can be calculated from equations (6.12) and (6.15b) to be 94 kJ/mol ATP. Relative to glucose oxidised directly, the efficiency of the indirect oxidations are 0.95 for glucose oxidised after synthesis of glycogen and 0.84 after synthesis of palmitic acid.

Flatt (1978, 1985) attributes the specific dynamic effect, or perhaps more correctly the immediate effect of nutrients on heat production, firstly, to the metabolic costs of transport of nutrients (see Section 12.2.1.c) and secondly, to their intermediate storage before complete oxidation. He assumed that the heat produced per mole of ATP formed could on average be taken to be 84 kJ. On this basis he computed that the 'dissipation' of energy due to the storage component alone was 4% for glucose stored temporarily as glycogen and 28% when it was stored temporarily as fat. These values are in accord with the relative efficiencies computed above. When fats are hydrolysed in the gut, absorbed and then stored temporarily as triacyglycerides before complete oxidation he estimated the cost of storage to be 7%, Additionally, he recognised that the yield of ATP on the oxidation of protein was low, largely attributed to the ATP required for synthesis of urea.

g. Heat associated with net synthesis of fat and protein

When young animals grow or adult ones accrete new tissue, energy retention is positive and the energy supplied by nutrients is in part stored fairly permanently as 'fat' and 'protein'. It is probable that temporary storage (in the sense employed in the preceding paragraphs) takes place during the course of a day, but over the whole day, or for a protracted period, a net gain of energy takes place. The efficiency of utilisation of metabolisable energy, measured above zero energy retention, is less and the increment of heat greater than it is below zero energy retention. A considerable component of these differences can be attributed to the endergonic reactions concerned with the synthesis of lipid and protein which were discussed in Sections 6.3.2 and 6.3.3.

Much of the earlier experimental work undertaken to measure efficiency of utilisation of metabolisable energy above maintenance was with domesticated animals. The reason is obvious; efficient use of food for the growth of meat animals has considerable economic significance. The studies were mostly made with mature animals in which there was little increase in protein deposition and the energy retention was predominantly as fat. These studies showed that dietary fat was used more efficiently than dietary carbohydrate and the latter was usually, but not invariably, used more efficiently than dietary protein in promoting energy retention. Some representative values of k_{f+p} are included in Table 12.1. It was energy retention that was measured in these experiments; there were few studies in which retention of energy as fat or protein were separately considered. The relevance of results with mature animals, which deposit much fat and little protein, to growing animals, which retain a considerable proportion of their body energy as protein, is open to question. Kielanowski (1965) working with pigs, used regression methods to separate the efficiency of utilisation of metabolisable energy for deposition of protein (k_p) from that for deposition of fat (k_f). Many similar investigations have since been made with pigs, rats and ruminants and those with pigs have been summarised in a publication of the Agricultural Research Council (1981). The usual method adopted in such studies has been to solve the following equation using statistical methods:

$$\dot{M}_E = \dot{H}_{(0)} + \frac{\dot{R}_P}{k_p} + \frac{\dot{R}_F}{k_f} \qquad (12.3)$$

where $\dot{M}_E$ is the metabolisable energy intake; and $\dot{H}_{(0)}$ is the intercept of the equation, representing the value of $\dot{M}_E$ when both protein energy retention ($\dot{R}_P$) and fat energy retention ($\dot{R}_F$) are zero. In simple-stomached species the values obtained for k_f differ very little from a mean value of 0.76. Values for the corresponding efficiency of protein deposition are much more variable and one usually taken as representative is 0.56.

The equation can be criticised. First, in most data sets on which the regression estimates of $\dot{H}_{(0)}$, k_p and k_f are based, there is massive autocorrelation thus raising statistical doubts about the validity of the separation of the components. Secondly, the intercept term does not coincide with zero energy retention because protein deposition and fat loss occur in growing animals when energy retention is zero. Lastly, the approach assumes that all processes contributing to the heat increment above maintenance can conveniently be ascribed to two processes – net synthesis of fat or net synthesis of protein. Increases in other metabolic processes associated with the ingestion of food are necessarily incorporated into the constants. Because of these criticisms a different approach has been employed to estimate k_p and k_f. An estimate is made of the metabolisable energy required to ensure zero energy retention and the actual metabolisable energy intake less this maintenance energy requirement is then related to the amounts of protein and fat energy retained. This approach while meeting one criticism does not meet the others. It does not affect the efficiency of utilisation of metabolisable energy for fat deposition very much but tends to increase the uncertainty associated with the corresponding efficiency of protein deposition.

The efficiency of the synthesis of fat and protein can be estimated from biochemical considerations alone. This was indeed done in the preceding section when considering temporary storage of glycogen and fat. Table 12.5 summarises these 'theoretical, biochemical efficiencies' for glycogen, fat and protein deposition in the body when dietary carbohydrates, lipids and proteins are the sources of energy. The biochemically-derived value for the efficiency of fat synthesis from carbohydrate (0.80) is higher than that found experimentally (0.78), but not outstandingly so. The theoretical value does not include the cost of absorption from the gut. The corresponding biochemical efficiency for the deposition of fat when dietary fat is the source (0.95) is also higher than experimentally observed values (about 0.85). The theoretical value, given in Table 12.5, for the efficiency with which fat is formed from protein is probably a maximal one; it is of the same order as the experimentally-derived values in Table 12.1; agreement between observation and theory is reasonable. There is however massive disagreement between the theoretical efficiency of deposition of energy in the form of protein from a dietary protein source (0.85) and estimates made experimentally which probably average 0.56.

The theoretical biochemical efficiency for protein deposition from di-

Table 12.5. *The theoretical efficiency (J/J) with which the energy of nutrients is employed in synthesis, calculated from the stoichiometry of transport and synthesis*

Dietary substrate	Product	Estimated efficiency	Heat increment
Carbohydrate	Glycogen	0.95	0.05
	Body fat	0.80	0.20
Lipid	Body fat	0.96	0.04
Protein	Body fat	0.66	0.33[a]
	Body protein	0.86	0.14[b]

Notes:

[a] The value depends on the amino acid content of the protein; minimal glucogenesis is assumed.

[b] The value is that for the formation of the peptide bond, assumed to be 5 ATP and the enthalpy associated with the formation of ATP is assumed to be 81 kJ/mol. Dietary protein is assumed to have the same composition as body protein.

The values above differ slightly from those given by Blaxter (1962b) and by Millward, Garlick & Reeds (1976).

etary protein, given in Table 12.5, was computed as the enthalpy of combustion of the protein divided by the enthalpy of combustion of its constituent amino acids plus the energy cost of providing the ATP necessary to synthesise the peptide bonds of the protein. The number of moles of ATP required per bond formed was assumed to be five, the enthalpy/mole of ATP was estimated to be 81 kJ and the mean molecular weight of the constituent amino acids of the casein was taken to be 100. Calculation shows that to obtain an efficiency of 0.56, 24 moles of ATP would be required to form a peptide bond. This has suggested to many that the net deposition of protein involves the synthesis of about five times as much protein as is eventually deposited and that the reason for the low efficiency of protein synthesis resides in the extent of turnover. Superficially this appears plausible; protein turnover in the whole body commonly exceeds the rate of deposition by a factor of five. As Fuller *et al.* (1987a) have pointed out, however, a large component of this turnover of body protein occurs even when there is no net deposition of protein, and this is irrelevant to the interpretation of the efficiency of net protein deposition. In their experiments they changed the rate of protein deposition in the body, either by increasing the amount of a lysine-poor protein in the diet, or by supplementing that protein with lysine. They calculated that the increase in heat production was greater (and efficiency lower) when additional protein in the diet led to increased protein retention than it was when the amino acid which limited net deposition was supplied. They further showed (Fuller *et al.* 1978b) that a doubling of protein deposition brought about by improving the quality of the protein in a diet of fixed metabolisable energy content, was not associated with an increase in protein turnover. There thus appears to be no obligatory link between protein deposition and protein turnover. Their experiments also illustrate the difficulties which are involved in trying to isolate experimentally the energy cost of depositing protein. Addition of lysine to the constant diet, which resulted in an increase in protein deposition, necessarily resulted in a reduction in the amount of protein oxidised. Concomitantly there was a fall in the amount of lipid deposited. Quite apart from the assumption that the only energy-requiring process in protein synthesis is that of forming the peptide bond in the accreted protein, an assumption discussed in Section 6.3.3, there are real difficulties in separation of that part of the increment of heat which is specifically concerned with the synthesis of protein from other processes related to growth.

h. The cause of the heat increment

From the above analysis, the wisdom of Mitchell's remark, quoted earlier, about the multiple causes of the increase in heat production following food is very evident. There are components of the heat increment due to increased muscular activity, more evident in some types of investigation than in others, and certainly a component related to the actual act of eating, although this is probably small in animals other than herbivores. The heat increment includes fermentation heat, which in ruminants is particularly large, but in omnivora with limited hind-gut fermentation is quite small and in carnivores is likely to be non-existent. Small amounts of heat arise from hydrolyses in the lumen of the gut, absorption processes are endergonic in many instances and so too are those concerned with transport of the absorbed nutrients, the provision of oxygen at the cell level and the disposal of carbon dioxide and other products of the oxidations involved. Food also affects metabolism by changing neural and endocrinal status, the activity of ion pumps and the rates at which protein synthesis and degradation occur. Other effects of food which are also relevant relate to pharmacological effects of food constituents. Caffeine, for example increases heat production (Acheson *et al.* 1980).

Even so, it does appear that the dominant causal factor is to be found in the heat associated with the provision of ATP and reduced coenzymes from food substrates. Although there is a quantitative uncertainty related to the net synthesis of protein, there is no doubt that the high increments of heat associated with the retention of energy in the body can be explained by the stoichiometry of the syntheses of the compounds which are stored, and that the endergonic processes involved are expensive in terms of ATP and reducing power. The replacement values of nutrients at and below maintenance also accord with the contention that the value of nutrients as energy sources are proportional to the ATP formed from them when they are completely oxidised. The initial increases in heat production associated with meal consumption are again in accord with the hypothesis that temporary storage occurs with consequent increased requirement for ATP. While there is a number of factors which influence the increment of heat, the major one relates to the inevitable enthalpy change associated with the generation of ATP.

12.2.2 *Heat increments in ruminant animals*

More studies have probably been conducted to measure heat increments (or, alternatively, efficiencies) in ruminants than in other animal groups. These studies include the systematic measurement of the increments associated with a wide variety of feeds – studies which include those conducted at the turn of the century by G. Kuhn (1894) and Kellner & Köhler (1896) – as well as investigations designed to elucidate the determining factors. Blaxter and Boyne (1978) summarised the results of 80 calorimetric experiments with natural diets to show that efficiency of utilisation of the total energy of a diet could be represented by a simple limiting exponential equation from which could be derived the efficiencies of utilisation of its metabolisable energy both for maintenance (k_m) and for fat and protein deposition (k_{f+p}). Statistical analysis of the results of the 80 experiments showed that these efficiencies could be predicted from attributes of the diet:

$$\left.\begin{aligned} k_m &= 0.947 - 0.000\,10\,(P/q) - 0.128/\mathrm{g} \\ k_{f+p} &= 0.951 + 0.000\,37\,(P/q) - 0.336/q \end{aligned}\right\} \qquad (12.4)$$

where P is the protein content of the organic matter of the diet (g/kg); and q the metabolisable energy as a proportion of its total energy. The residual standard deviation associated with these regressions was ± 0.03 and this is the error which attaches to the two efficiencies and to the corresponding heat increments. The equations show that both efficiencies increase with increase in metabolisability (q), and that below maintenance efficiency declines while above maintenance it increases with increase in the protein content of the diet. The equations, based on about 1000 complete energy budgets determined in respiration chambers, provide a good basis for predicting the efficiencies with which the energy provided by different diets is used by *adult* cattle and sheep in which most of the energy is retained as

Table 12.6. *Experimentally determined increments of heat in ruminants used in the computations in Table 12.7*

Nutrient	Heat increment (J/J)	
	Below maintenance	Above maintenance
Glucose	0.0	0.27
Protein	0.19	0.35
Acetic acid	0.15	0.67
Propionic acid	0.14	0.44
n-butryic acid	0.09	0.38
Linoleic acid	—[a]	0.38

Note:
The experiments involved continuous infusion of the nutrients into the gut. Thus the heat increments do not include any component referrable to the act of eating. Glucose and protein were given by infusion into the abomasum thus avoiding rumenal fermentation; the heat increments therefore exclude any component related to fermentation heat.
[a] No data available.

fat. Probably – though this has not been tested explicitly – the efficiencies also apply to other ruminant species.

In addition to these essentially practical estimates, systematic studies have been made of the factors which determine the heat increment. Those related to the act of eating and to fermentation heat (undertaken by Webster and his colleagues) and those related to the utilisation of the energy of the end products of digestion (made by Blaxter and his colleagues) have already been mentioned. Table 12.6 summarises the latter. In Table 12.7 these data are assembled to ascertain if they do indeed account for the heat increments predicted from the large compilation of trials with practical diets. To make the comparison, extreme diets were selected; one entirely of poor roughage and the other of grain. These diets were characterised by metabolisable energies of 0.40 and 0.70 J/J enthalpy of combustion, respectively. Each had a protein content of 150 g/kg and their heat increments were predicted from equations (12.4) to give the 'observed' values. The value estimated from the components of the heat increment involved assessing the time spent eating the two diets to arrive at the energy cost of eating and ruminating and calculating fermentation heat from methane production. It was assumed that with roughage the enthalpy of combustion

Table 12.7. *Components of the heat increment in ruminants at maintenance and above maintenance when fat and protein is being deposited. The comparison is made with respect to a roughage diet and a grain diet which can be regarded as extreme types of diet consumed by these animals*
The roughage diet was assumed to have a metabolisability of 0.4 from which the equations of Blaxter & Boyne (1979) predict that the value of k_m would be 0.59 and of k_{f+p} 0.25. The grain diet was assumed to have a metabolisability of 0.7 and the corresponding efficiencies are: k_m=0.74 and k_{f+p}=0.55. The heat increments calculated from these efficiencies are taken as the observed values.

Diet	Increment below or above maintenance	Components of heat increment (J/J)				Observed heat increment
		Cost of eating	Fermentation heat	Nutrient metabolism	Total	
Roughage	Below	0.07	0.13	0.16	0.36	0.41
	Above	0.07	0.13	0.53	0.73	0.75
Grain	Below	0.01	0.08	0.12	0.21	0.26
	Above	0.01	0.08	0.44	0.53	0.45

Note:
The nutrients absorbed as a proportion of the total metabolisable energy were for the roughage diet: protein 0.15, acetic acid 0.51, propionic acid 0.18, and n-butyric acid 0.17. For the grain ration they were: protein 0.15, hexose 0.05, acetic acid 0.23, propionic acid 0.31 and n-butyric acid 0.26.
Source: Blaxter & Boyne (1979).

of the nutrients absorbed consisted of: protein 15%, acetic acid 50%, propionic acid 18% and n-butyric acid 17%. For grain the corresponding figures were protein 15%, hexose 5%, acetic acid 23%, propionic acid 31%, and n-butyric acid 26%. For the roughage these correspond to a fermentation producing 75% acetic acid on a molar basis. For the grain diet the assumption is that propionic acid makes up 35% on a molar basis of the total steam-volatile acids produced. The factors in Table 12.6 were then used to calculate the heat arising from the metabolism of the absorbed nutrients.

The heat increments below maintenance, calculated by summation of its components, agree reasonably well with those which are expected from the results of the practical experiments. The same is true for the roughage diet when efficiency is measured above maintenance. In all three cases the differences are within the bounds of likely experimental error. For the grain diet given above maintenance, however, the discrepancy between the summated value for the components and expectation, is very large and cannot be explained by errors of measurement. The practical experiments show that less heat is produced when a grain diet promotes fat and protein retention than can be calculated from the components of the heat increment. The component factors predict the corresponding heat increment of roughage with acceptable precision.

a. The utilisation of acetic acid as an energy source

To explain the above anomaly, attention has focussed on the factors which determine the efficiency with which acetic acid is used for net retention of energy in the body. Comparative slaughter experiments in which rations high in grain were given, suggested that supplements of acetic acid were utilised with the evolution of considerably less heat than the 0.67 given in Table 12.6 (Ørskov & Allen 1966 a,b). Calorimetric experiments in which sheep were sustained entirely by the intragastric infusion of mixtures of steam-volatile fatty acids and casein, supported this contention (Ørskov *et al.* 1979). Mixtures of fatty acids with a high molar proportion of propionic acid admittedly gave the smallest increments of heat but there was little increase in heat increment as the molar proportion of acetic acid in the mixture of steam-volatile acids infused, was increased. These results suggest a synergism between the metabolism of two- and four-carbon fatty acids and glucogenic components of the diet such as propionic acid, glucose and glucogenic amino acids. For the formation of fat from acetic acid, glyceryl phosphate and reduced NADP are required. These arise from the pentose phosphate pathway and from the isocitrate dehydrogenase shuttle. The value obtained for the heat increment of acetic acid given in Table 12.6 refers to its value when added to a diet in which glucogenic compounds were limiting. With grain diets they are not; nor were they limiting in the infusion mixtures used by Ørskov because casein, which provides glucogenic amino acids, was in excess. Supporting evidence for the hypothesis that glucose or its precursors is necessary to ensure high efficiency of utilisation of the energy of acetate, first proposed by Armstrong & Blaxter (1961), comes from experiments in which the increment of heat of acetic acid has been measured when it is added to grain rations and to roughage rations, showing that the increment is low in the former and high in the latter case (Tyrell, Reynolds & Moe 1976).

12.2.3 *Biochemical models of metabolism*

With the growth of knowledge about intermediary metabolism, there have been many attempts to integrate this information to provide explanations of phenomena such as heat increments. Many of these attempts have taken the form of simulation models based on enzyme kinetics and the stoichiometry of the biochemical pathways involved. They mostly involve the definition of a number of state variables which define the system, driving variables (which in the instance of nutritional models are inputs of nutrients) and the description of reaction rates which determine the inter-relationships between variables. The latter are usually represented by the Michaelis–Menten formulation, often modified to provide sigmoidal types of response. Commonly many differential equations are involved and these are then numerically integrated over short intervals of time to provide a series of outputs in response to prescribed patterns of input. The methods employed are described by France & Thornley (1984).

Several simulation models have been constructed to examine whether the results of experiments on the heat increment in ruminants can be explained in biochemical and kinetic terms. The model of Gill *et al.* (1984) is of particular interest since two of the simulations dealt with grain and roughage diets. The model is depicted in Figure 12.5 and was described by about 72 equations. The results showed that the efficiency of utilisation of the metabolisable energy of grain was always in excess of that for roughage and that more generally low efficiencies of energy utilisation when the supply of glucose or glucose precursors is low relative to the supply of acetate. Gill and her associates are careful to point out that the model does not provide a

Fig. 12.5. The basic model employed by Gill *et al.* (1984). The reactions are coded: ■ represents reactions which result in the synthesis of ATP, □ those requiring ATP, ● represents reactions producing NADPH and ○ those which require participation of NADPH. This model predicts the increments of heat associated with the metabolism of absorbed nutrients.

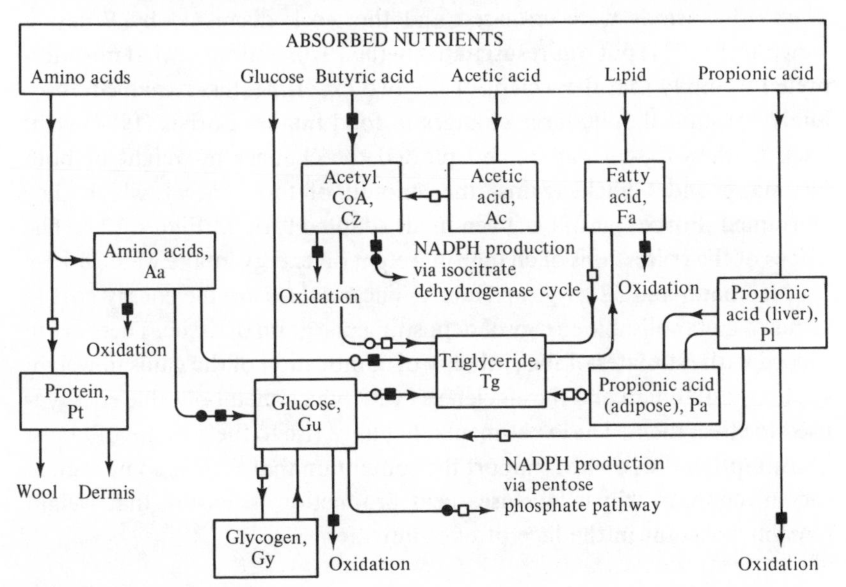

unique test of the hypothesis since alternative representations of metabolism have not been tested. The more sophisticated approach agrees, however, with experimental results and with the hypothesis reached on other grounds.

12.3 Regulatory heat production

12.3.1 *Luxuskonsumption*

The discussion of the components of the heat increment in animals generally, and more particularly the discussion of heat increments in ruminants suggest that the listing of the components is complete; the heat increment can be accounted for without postulating other factors. There is an opinion, however, that, given excess food, man can certainly – and some animals can possibly – increase their heat production to the extent that no retention of energy occurs. Casual observation might suggest that the relative stability of body weight in man over long periods of time despite variation in food intake, implies that there is a regulatory component. Such a regulation would involve compensation by an increase in energy loss for the increase in food energy intake.

This idea was first proposed by Neumann (1902) and Gulick (1922) who conducted experiments upon themelves in which they varied their food intake and measured their body weight. They concluded that there was some energy dissipating mechanism which enabled them to maintain a reasonably constant body weight and this was termed 'luxuskonsumption'. Their conclusions were criticised and the work dismissed by Wiley & Newburgh (1931) but was resuscitated in the 1970s and somewhat misinterpreted to imply that the weight of the two experimenters remained absolutely constant despite large changes in food intake. Forbes (1984) went back to the original papers and plotted the changes in weight of both Neumann and Gulick against the amount of food energy which they consumed. Forbes' graph is given in an adapted form in Figure 12.6. The slopes of the regressions of change in weight on energy intake were 38 kJ/g for Neumann and 32 kJ/g for Gulick. These values are the energy cost in terms of metabolisable energy of depositing one gram of tissue. These are in accord with estimates of the enthalpy of combustion of the gains in weight made by adult man and the efficiencies with which metabolisable energy is used to effect them. The experiments that gave rise to the concept of luxuskonsumption thus do not support the contention that there is some regulatory mechanism which increases heat production to ensure that weight remains constant in the face of over-nutrition.

12.3.2 *Low protein diets and heat production*

An experiment much quoted as evidence for a regulatory component of heat production is that of Miller & Payne (1962). It consisted of feeding one young pig a very low protein diet and another similar pig a diet with normal protein content. Both were given their respective diets in amounts which were adjusted so that they maintained body weight. After a period the diets were reversed; the pig which had received the low protein diet then received the high protein diet and its companion did the opposite. When given the low protein diet the pigs consumed several times as much energy to maintain weight as they did when given the normal diet, suggesting that because at energy equilibrium heat production is equal to the metabolisable energy intake, heat production must have been massively increased in the pigs when they were given the low protein diet. What was measured, however was not energy equilibrium but stability of body weight. When young growing animals which are in energy equilibrium are given diets adequate in protein content protein accretion occurs, and this is balanced by an equivalent loss of body energy as fat. As discussed in Section 5.1.2, protein accretion is accompanied by accretion of water. Energy

Fig. 12.6. The relation between the intake of metabolisable energy as food and associated changes in body weight. The data refer to the experiments conducted on themselves by Neumann (1902) and by Gulick (1922). These experiments led to the spurious idea that in man there was a luxuskonsumption, namely an ability to oxidise excess food and so maintain weight stability. The reanalysis of these old experiments was done by Forbes (1984).

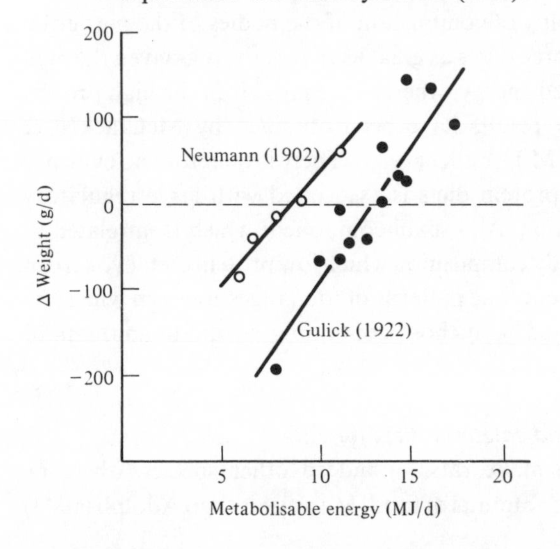

Table 12.8. *The metabolisable energy necessary to maintain body weight in young pigs when given diets either low or high in protein*

	Protein content of diet	
	High (268 g/kg)	Low (20 g/kg)
Metabolisable energy found to be required for maintenance of body weight (MJ/d)	4.34±0.6	13.95±17
Final enthalpy of combustion of the body (MJ)	169±0.1	453±1.7
Gain or loss of energy from the body in 42 d (MJ)	−31	238

Note:
Calculated from the results of Gurr *et al.* (1980).

equilibrium in young animals given normal protein diets is thus associated with a net increase in weight. Weight equilibrium in such animals is associated with loss of energy from the body. Protein deficiency due to consumption of a low protein diet results in loss of protein and water from the body. Weight equilibrium is achieved by balancing this loss by the deposition of an equal amount of fat. There is no doubt that the observations of Miller & Payne (1962) were correct; but their interpretation is erroneous. This is illustrated by the results of a similar experiment by Gurr *et al.* (1980) which are summarised in Table 12.8. The metabolisable energy for maintenance of body weight was more than three times greater for pigs given the low protein diet compared with the high protein diet. At the end of the experiment the enthalpy of combustion of the bodies of the pigs given low protein was almost three times as great as that of the pigs given the high protein diet. They retained energy, whereas the pigs given the high protein diet lost energy. Similar results have been obtained by McCracken & McAllister (1984) and by M.F. Fuller (1983). There is therefore no evidence that the feeding of low protein diets is associated with large regulatory increases in heat production. Any small component which is unrelated to the marked changes in body composition which low protein diets evoke, can be explained by the different time patterns of food ingestion seen when low protein diets are compared with those containing adequate amounts of protein.

12.3.3 *Over-nutrition and cafeteria diets for rats*

It is difficult to make rats, or indeed other species, obese by inducing them to eat large amounts of food. Many years ago Adolph (1943) showed that rats 'eat for energy', that is, they adjust their food energy intakes to meet their energy requirements for basal metabolism, muscular work, thermoregulation and growth. If their diets are diluted with an inert material they eat a greater weight of food and keep constant their total energy intake. This normal control of food intake can be over ridden by feeding diets very high in fat content, by substituting a sucrose solution for their drinking water, by giving food by gastric intubation or by offering a varied diet. High fat diets and feeding by intubation achieve higher than normal intakes and the efficiency with which the metabolisable energy of food is used to promote energy retention is also high when results are compared with those noted with more normal diets. With high fat diets a high efficiency is consistent with direct incorporation of dietary fatty acids into the lipids of the body and with diets fed by tube the results are consistent with other observations that ingestion of food at long intervals (meal eating) results in somewhat greater energy retention than occurs when food is given at short intervals and the rats 'nibble'.

Studies by Sclafani & Springer (1976) show that by offering rats a variety of palatable foods in addition to a basal diet, total food intake could be increased and obesity induced. They called this diet a 'supermarket diet' since most of the foods offered were processed ones purchased in supermarkets; the term now employed to describe such a process for enhancing food intake is 'cafeteria feeding'. Rothwell & Stock (1979) used this process to show that cafeteria feeding induced increases in energy intake of 80% yet rats fed in this way did not become obese. They claimed from the results of comparative slaughter experiments (see Section 5.1.1) that heat production had been massively increased and that this response of heat production to overfeeding, which they termed 'diet-induced thermogenesis', was a mechanism to maintain zero energy balance in the face of over-nutrition. They further suggested that the physiological mechanism involved was through sympathetic activation of the mitochondrial proton conductance pathway in brown adipose tissue in a manner analogous to that seen in non-shivering thermogenesis.

This was an attractive hypothesis. If, as they suggested, it applied to man, then one could postulate that people who ate large amounts of food and yet remained thin possessed the mechanism, whereas in obese individuals the mechanism was faulty. Obesity could then be regarded as a metabolic defect. Attempts to confirm the results which Rothwell & Stock obtained have not been uniformly successful. Increases in food intake can be achieved by cafeteria feeding, but the diet-induced thermogenesis, that is, a large increase in heat production not explained by accepted views about the nature of the heat increment, has not been an invariable effect. The increase in heat production in response to cafeteria feeding appears from later work by Rothwell & Stock to vary with strain of rat, with particular colony of rat within a strain and with the age of rat. It does not occur if the rats are kept at temperatures above the critical temperature (see Section 10.3.1; and Fig. 10.2). The phenomenon of a regulatory increase in heat production has been reviewed by those who deny and those who accept its existence (Hervey & Tobin 1983; Rothwell & Stock 1983). There is no evidence that a comparable regulation of heat production occurs in man and the fact that the increase in heat production can only be elicited at, what for the laboratory rat, are low environmental temperatures (21–25 °C) and not at 29 °C or above (Barr & McCracken 1984; Rothwell & Stock 1986) suggests a confounding of the effect of overfeeding with thermoregulatory responses.

12.3.4 *Individual differences in metabolism*

Several other hypotheses involving a regulatory increase in heat production have been proposed to account for the apparent stability of body weight in some individuals and the development of obesity in others. They include differences between individuals in the amount of 'futile cycling' – the regulatory process discussed in Section 6.5 – and differences between individuals in the hydrogen acceptor in the oxidation of glycerophosphate with the transfer of hydrogen to FAD in normal subjects and to NAD in the obese. This would entail generation of more ATP in obese subjects than in the normal subject. Other hypotheses have included an effect of exercise and activity on the heat increment of food, implying that exercise in the fed state has different effects than it does when the subject is post-absorptive (see p. 173). None of these hypotheses has been substantiated in any acceptable way.

Nevertheless, there are differences in the amounts of food consumed over long periods of time by individuals of the same size; some people state that they eat a lot and yet remain thin while others aver that they eat very little and yet tend to gain weight unless they reduce their food intake still further. Sometimes such remarks are greeted with scepticism but Warwick's studies with two normal young female students demonstrate that the contention is true. His results, as reported by Garrow (1985), are presented in Table 12.9. They show that subject NB, selected because she was known to be a large eater, was estimated to consume about 50% more energy as food than did subject ET who was known from observations made by her fellow students to eat very little. These estimates of intake were made by diet assessment methods, and as indicated in Section 3.5, are open to error. The heat production of the two women measured during confinement in a direct

Table 12.9. *A comparison of the food intakes and heat productions of two young women selected as examples of two people of the same body size and adiposity who habitually consume widely differing amounts of food*

Attribute	Large-eater (subject NB)	Small-eater (subject ET)	Ratio (NB/ET)
Age (y)	23	23	
Body weight (kg)	54.2	52.7	
Lean body mass (kg)	45.1	43.6	1.03
Average food intake during normal life as determined by diet assessment (MJ/d)	9.9	6.5	1.54
24-h heat production determined on second day of confinement in a calorimeter (MJ/d)	9.1	5.8	1.55
Rate of heat production in the calorimeter during the night hours (MJ/d)	7.2	4.6	1.56

Note:
From the results of Warwick as quoted by Garrow (1985).

calorimeter showed a similar difference in their 24-h heat production. These results show that the two methods of assessment agree. The fact that the difference was discernible during sleep in the night hours suggests that most of the difference in the 24-h metabolism arose from differences in minimal metabolism although some could have been due to the continuing effect of food.

12.4 Heat increments of food in reproduction

12.4.1 *The problem of estimation*

It is simple to estimate the efficiency of utilisation of food in animals which are not reproducing; it is the increment of energy retention divided by the increment of metabolisable energy that results in the net retention of energy. This applies both below and above maintenance, and in the latter case, can be interpreted as the efficiency with which food is used to promote growth or accretion. The effects of an increase in the metabolisable energy supplied to a reproducing animal are not so easily interpreted. For example, if the energy intake of a pregnant animal is increased, energy retention will occur but the retention is not necessarily in the reproductive structures; the major proportion of the deposition will occur in the maternal body. Again, in lactation an increase in food intake may lead to a small increase in energy secretion in milk, but it also results in an increase in the fat and protein stored in the maternal body. The same is true of egg production.

An increment of metabolisable energy given to a laying bird does not result in a consequential increase in the enthalpy of combustion of the egg or an increase in the rate at which they are produced; most is stored by the bird in her own tissues. Various devices have been employed to try to isolate efficiency terms which specifically relate to the various aspects of reproduction and these are discussed below.

a. Efficiency of utilisation of energy in pregnancy

Several approaches have been used to estimate the amount of metabolisable energy which a mammal requires to support pregnancy and some of them are discussed here. The approach used for pregnancy in woman is that employed by Hytten & Leitch (1971) and modified by Prentice & Whitehead (1987). The primary requirement is set by the rate at which energy is retained in the gravid uterus and in the developing breasts. For the whole of pregnancy these rates total 43 MJ. To this is added an assumed ideal amount of maternal fat storage which Hytten & Leitch estimated to be 4 kg for the whole of pregnancy, or 131 MJ. Finally, a further 149 MJ is added to allow for the theoretical increase in maternal metabolism which occurs in pregnancy. The total additional energy required for the whole of pregnancy is thus estimated to be 323 MJ. The metabolisable energy needed to meet these necessary energy costs is then computed assuming that the efficiency of utilisation of metabolisable energy is 0.9, that is, the same as that noted for the maintenance of man.

This approach is admittedly one concerned with estimation of metabolisable energy requirements for pregnancy rather than with estimation of the way in which the utilisation of dietary energy might be affected by pregnancy. It predicates a norm of maternal fat storage, which may not be necessary or achieved, and assumes an efficiency for the joint processes – synthesis of this fat, synthesis of the products of conception, maintenance of the foetus and adjustment of maternal metabolism to provide nutrients to the gravid uterus. It seems unlikely that this overall efficiency would be the same as that for maintenance.

The approach with animals has been to relate the energy stored in the products of conception to the amount of metabolisable energy consumed above that necessary to maintain zero energy retention in the non-pregnant animal. To make such a calculation, ideally, the energy retention in the maternal body during pregnancy should be zero. This may not be so; in many instances, and particularly in late pregnancy, maternal tissue is lost. Then the source of energy for foetal development, its continuing metabolism and the augmented metabolism of the mother is met in part from maternal body tissues. This approach in effect states that any increase in heat production of the mother during pregnancy is an indication of inefficiency in using the energy derived from food or her own tissues in supporting the growth of the products of conception.

Robinson *et al.* (1980) summarised their careful studies with sheep and those of others in which this approach has been employed. They showed that, when the maternal body is not depleted, the efficiency of utilisation of metabolisable energy for the growth of the conceptus averaged about 0.13. The calculation entailed dividing the accretion in the uterus (determined by comparative slaughter methods) by the amount of metabolisable energy consumed above that necessary to maintain the observed *maternal* weight in the absence of pregnancy. The latter was the weight of the animal less that of its gravid uterus. The results also permitted estimation of efficiency when maternal tissues alone were the source of energy for conceptus growth. The efficiency of energy utilisation for conceptus energy gain were then higher at 0.18.

Studies in pregnant sows have been undertaken in which energy retention in their bodies has been measured at different stages of pregnancy and regressed on their concomitant intakes of metabolisable energy. Both metabolisable energy intakes and energy retentions were scaled by dividing them by the weight of the sow (measured at each stage of pregnancy) raised to the power 0.75 (Close, Noblet & Heavens 1985). The regression analysis was carried out with these scaled values and without separating different stages of pregnancy. The slope of the regression, that is, the efficiency, was found to be 0.74. It has mixed significance because it relates to both the rate of gain of energy by the conceptus and by the maternal body. In addition, the maintenance term includes a component which relates to the weight of the conceptus and hypertrophied uterus. It is thus a different calculation to that made by Robinson's group; in the sheep maternal metabolism alone was deducted from the metabolisable energy intake and gains of maternal tissue were either zero or negative.

Similar experiments to these with pigs have been made with rabbits (Partridge, Lobley & Fordyce 1986). These showed an efficiency of utilisation of metabolisable energy of 0.67 in pregnant rabbits, a value which was the same as that found when the same diet supported growth in non-pregnant rabbits. The proportion of the positive energy retentions in these studies which was due to foetal growth was not determined; it must have varied considerably since data for three successive eight-day periods of pregnancy were pooled to arrive at the regression estimate of efficiency.

Studies of the efficiency of energy utilisation during pregnancy are thus complicated by differences in approach and methods of expression. The alternative approaches may not affect any estimates made from them of the amounts of metabolisable energy required to support pregnancy. Nevertheless, there is not a very good basis on which to extrapolate from those species which have been most studied to those for which there is little direct information.

b. Energy utilisation in lactation

The constituents of milk are synthesised within the epithelial cells of the alveoli of the mammary gland. The precursor of lactose is the glucose of blood plasma, for the proteins the plasma amino acids and the fatty acids of the milk lipid arise either by *de novo* synthesis from acetyl CoA or directly from the free fatty acids of the plasma. The acetyl CoA in non-herbivores arises from the oxidation of glucose via the Embden–Meyerhof pathway. In ruminants it arises from acetate for there is an active acetyl CoA synthetase in the cytosol. Calculations of the energy cost of synthesis of milk can be made; these are not complicated by turnover of the proteins as in the instance of tissue growth since the protein is an 'export' protein (Rook & Thomas 1983). These calculations suggest that, depending on the composition of the milk secreted, efficiency of utilization of the energy derived from maternal tissues should be about 0.85. When the metabolisable energy of the diet is the energy source, efficiencies may be expected to be lower, particularly in ruminants for the reasons given earlier.

Many experiments have been made with several species to estimate the efficiency of utilisation of the energy of both food and body tissues in lactation. These show a good concordance and some of the results are given in Table 12.10. All the results show that the efficiency with which the energy of food supports lactation is less than it is when lactation is supported from energy derived from maternal tissues. In simple-stomached species an efficiency of 0.70 for the efficiency of utilisation of metabolisable energy appears generally applicable. This value is supported by other studies not depicted in Table 12.10. For ruminants efficiency varies slightly with the composition of the diet (Agricultural Research Council 1980) around a value of about 0.65. The efficiency with which the energy of body tissues is used to support lactation can be taken to be 0.85, irrespective of species, a conclusion which agrees with the theoretical calculation given above.

c. Energy utilisation of egg production

As with lactation and pregnancy, most of the studies designed to assess the efficiency with which metabolisable energy is used in egg production have relied on regression techniques to partition the metabolisable energy intake between maintenance needs, those for growth and fat deposition and those for the production of the egg. Understandably such studies have largely

Table 12.10. *The efficiencies with which the metabolisable energy of food or of body tissues are used to support lactation*

Species and authority	Efficiency when energy source is	
	Metabolisable energy of food	Body tissues
Bovine (Agricultural Research Council 1980)	0.63	0.84
Bovine (Vermorel *et al.* 1982)	0.59	0.795
Pig (Noblet & Etienne 1987)	0.72	0.886
Rabbit (Partridge *et al.* 1983)	0.735	0.938

Note:
The data ascribed to the Agricultural Research Council (1980) consist of the results of studies made throughout the world up to that time. In later studies with rabbits Partridge *et al.* (1986) found the same value for the efficiency with which body tissues were used for lactation as that given in the table. The value for utilisation of metabolisable energy for lactation was, however, higher, a finding thought to be due to the high fat content of the diet used.

been made with laying hens, not only because of their economic importance but because their rate of egg production is extremely high. Earlier work on these lines related weight of food ingested to body weight, change in weight and weight of egg, rather than in terms of metabolisable energy and the enthalpy of combustion of gains in weight and of the egg. Making reasonable assumptions about the metabolisable energy of the diets used and assuming that the egg has an enthalpy of combustion of 6.7 kJ/g and body gain one of 16.7 kJ/g, these earlier data have been summarised (Vohra, Wilson & Stopes 1975). The results showed considerable variation about a mean value from eight studies of 0.69. More recent calorimetric work confirms this rather imprecise mean estimate. Kirchgessner (1982) found the efficiency of utilisation of metabolisable energy for egg production to be 0.68 and the energetic efficiency with which body tissues could sustain egg production to be 0.83. Chwalibog (1982) found values of 0.65 for the efficiency of utilisation of metabolisable energy and 0.82 for the efficiency when the body tissues of the bird was the energy source for egg production. As with milk secretion, the value for the efficiency with which body reserves are employed to produce an egg agrees with stoichiometric calculation.

13

The application of studies of the energy exchange

From the preceding chapters it is perhaps evident that it is an over-simplification to express the many metabolic events which take place in animals solely in terms of overall energy transductions. Extremely complex biochemical and biophysical processes are involved and their subtleties are glossed over when all that is considered is the overall changes in heat production or the enthalpy of the body. However precise calorimetric observations on whole animals may be, they are gross in the sense that they represent a massive summation of a host of reactions at cellular and sub-cellular levels and only rarely provide an unequivocal insight into what these reactions might be. Admittedly, explanations of phenomena in the whole animal can often be satisfactorily resolved in terms of some underlying biochemical events, an example being the role of brown adipose tissue and proton conductance in explaining non-shivering thermogenesis. Similarly explanations can be given of the varying effects of different physical environments on the overall heat loss of animals in terms of the physics of radiative and convective heat transfers. Again the increase in heat production that accompanies physical work can be explained, albeit not in an entirely satisfactory way, in terms of the bioenergetics of isolated muscle and elementary kinetics. In many instances, however, the description of the overall exchanges of energy by an animal simply represents the end result of a series of complex metabolic integrations in which the underlying processes that are integrated are only partly characterised and understood. Such a situation constitutes a challenge. Studies of the bioenergetics of whole animals have value in that they pose questions which demand answers in more fundamental terms. These questions include those related to the causes of the post-prandial rise in heat production, the nature of the circadian and seasonal rhythms of metabolism, why metabolic rate falls in under-nutrition, why environmental cold affects digestion, and what constitute the causal factors which account for differences in the heat productions of individuals within a species. It can equally be stated that advances in other areas of biological science pose questions relating to the metabolism of the organism as a whole. What, for example, is the likely effect of alteration of hormonal status through immunological or genetic engineering techniques on the utilisation of the energy of food, what are the consequences of pharmacological manipulation of neural activity, and what is the implication of ion fluxes or indeed of species differences in intermediary metabolism on the production of heat by animals?

Future studies of the interaction between observations of the energy exchanges of whole animals under a variety of circumstances on the one hand, and, on the other, experiments which elucidate biochemical and biophysical processes at the cellular and sub-cellular level, will obviously enhance our understanding of the former and place the latter in a wider context. Such studies can be regarded as continuing the endeavours begun two centuries ago by Lavoisier and by Crawford, who, led by curiosity, sought solutions to the ancient problem of the origin of animal heat.

In addition to this interactive and integrative aspect of studies of the energy exchanges of whole animals, bioenergetic studies have a more immediate application. Some applications have an obvious utility – the nutritional needs of man and his domesticated animals and the design of shelter and clothing are examples. Other applications are more concerned with the provision of concepts which provide a wide perspective about the living world. The energetics of populations and of ecosystems and comparative aspects of bioenergetics are examples of the extension of more detailed studies of the energy exchanges of individuals.

In domesticated animals knowledge of the amounts of energy required to support growth and reproduction enables a rationalisation of their feeding and management to be undertaken. This rationalisation involves the planning of supplies of feeds, including the design of cropping policies to provide them, and the application of econometric methods to arrive at optimal dispositions of land and other resources. Indeed, considered purely from an economic stance, the cost of food is the single most important determinant of output from an animal production enterprise, accounting for 70% or more of total input costs in most situations. Primary consideration of the energy requirements of different classes of animals in such planning has additional advantage in that requirements of other nutrients are closely linked to the energy supply. Thus requirements of most of the B-complex vitamins are proportional to metabolic rate – a matter which is understandable since most vitamins are components of the enzyme systems concerned with the utilisation of the nutrients which are sources of energy for the cell.

Furthermore, requirements for amino acids cannot be considered separately from the concomitant utilisation of energy-yielding constitutents of the diet.

Knowledge about the energy requirements of man is applied in a rather different way to that adopted with domesticated animals. Admittedly, in a very broad sense, the estimation of a country's food supplies in relation to the estimated need for food by its population is usually undertaken in terms of energy, and it is recognised that in the prediction of serious food shortages an energy accounting system is useful. However, it is doubtful whether such an accounting is used in the direct and immediate way in which it is employed with domesticated animals. The value of knowledge about the energy requirements of man lies mostly in its application to the individual, and an optimalisation process which is concerned with the long-term maintenance of health, rather than, as with agricultural animals, an economically efficient use of food resources. The design of optimal dietary regimes to be followed in pregnancy, lactation, during juvenile growth and for hospitalised patients are examples. Knowledge about energy metabolism is obviously also of considerable importance in dealing with its so-called disorders. Obesity, anorexia, cachexia, infection and trauma are again all matters which relate to individuals and all involve the energy exchange.

The modification of the physical environment and hence of the heat loss from animals and man by the provision of housing and shelter is of considerable antiquity! Design of buildings to accommodate animals or people necessarily takes into account factors other than the thermal comfort of their inhabitants. In recent years the design of buildings to accommodate people has changed considerably in that it is recognised that, firstly fuel is no longer cheap and heating costs can be considerable, and secondly people no longer wish to wear thick cumbersome clothing. Recognition of the decline in the metabolic rates of people with advancing age has led to a deeper concern about the definition of optimal environments for them. Certainly increased recognition is now being given, in regions where winters are cold, to the biological factors which define the optimal conditions for occupants of buildings.

Ecological energetics is concerned with populations rather than with individuals. Central to this area of study is the idea of energy flow, that is the transfer of energy, in terms of enthalpy of combustion, from one population component of the ecosystem to another by grazing, predation, ingestion of excretory products or consumption of the whole or parts of dead organisms. It is obvious that in a stable ecosystem the input of energy into the system as a whole is balanced by the output from the system. The energy input may be of solar radiation (where green plants are involved) or it may be of organic matter (as in systems which do not have photosynthetic components). The only output from such a system in equilibrium is heat. Studies of these gross aspects of energy flow in an ecosystem give no information on the number of species within the system, their population sizes, their spatial distribution, their growth rates or their death rates. The situation is precisely that given by the first law of thermodynamics; the enthalpy change in a system is determined only by its initial and final states and is independent of what happens in between. In natural ecosytems, much happens in between. There are considerable and complex interactions between components; grazing by herbivorous species, for example, alters the productivity of the species grazed. Furthermore no natural ecosystems can be regarded as completely isolated in that they are subject to random effects, notably those due to variation of climate and of season. Considerable effort has been expended in devising ways of integrating population dynamics and studies of the bioenergetics of individual organisms with the objective of achieving a greater understanding of the working of ecosystems. Because of the inherent complexity of natural communities of plants and animals it is perhaps understandable that no simple, quantitative generalisations have emerged about the determinants of energy flow within them. Even so, the supply of energy as food is axiomatically a determinant of population survival and the delineation of the relationships between supply of energy as food and the stability of populations of a single species is a subject of considerable interest. Here extensions of studies with the individual animal, describing its reactions to changes in its food energy supply, to whole populations brings in additional dimensions related to food allocation between classes within the species and behavioural responses as well as physiological ones.

Zoologists, confronted with the enormous wealth of animal species existing in a wide variety of natural environments have, understandably, been interested in the formulation of integrating hypotheses to generalise their observations. The Brody–Kleiber relationship, which relates the metabolic rates of animals of different species to their body weight is an example of such a generalisation and so too is its corollary, the concept of biological time. These relationships have done much to provide general statements about a number of biological phenomena – for example the relation of pulse frequency to body weight, the speeds of locomotion of different species and the relative magnitude of their summit metabolisms. Much work has been done to relate other physiological attributes to body weight using allometric methods, and many of these reveal a proportionality to body weight raised either to the powers 0.75 or -0.25, that is, they accord with an expectation based on the Brody–Kleiber energy relationship. However, and as discussed in earlier chapters, as more information has accrued, there is little doubt that these generalisations have poor predictive power when applied to many species of animal. Species are not all designed to accord with a single generalisation relating to their energy metabolism. This does not mean that the initial generalisation is not of value but rather that it is but a general statement providing perhaps a baseline from which to assess the magnitude and nature of the adaptations that species, genera or other taxonomic groups have made during the course of evolution. Certainly the large amount of descriptive work presently being undertaken on the bioenergetics of wild species will provide the basis for more refined general statements about the determinants of their size, form and function.

Studies of the bioenergetics of animals and man are of considerable interest and fascination. Some of them have an immediate utility and application while others are of value in that they raise new concepts and provide the basis of observation for problems which can be explored using more sophisticated approaches. The current studies being made with man, with his domesticated animals and with wild species of bird and mammal represent a continuance of the studies made two centuries ago when, on that February day, Lavoisier and Laplace first placed a guinea pig in a calorimeter.

APPENDIX
Enthalpies (heats) of combustion

There are many listings of the enthalpies of combustion of compounds of biological interest. Examples of such listing are the tabulations of values for amino acids by Hutchens (1975), for fatty acids by Kharasch (1929) and for carbohydrates in the International Critical Tables (1929). Fuller listings are given in the Handbook of Chemistry and Physics (1985) and in Kaye & Laby (1973). In these compilations the precision of the measurements, the conditions under which they were determined and the conventions used in their presentation are rarely stated. Some of the data can be traced to determinations made a century ago and their accuracy is not known. There is doubt about the purity of the compounds examined and certainly some enthalpies of combustion were determined at a temperature of 20 °C rather than the standard 25° C. Because of problems related to the definition of the International Joule at that time – and consequent uncertainty about the enthalpy of combustion of the thermochemical reference standard employed – the constants used to convert values expressed in calories to joules are open to conjecture. A further problem relates to the convention employed with respect to the oxidation of compounds containing sulphur. Thus in the tabulation of Hutchens (1975) the end-product of sulphur oxidation in sulphur-containing compounds is elemental sulphur; other determinations (Tsuzuki *et al.* 1958) state that the end-product is gaseous sulphur trioxide. The accepted convention is that the end-product is H_2SO_4 $115H_2O$. The differences between the enthalpies of combustion of sulphur-containing compounds are not small. Thus Hutchens (1975) lists the enthalpy of combustion of methionine as −2781.5 kJ/mol, the value obtained by Tsuzuki *et al.* (1958) is −3176 kJ/mol and the value for the conventional oxidation is −3386.9 kJ/mol. The values listed in Table A1.1 below all relates to standard conditions and mostly stem from a critical interpretation and assessment of the relevant published data by Dr E.

Domalski of the US Bureau of Standards (Domalski 1972). They can be regarded as the most accurate values presently available. There is doubt about some of the values central to the methods of estimating heat production by indirect calorimetry. For example, it is not clear whether corrections for pressure changes were made when the enthalpies of combustion of body fats were determined or if the data were corrected to the standard temperature.

Enthalpies of formation can be computed from enthalpies of combustion using a value of −393.51 kJ/mol for the enthalpy of formation of gaseous carbon dioxide, and −285.84 kJ/mol for the enthalpy of formation of liquid water. Methods for calculating enthalpies of formation are given in Chapter 2.

Table A1.1. *Enthalpy of combustion for compounds of biological interest*

Compound	ΔH_c (kJ/mol)
Carbohydrates	
Arabinose (β-D, β-L)	−2338.8
Xylose (α-D)	−2338.9
Galactose (α-D)	−2803.7
Glucose (α-D)	−2803.0
Glucose (β-D)	−2809.3
Fructose (β-D)	−2811.6
Sorbose (L)	−2805.8
Lactose (β)	−5648.4
Maltose	−5645.5
Sucrose	−5640.9
Cellobiose	−5638.4
Raffinose	−8472.6
Glycerol	−1655.4
Starch (MJ/kg)	−17.48
Inulin (MJ/kg)	−17.28
Dextrin (MJ/kg)	−17.20
Glycogen (MJ/kg)	−17.51
Cellulose (MJ/kg)	−17.49
Xylan (MJ/kg)	−17.82
Lignin (MJ/kg)	−27.28
Fatty acids	
Formic	−254.6
Acetic	−874.5
Propionic	−1527
n-Butyric	−2183.5
Pentanoic	−2837.3
Hexanoic	−3491.5
Heptanoic	−4145.5
Decanoic	−6109.1
Undecanoic	−6764
Dodecanoic (lauric)	−7414
Tetradecanoic (myristic)	−8722

Table A1.1. (*cont.*)

Compound	ΔH_c (kJ/mol)
Hexadecanoic (palmitic)	−10031
Octadecanoic (stearic)	−11339
Cis-9-octadecanoic (oleic)	−11194
Lipids	
Cattle depot fat (MJ/kg)	−39.3
Human liver and muscle fat (MJ/kg)	−39.4
Human depot fat (MJ/kg)	−39.8
Cattle mixed phospholipids (MJ/kg)	−34.0
Milk fat (MJ/kg)	−38.7
Grain lipids (MJ/kg)	−38.9
Amino acids	
Alanine (L)	−1620.5
Arginine (L)	−3739.9
Asparagine (L)	−1928.7
Cysteine (L)	−2261.5
Cystine (L)	−4252.2
Glutamic acid (L)	−2243.5
Glycine	−973.5
Histidine	*
Isoleucine (D, L)	−3583.6
Leucine (D, L)	−3583.1
Lysine (DL)	−3683.2
Methionine	−3386.9
Proline (DL)	−2728
Phenylalanine (DL, L)	−4645.5
Serine (L)	−1454.8
Threonine (DL, L)	−2102.2
Tyrosine (L)	−4430.0
Valine (DL, L)	−2919.6
Tryptophan	−5628.3
Proteins	
Myosin (MJ/kg)	−24.89
Gelatin (collagens) (MJ/kg)	−22.05
Mammalian muscle (MJ/kg)	−23.70
Other compounds	
Acetone	−1790.4
Ethanol	−1366.8
Methane	−890.3
Uric acid	−1921.1
Lactic acid (DL)	−1367.3
Creatine	−2324.0
Creatinine	−2336.9
Urea	−631.8
Hippuric acid	−4218.5
β-Hydroxy butyric acid (DL)	−2038.4

Note:
*No value traced.

REFERENCES

Abderhalden, A. (1924). *Handbuch der Biologischen Arbeitsmethoden*. Berlin: Urban & Schwartzenberg.

Acheson, K.J., Zahorska-Markiewicz, B., Anantharum, K. & Jequier, E. (1980). Caffeine and coffee: their influence on metabolic rate and substrate utilization in normal weight and obese individuals. *American Journal of Clinical Nutrition*, **33**, 989–97.

Acheson, K.J., Schutz, Y., Bessard, E., Ravussin, E. & Jequier, E. (1984). Nutritional influences on lipogenesis and thermogenesis after a carbohydrate meal. *American Journal of Physiology*, **246**, E62–70.

Adam, I., Young, B.A., Nicol, A.M. & Degen, A.A. (1984). Energy cost of eating in cattle given diets of different forms. *Animal Production*, **38**, 53–6.

Adams, N.J., Brown, C.R. & Nagy, K.H. (1986). Energy exchanges of free-ranging albatrosses (*Diomedea exulans*). *Physiological Zoology*, **89**, 583–91.

Adolph, E.F. (1943). *Physiological Regulations*. Lancaster, Pa.: Jaques Cattell Press.

Agricultural Research Council (ARC) (1980). *The Nutrient Requirements of Ruminant Livestock*. Farnham Royal, UK: Commonwealth Agricultural Bureau.

Agricultural Research Council (ARC) (1981). *The Nutrient Requirements of Pigs*. Farnham Royal, UK: Commonwealth Agricultural Bureau.

Agricultural Research Council (ARC) & Medical Research Council (MRC) (1974). *Food and Nutrition Research*. London: HMSO.

Alexander, G. (1961). Temperature regulation in the new-born lamb. 3. Effect of environmental temperature on metabolic rate, body temperatures and respiratory quotient. *Australian Journal of Agricultural Research*, **12**, 1152–74.

Alexander, G. (1962). Temperature regulation in the new-born lamb. 5. Summit metabolism. *Australian Journal of Agricultural Research*, **13**, 100–21.

Alexander, G. (1974). Heat loss from sheep. In *Heat Loss from Animals and Man*, ed. J.L. Monteith & L.E. Mount, pp. 173–203. London: Butterworth.

Alexander, R. McN. (1968). *Animal Mechanics*. London: Sidgwick & Jackson.

Alexander, R. McN. (1977). Mechanics and scaling of terrestial locomotion. In *Scale Effects of Terrestial Locomotion*, ed. T.J. Pedley, pp. 93–110. London: Academic Press.

Alexander, R. McN. (1980). Optimum walking techniques for quadrupeds and bipeds. *Journal of Zoology (London)*, **192**, 97-117.

Alexander, R. McN. (1981). *The Chordates*, 2nd edn. Cambridge: Cambridge University Press.

Alexander, R. McN. & Goldspink, G. (1977). *Mechanics and Energetics of Animal Locomotion.* London: Chapman & Hall.

Apfelbaum, M. (1973). Influence of energy intake on energy expenditure in man. In *Obesity in Perspective*, ed. G.A. Bray, pp. 145–55. USA: Publication No 75–708 National Institutes of Health.

Argo, C.M. & Smith, J.S. (1983). Relationship of energy requirements and seasonal cycles of food intake in Soay rams. *Journal of Physiology*, **343**, 23–4.

Arieli, A. (1986). Effects of glucose on fermentation heat in sheep rumen fluid *in vitro*. *British Journal of Nutrition*, **56**, 305–11.

Armstrong, D.G. (1969). Cell bioenergetics and energy metabolism. In *Handbuch der Tierernährung*, Band 1, ed. W. Lenkeit, K. Breirem & E. Crasemann, pp. 385–414. Hamburg: Paul Parey.

Armstrong, D.G. & Blaxter, K.L. (1961). The utilization of the energy of carbohydrate by ruminants. In *2nd Symposium on Energy Metabolism*, ed. E. Brouwer & A.J. Van Es, pp. 187–97. Wageningen: European Association for Animal Production.

Aschoff, J. (1981). Thermal conductance in mammals and birds: its dependence on body size and circadian phase. *Comparative Biochemistry and Physiology*, **69A**, 611–9.

Aschoff, J. & Pohl, H. (1970). Die rüheumsatz von vogeln als funktion der tageszeit under der korpergrosse. *Journale für Ornithologie*, **111**, 38-47.

Atwater, W.O. (1902). *On the Digestibility and Availability of Food Materials*. Agricultural Experiment Station 14th Annual Report. Storrs, Connecticut.

Atwater, W.O. (1910). *Principles of Nutrition and Nutritive Value of Food.* Farmer's Bulletin No. 142. Washington, DC: US Department of Agriculture.

Atwater, W.O. & Benedict, F.G. (1903). *Experiments on the Metabolism of Matter and Energy in the Human Body*. Agricultural Bulletin No. 136. Washington DC: US Department of Agriculture.

Atwater, W.O. & Bryant, A.P. (1900). *The Availability and Fuel Value of Food Materials.* Agricultural Experiment Station 12th Annual Report. Storrs, Connecticut.

Austin, C.R. & Short, R.V. (1984). *Reproduction in Mammals, Book 4: Reproductive Fitness*, 2nd edn. Cambridge: Cambridge University Press.

Bachmair, A., Finley, D. & Varshavsky, A. (1986). *In vivo* half-life of a protein is a function of its amino-terminal residue. *Science*, **234**, 179–86.

Bagshaw, C. (1982) *Muscular Contraction*. London: Chapman & Hall.

Bahr, R. & Maehlum, S. (1986). Excess post-exercise oxygen consumption. *Acta Physiologica Scandinavia*, **128, suppl. 556**, 99–104.

Bakken, G.S., Buttemer, W.A., Dawson, W.R. & Gates, D.M. (1981). Heated taxidermic mounts: a means of measuring the standard operative temperature affecting small endothermic animals. *Ecology*, **62**, 311–8.

Baldwin, R.L., Smith, N.E., Taylor, J. & Sharp, M. (1980). Manipulating metabolic parameters to improve growth rate and milk secretion. *Journal of Animal Science*, **51**, 1416–28.

Barr, H.G. & McCracken, K.J. (1984). High efficiency of energy utilization in cafeteria and force-fed rats kept at 29 °C. *British Journal of Nutrition*, **51**, 379–87.

Bartholomew, G. (1977). Body temperature and energy metabolism. In *Animal Physiology: Principles and Applications*, 3rd edn, ed. M. Gordon, pp. 364–449. New York: McMillan.

Beamish, F.W.H. (1978). Swimming capacity. In *Fish Physiology*, VIII, ed. W.S. Hoar & D.J. Randall, pp. 101-87. New York: Academic Press.

Behnke, J. (1965). *Human Body Composition*. Oxford: Pergamon Press.

Behnke, A.R. & Yaglou, C.P. (1951). Physiological responses of men to chilling in ice water and fast and slow rewarming. *Journal of Applied Physiology*, **3**, 591–5.

Benedict, F.G. (1915). *A Study of Prolonged Fasting*. Publication No. 203, Carnegie Institute. Washington DC: Carnegie Institute.

Benedict, F.G. (1932). *The Physiology of Large Reptiles*. Publication No. 425, Carnegie Institute. Washington DC: Carnegie Institute.

Benedict, F.G. (1936). *The Physiology of the Elephant*. Publication No. 474, Carnegie Institute. Washington DC: Carnegie Institute.

Benedict, F.G. (1938). *Vital Energetics. A Study in Comparative Basal Metabolism*. Publication No. 503, Carnegie Institute. Washington DC: Carnegie Institute.

Benedict, F.G., Horst, K. & Mendel, L.B. (1932). The heat production of unusually large rats during prolonged fasting. *Journal of Nutrition*, **5**, 581–97.

Benedict, F.G. & Lee, R.C. (1937). *Lipogenesis in the Animal Body with Special Reference to the Physiology of the Goose*. Publication No. 489, Carnegie Institute. Washington DC: Carnegie Institute.

Benedict, F.G., Miles, W.R., Roth, P. & Smith, H.M. (1919). *Human Vitality and Efficiency under Prolonged Restricted Diet*. Publication No. 280, Carnegie Institute. Washington DC: Carnegie Institute.

Benedict, F.G. & Mürchhauser, H. (1915). *Energy Transformations during Horizontal Walking*. Publication No. 126, Carnegie Institute. Washington DC: Carnegie Institute.

Benedict, F.G. & Ritzman, E.G. (1923). *Undernutrition in Steers. Its Relation to Metabolism and Subsequent Realimentation*. Publication No. 324, Carnegie Institute. Washington DC: Carnegie Institute.

Benedict, F.G. & Ritzman, E.G. (1927). *The Metabolism of the Fasting Steer*. Publication No. 377, Carnegie Institute. Washington DC: Carnegie Institute.

Benedict, F.G. & Root, H.F. (1926). Insensible perspiration: its relation to human physiology and pathology. *Archives of Internal Medicine*, **38**, 1–35.

Bennett, J.W. (1972). The maximal metabolic response of sheep to cold: effects of rectal temperature, shearing, feed composition, body posture and body weight. *Australian Journal of Agricultural Research*, **23**, 1045–58.

Ben-Porat, M., Sideman, S. & Bursztein, S. (1983). Energy metabolism rate equations for fasting and post-absorptive subjects. *American Journal of Physiology*, **244**, R764–9.

Ben Shaul, D.M. (1962). The composition of the milk of wild animals. *International Zoological Yearbook*, **4**, 333–42.

Benzinger, T.H. & Kitzinger, C. (1949). Direct calorimetry by means of the gradient principle. *Review of Scientific Instruments*, **20**, 849–60.

Bergmann, C. (1847). Über die verhaltnisse der warme okonomie der thiere zu ihre grosse. *Gottingen Studien*, **1**, 595–708.

Berraud, J., -P. (1975). *La Vie et l'Ouvre de F. Laulanie (1850–1906)*. Toulouse: Cepadons.

Bertin, L. (1973). *La Vie des Animaux*. English version *Larousse Encyclopaedia of Animal Life*, 3rd edn. London: Paul Hamlyn.

Bessman, S.P. (1985). The creatine-creatinine phosphate energy shuttle. *Annual Reviews of Biochemistry*, **54**, 831–62.

Bewley, D.K. (1986). Anthropomorphic models for the calibration of equipment for *in-vivo* neutron activation analysis. United States Department of Energy, Brookhaven National Laboratory 1986 conference p1.

Bisdee, J.T. & James, W.P.T. (1983). Whole body calorimetry studies in the menstrual cycle. *Proceedings of the 4th International Congress on Obesity*, 52A .

Blake, R.W. (1983). *Fish Locomotion*. Cambridge: Cambridge University Press.

Blaxter, K.L. (191b). Energy utilization in the ruminant. In *Digestive Physiology and Nutrition of the Ruminant*, ed. D. Lewis, pp. 183–97. London: Butterworths.
Blaxter, K.L. (1961a). Lactation and the growth of the young. In *Milk: the Mammary Gland and its Secretion*, ed. S.K. Kon & A.T. Cowie, pp. 305–61. London: Academic Press.
Blaxter, K.L. (1962a). The fasting metabolism of adult wether sheep. *British Journal of Nutrition*, **16**, 615–26.
Blaxter, K.L. (1962b). *The Energy Metabolism of Ruminants*. London: Hutchinson.
Blaxter, K.L. (1964). Protein metabolism and requirements in pregnancy and lactation. In *Mammalian Protein Requirements*, ed. H.N. Munro & J.B. Alison, pp. 173–223. New York: Academic Press.
Blaxter, K.L. (1967). Techniques in energy metabolism studies and their limitations. *Proceedings of the Nutrition Society*, **26**, 86–96.
Blaxter, K.L. (1968). The effect of the energy supply on growth. In *Growth and Development of Mammals*, ed. G.A. Lodge & G.E. Lamming, pp. 329–44. London: Butterworths.
Blaxter, K.L. (1972). Fasting metabolism and the energy requirements for maintenance. In *Festskrift til Knut Breirem*, pp.19–36. Gjovik: Mariendals Boktrykkerie.
Blaxter, K.L. (1977). Environmental factors and their influence on the nutrition of farm livestock. In *Nutrition and the Climatic Environment*, ed. W. Haresign, H. Swan & D. Lewis, pp. 1–16. London: Butterworth.
Blaxter, K.L. (1980). Energy metabolism: some lessons from studies with animals. In *Assessment of Energy Metabolism in Health and Disease*, ed. S. Calvert, pp. 6–15. Columbus Ohio: Ross Laboratories.
Blaxter, K.L. & Boyne, A.W. (1978). The estimation of the nutritive value of feeds as energy sources for ruminants and the derivation of feeding systems. *Journal of Agricultural Science (Cambridge)*, **90**, 47–68.
Blaxter, K.L. & Boyne, A.W. (1979). The estimation of the nutritive value of feeds as energy sources for ruminants and the derivation of feeding systems. *Journal of Agricultural Science, (Cambridge)*, **90**, 47–68.
Blaxter, K.L. & Boyne, A.W. (1982). Fasting and maintenance metabolism of sheep. *Journal of Agricultural Science (Cambridge)*, **99**, 611–20.
Blaxter, K.L. & Clapperton, J.L. (1965). Prediction of the amount of methane produced by ruminants. *British Journal of Nutrition*, **19**, 511–22.
Blaxter, K.L., Clapperton, J.L. & Martin, A.K. (1966). The heat of combustion of the urine of sheep and cattle in relation to its chemical composition and to diet. *British Journal of Nutrition*, **20**, 449–60.
Blaxter, K.L., Fowler, V.R. & Gill, J.C. (1982). A study of the growth of sheep to maturity. *Journal of Agricultural Science (Cambridge)*, **98**, 405–20.
Blaxter, K.L. & Graham, N.McC. (1956). The effect of the grinding and cubing process on the utilization of the energy of dried grass. *Journal of Agricultural Science (Cambridge)*, **47**, 207–17.
Blaxter, K.L., Graham, N.McC. & Wainman, F.W. (1956). Some observations on the digestibility of food by sheep and on related problems. *British Journal of Nutrition*, **10**, 69–91.
Blaxter, K.L., Graham, N.McC. & Wainman, F.W. (1959). Environmental temperature, energy metabolism and heat regulation in sheep. 3. The metabolism and thermal exchanges of sheep with fleeces. *Journal of Agricultural Science (Cambridge)*, **52**, 41–9.
Blaxter, K.L. & Joyce, J.P. (1963a). The accuracy and ease with which measurements of metabolism can be made with tracheostomized sheep. *British Journal of Nutrition*, **17**, 523–37.

Blaxter, K.L. & Joyce, J.P. (1963b). Artificial sheep. *Animal Production*, **5**, 216.

Blaxter, K.L. & Rook, J.A.F. (1953). The heat of combustion of the tissues of cattle in relation to their chemical composition. *British Journal of Nutrition*, **7**, 83–91.

Blaxter, K.L. & Wainman, F.W. (1961). Environmental temperature and the energy metabolism and heat emission of steers. *Journal of Agricultural Science (Cambridge)*, **56**, 81–90.

Blaxter, K.L. & Wainman, F.W. (1964a). The utilization of the energy of different diets by sheep and cattle for maintenance and for fattening. *Journal of Agricultural Science (Cambridge)*, **63**, 113–28.

Blaxter, K.L. & Wainman, F.W. (1964b). The effect of increased air movement on the heat production and emission of steers. *Journal of Agricultural Science (Cambridge)*, **62**, 207–14.

Bleibtreu, M. (1901). Fettmast und respiratorischer quotient. *Pflugers Archive*, **85**, 345–82.

Bligh, J. & Johnson, K.G. (1973). Glossary of terms for thermal physiology. *Journal of Applied Physiology*, **35**, 941–61.

Bligh, J. & Robinson, S.G. (1965). Radiotelemetry in a veterinary research project. *Medical and Biological Illustration*, **15**, 94–9.

Blueweiss, L., Fox, H., Kudzma, V., Nakashima, D., Peters, R. & Sams, S. (1978). Relationships between body size and some life history parameters. *Oecologia (Berlin)*, **37**, 257–72.

Bonjour, J.P., Welti, H.J. & Jequier, E. (1976). Etude calorimetrique des consignes thermoregulatrices au declenchement de la sudation et au cours du cycle menstruel. *Journal Physiologie (Paris)*, **72**, 181–204.

Bonner, W.N. (1984). Lactation strategies in Pinnipeds: problems for a marine mammalian group. *Symposium of the Zoological Society of London*, **51**, 253–72.

Boyne, A.W., Brockway, J.M., Ingram, J.F. & Williams, J. (1981). Modification of tractive loading on the energy cost of walking in sheep, cattle and man. *Journal of Physiology (London)*, **315**, 303–16.

Brackenbury, J. (1984). Physiological reactions of birds to flight and running. *Biological Reviews*, **59**, 559–75.

Bradley, S.R. & Deavers, D.R. (1980). A re-examination of the relationship between thermal conductance and body weight in mammals. *Comparative Biochemistry and Physiology*, **65A**, 465–76.

Brafield, A.E. & Llewellyn, M.J. (1982). *Animal Energetics*. Glasgow: Blackie.

Brand, M.D. & Murphy, M.P. (1987). Control of electron flux through the respiratory chain in mitochondria and cells. *Biological Reviews*, **62**, 141–93.

Breirem, K. (1936). Erhaltungs-stoffwechsels des wachsenden schweiner. *Tierernährung*, **8**, 463–98.

Brett, J.R. (1965). The relation of size to rate of oxygen consumption and sustained swimming speed of sockeye salmon (*Oncorhynchus nerka*). *Journal of the Fisheries Research Board of Canada*, **22**, 1491–7.

Briedis, D. & Seagrave R.C. (1984). Energy transformation and entropy production in living systems. 1. Applications to embryonic growth. *Journal of Theoretical Biology*, **110**, 173–93.

Brockway, J.M. (1987). Derivation of formulae used to calculate energy expenditure in man. *Human Nutrition: Clinical Nutrition*, **41C**, 463–71.

Brockway, J.M., McDonald, J.D. & Pullar, J.D. (1965). Evaporative heat loss mechanisms in the sheep. *Journal of Physiology (London)*, **179**, 554–68.

Brockway, J.M. & McEwan, E.H. (1969). Oxygen uptake and cardiac performance in sheep. *Journal of Physiology (London)*, **202**, 661–9.

Broady, S. (1945). *Bioenergetics and Growth*. New York: Reinhold Publishing Co.

Brody, S. & Nisbet, R. (1938). Growth and development with special reference to domestic animals. 47. A comparison of the amounts of energetic efficiencies of milk production in rat and dairy cow. *University of Missouri Agricultural Experiment Station*, Research Bulletin No. 285.

Brody, S. & Procter, R.C. (1932). Growth and development with special reference to domestic animals: further investigations of surface area in energy metabolism. *University of Missouri Agricultural Experiment Station*, Research Bulletin No. 116.

Brody, S., Procter, R.C. & Ashworth, U.S. (1934). Basal metabolism, endogenous metabolism and neutral sulphur excretions as functions of body weight. *University of Missouri Agricultural Experiment Station*, Research Bulletin No. 220.

Brook, A.H. & Short, B.F. (1960). Sweating in sheep. *Australian Journal of Agricultural Research*, **11**, 557–69.

Broster, W.H. & Broster, V.J. (1984). Long-term effects of plane of nutrition on the performance of the dairy cow. *Journal of Dairy Research*, **51**, 149–96.

Brouwer, E. (1958). On simple formulae for calculating the heat expenditure and the quantities of carbohydrate and fat metabolized in ruminants from data on gaseous exchange and urine N. In *1st Symposium on Energy Metabolism*, pp. 182–94. Rome: European Association for Animal Production.

Brouwer, E. (1965). Report of subcommittee on constants and factors. In *Energy Metabolism*, ed. K.L. Blaxter, pp. 441–3. London: Academic Press.

Browne, T. (1646). *Pſeudodoxia Epidemica: or Enquiries into Very Many Received Tenets and Commonly Preſumed Truths*. London: Edward Dod.

Bruce, J.M. & Clark, J.J. (1979). Models of heat production and critical temperature for growing pigs. *Animal Production*, **28**, 353–69.

Bull, N.L. & Wheeler, E.F. (1986). A study of different dietary survey methods among 30 civil servants. *Human Nutrition: Applied Nutrition*, **40A**, 60–6.

Burkinshaw, L. (1982). The contribution of nuclear activation techniques to medical science. *Journal of Radioanalytical Chemistry*, **69**, 27–45.

Burkinshaw, L. (1985). Measurement of human body composition *in vivo*. *Progress in Medical Radiation Physics*, **2**, 113–37.

Burkinshaw, L. & Oxby, C.B. (1986). New methods of measuring body composition. In *Proceedings of the 13th International Congress of Nutrition*, ed. T.G. Taylor & N.K. Jenkins pp. 259–61. London: John Libbey.

Burnett, G.H. & Bruce, J.M. (1978). Thermal simulation of a suckler cow. *Farm Buildings Progress*, **54**, 11–13.

Burton, A.C. (1934). The application of the theory of heat flow to the study of energy metabolism. *Journal of Nutrition*, **7**, 497–533.

Burton, A.C. (1935). Human calorimetry. 2. The average temperature of the body. *Journal of Nutrition*, **9**, 261–79.

Burton, A.C. & Bronk, D.W. (1937). The motor mechanism of shivering and thermal muscular tone. *American Journal of Physiology*, **119**, 284–90.

Burton, A.C. & Edholm, O. (1955). *Man in a Cold Environment*. London: Edward Arnold.

Burton, K. (1974). The enthalpy change for the reduction of nicotinamide-adenine dinucleotide. *Biochemical Journal*, **143**, 365–8.

Buttemer, W.A., Hayworth, A.M., Weathers, W.W. & Nagy, K.A. (1986). Time budget estimates of avian energy expenditure: physiological and meteorological considerations. *Physiological Zoology*, **59**, 131–49.

Calder, W.A. 3rd. (1981). Scaling of physiological processes in homeothermic animals. *Annual Reviews of Physiology*, **43**, 301–22.

Calder, W.A. 3rd. (1987). Scaling energetics of homeothermic variates: an operational allometry. *Annual Reviews of Physiology*, **49**, 107–20.

Calder, W.A. 3rd. & King, J.R. (1974). Thermal and calorie relations of birds. In *Avian Biology*, vol. IV, ed. D.S. Farmer & J.R. King, pp. 259–413. London: Academic Press.

Calloway, D.H., Colastio, D.J. & Mathew, R.D. (1966). Gases produced by human intestinal flora. *Nature (London)*, **212**, 1238.

Campbell, R.G. & Dunkin, A.C. (1983). The effects of energy intake and dietary protein on nitrogen retention, growth performance, body composition and some aspects of energy metabolism in baby pigs. *British Journal of Nutrition*, **49**, 221–30.

Carlsson, F.D., Hardy, D. & Wilkie, D.R. (1967). The relation between the heat produced and phosphorylcreatine split during isometric contraction of frog muscle. *Journal of Physiology (London)*, **189**, 209–35.

Carnot, S. (1824). *Reflexions sur la Puissance motrice du Feu*. Paris.

Cathcart, E.P. & Cuthbertson, D.P. (1931). The composition and distribution of the fatty substance of the human subject. *Journal of Physiology (London)*, **72**, 349–60.

Causton, D.R. (1977). *A Biologist's Mathematics*. London: Edward Arnold.

Cavagna, G.A., Heglund, N.C. & Taylor, C.R. (1977). Walking, running and galloping: mechanical similarities between different animals. In *Scale Effects in Animal Locomotion*, ed. T.J. Pedley, pp. 111–25. London: Academic Press.

Cena, K. (1974). Radiative heat loss from animals and man. In *Heat Loss from Animals and Man*, ed. J.L. Monteith & L.E. Mount, pp. 33–58. London: Butterworths.

Cena, K. & Clark, J.A. (1978). Thermal resistance units. *Journal of Thermal Biology*, **3**, 173–4.

Cena, K. & Monteith, J.L. (1975a). Transfer processes in animal coats. 1. Radiative transfer. *Proceedings of the Royal Society of London, Series B*, **188**, 377–93.

Cena, K. & Monteith, J.L. (1975b). Transfer processes in animal coats. 2. Conduction and convection. *Proceedings of the Royal Society of London, Series B*, **188**, 395–411.

Charlet-Lery, G. & Szylit, O. (1980). Energy and protein efficiency of some diets in axenic and holoaxenic growing chickens. In *Energy Metabolism*, ed. L.E. Mount, pp. 81–4. London: Butterworths.

Chwalibog, A. (1982). Energy efficiency for egg production. In *Energy Metabolism of Farm Animals*, ed. A. Ekern & B. Sundstol, pp. 270–3. Aas, Norway: Agricultural University of Norway.

Clapperton, J.L. (1964a). The effect of walking on the utilization of food by sheep. *British Journal of Nutrition*, **18**, 39–46.

Clapperton, J.L. (1964b). The energy metabolism of sheep walking on the level and on gradients. *British Journal of Nutrition*, **18**, 47–54.

Clapperton, J.L., Joyce, J.P. & Blaxter, K.L. (1965). Estimates of the contribution of solar radiation to the thermal exchanges of sheep at a latitude of 55 °N. *Journal of Agricultural Science (Cambridge)*, **64**, 37–49.

Close, W.H. (1971). The influence of environmental temperature and plane of nutrition on heat losses from individual pigs. *Animal Production*, **13**, 295–302.

Close, W.H. & Mount, L.E. (1978). The effects of plane of nutrition and environmental temperature on the energy metabolism of the growing pig. 1. Heat loss and critical temperature. *British Journal of Nutrition*, **40**, 413–21.

Close, W.H., Noblet, J. & Heavens, R.P. (1985). Studies on the energy metabolism of the pregnant sow. 2. The partition and utilization of metabolizable energy in pregnant and non-pregnant animals. *British Journal of Nutrition*, **53**, 267–79.

Close, W.H. & Stanier, M.W. (1984). Effects of plane of nutrition and environmental temperature on the growth and development of the early weaned piglet. 2. Energy metabolism. *Animal Production*, **38**, 221–31.

Costhill, D.L., Gollnick, E.D., Jansson, E.D., Saltin, B. & Stein, E.M. (1973). Glycogen depletion in human muscle fibres during distance running. *Acta Physiologica Scandinavica*, **89**, 374–83.

Cottle, M. & Carlson, L.D. (1956). Regulation of heat production in cold adapted rats. *Proceedings of the Society for Experimental Biology and Medicine*, **92**, 845–9.

Crawford, A. (1779). *Experiments and Observations on Animal Heat and the Inflammation of Combustible Bodies. Being an Attempt to Resolve these Phenomena into a General Law of Nature*. London: J. Murray.

Crawford, E.C. (1962). Mechanical aspects of panting in dogs. *Journal of Applied Physiology*, **17**, 249–51.

Crichton, G.W. & Pownall, R. (1974). The homeothermic status of the neonatal dog. *Nature (London)*, **251**, 142-4.

Croxall, J.P. (1982). Energy costs of incubation and moult in petrels and penguins. *Journals of Animal Ecology*, **51**, 177–94.

Cummings, J.H. (1978). Diet and transit time through the gut. *Journal of Plant Foods*, **3**, 83–95.

Curtin, N.A. & Woledge, R.C. (1978). Energy changes and muscular contraction. *Physiological Reviews*, **58**, 690–761.

Czerkawski, J.W. (1986). *An Introduction to Rumen Studies*. Oxford: Pergamon Press.

Dallosso, H.M. & James, W.P.T. (1984). Whole body calorimetry studies with adult man. 2. The interaction of exercise and over-feeding on the thermic effect of a meal. *British Journal of Nutrition*, **52**, 65–72.

Danforth, E. (1983). The role of thyroid hormones and insulin in the regulation of energy metabolism. *American Journal of Clinical Nutrition*, **38**, 1006–17.

Danforth, E. (1985). Hormonal adaptation to over- and under-feeding. In *Substrate and Energy Metabolism in Man*, ed. J.S. Garrow & D. Halliday, pp. 155–68. London: J. Libbey.

Datta, S.R. & Ramanthan, N.L. (1968). Energy expenditures in work predicted from heart rate and pulmonary respiration. *Journal of Applied Physiology*, **26**, 297–302.

Dauncey, M.J. (1980). Metabolic effects of altering the 24-hour energy intake in man using direct and indirect calorimetry. *British Journal of Nutrition*, **43**, 257–69.

Dauncey, M.J. (1981). Influence of mild cold on 24-hour energy expenditure, resting metabolism and diet induced thermogenesis. *British Journal of Nutrition*, **45**, 257–68.

Dauncey, M.J. (1988). Energy expenditure and activity. *Canadian Journal of Physiology and Pharmacology* (in press).

Dauncey, M.J., Ingram, D.L., Watts, D.E. & Legge, K.F. (1983). Evaluation of the effects of environmental temperature and nutrition on growth and development. *Journal of Agricultural Science (Cambridge)*, **101**, 291–9.

Davies, C.T.M. (1980). Effects of wind assistance and resistance on the forward motion of a runner. *Journal of Applied Physiology*, **48**, 702–9.

Dawes, G.S. & Mott, J.C. (1959). The increase in oxygen consumption of the lamb after birth. *Journal of Physiology (London)*, **146**, 295–315.

Dawson, T.J. & Dawson, W.R. (1982). Metabolic scope and conductance in response to cold of some dasyurid marsupials and Australian rodents. *Comparative Biochemistry and Physiology*, **71A**, 59–64.

Dawson, T.J. & Taylor, C.R. (1973). Energetic cost of locomotion in kangaroos. *Nature (London)*, **246**, 313–4.

Demment, M.W. & van Soest, P.J. (1985). A nutritional explanation for body size patterns of

ruminants and non-ruminant herbivores. *American Naturalist*, **125**, 641–72.
Drabkin, D.L. (1950). The distribution of chromoproteins, haemoglobin, myoglobin and cytochrome C in the tissues of different species and the relationship of the total of each chromoprotein to body mass. *Journal of Biological Chemistry*, **182**, 317–33.
Drobny, R.D. (1980). Reproductive energetics of wood ducks. *Auk*, **97**, 480–90.
Domalski, E.S. (1972). Selected values of heats of combustion and heats of formation of organic compounds containing the elements carbon, hydrogen, nitrogen, oxygen, phosphorus and sulphur. *Journal of Physical Chemistry Reference Data*, **1**, 221–77.
DuBois, E.F. (1936). *Basal Metabolism in Health and Disease*, 3rd edn. Philadelphia: Lea & Febiger.
Durnin, J.V.G.A. (1985). The energy cost of exercise. *Proceedings of the Nutrition Society*, **44**, 273–82.
Durnin, J.V.G.A. & Edwards, S.R.G. (1955). Pulmonary ventilations as an index of energy expenditure. *Quarterly Journal of Experimental Physiology*, **40**, 370–7.
Economos, A.C. (1981). The largest land mammal. *Journal of Theoretical Biology*, **80**, 445–50.
Economos, A.C. (1982). On the origin of biological similarity. *Journal of Theoretical Biology*, **94**, 25–60.
Egan, E.P. & Luff, B.B. (1966). Heat of solution, heat capacity and density of aqueous urea at 25 °. *Journal of Chemical and Engineering Data*, **11**, 192–198.
Eggum, B.O. & Chwalibog, A. (1982). Energy metabolism in rats with normal and reduced flora. In *Energy Metabolism of Farm Animals*, ed. E. Ekern & F. Sundstol, pp. 164–7. Aa, Norway: Agricultural University of Norway.
Elder, H.Y. & Trueman, E.R. (1980). *Aspects of Animal Movement*. Cambridge: Cambridge University Press.
Elkana, Y. (1974). *The Discovery of the Conservation of Energy*. London: Hutchinson Educational.
Else, P.L. & Hulbert, A.J. (1985). An allometric comparison of the mitochondria of mammalian and reptilian species. The implication for the evolution of endothermy. *Journal of Comparative Physiology B*, **156**, 3–11.
Else, P.L. & Hulbert, A.J. (1987). Evolution of mammalian endothermic metabolism: leaky membranes as a source of heat. *American Journal of Physiology*, **253**, R1–R7.
Erikson, H., Krog, J., Andersen, K.L. & Scholander, P.F. (1956). The critical temperature in naked man. *Acta Physiologica Scandinavia*, **37**, 35–9.
European Association for Animal Production (EAAP) (1958). *1st Symposium on Energy Metabolism*. Rome: EAAP.
Evans, E.E. & Miller, D.S. (1968). Comparative nutrition, growth and longevity. *Proceedings of the Nutrition Society*, **27**, 121–9.
Fancy, S.G. & White, R.G. (1985). Incremental cost of activity. In *Bioenergetics of Wild Herbivores*, ed. R.J. Hudson & R.G. White, pp. 143–60. Boca Raton, Florida: CRC Press.
Fanger, P.O. (1970). *Thermal Comfort*. Copenhagen: Danish Technical Press.
FAO/WHO/UNU. (1985). *Energy and Protein Requirements. Report of a Joint Expert Consultation*. WHO Technical Report Series No. 724. Geneva: World Health Organisation.
Farrell, D.J. (1974). General principles and assumptions of calorimetry. In *Energy Requirements of Poultry*, ed. T.R. Morris & B.M. Freeman, pp. 1–24. Edinburgh: British Poultry Science Ltd.
Farrell, D.J. (1980). The reduction in metabolic rate and heart rate of man during meditation. In *Energy Metabolism*, ed. L.E. Mount, pp. 279–82. London: Butterworths.
Farrell, D.J., Leng, R.A. & Corbett, J.L. (1972). Undernutrition in grazing sheep. 2.

Calorimetric measurements on sheep taken from pasture. *Australian Journal of Agricultural Research*, **23**, 499–509.

Feather, N. (1963). *Mass, Length and Time*. Harmondsworth: Penguin Books.

Feder, M.E. & Burggren, W.M. (1985). Cutaneous gas exchange in vertebrates: design patterns, control and implications. *Biological Reviews*, **60**, 1–45.

Feist, D. & Rosenmann, M. (1975). Seasonal sympatho-adrenal and metabolic responses to cold in the Alaskan snow-shoe hare *Lepus americanus*. *Comparative Biochemistry and Physiology*, **51A**, 449–55.

Ferguson, S.J. & Sorgato, M.C. (1982). Proton electro-chemical gradients and energy transduction processes. *Annual Reviews of Biochemistry*, **51**, 185–217.

Ferrell, C.L., Koong, L.J. & Nienaber, J.A. (1986). Effect of previous nutrition on body composition and maintenance energy costs of growing lambs. *British Journal of Nutrition*, **56**, 595–605.

Finch, V.A. (1985). Comparison of non-evaporative heat transfer in different cattle breeds. *Australian Journal of Agricultural Research*, **36**, 497–508.

Fisher, C. (1982). Energy evaluation of poultry rations. In *Recent Advances in Animal Nutrition – 1982*, ed. W. Haresign, pp. 113–39. London: Butterworths.

Fitzgerald, J.P. (1957). Cutaneous respiration in man. *Physiological Reviews*, **37**, 325–36.

Flatt, J.P. (1978). The biochemistry of energy expenditure. In *Recent Advances in Obesity Research*, ed. G.A. Bray, pp. 211–28. London: Newman Publications Ltd.

Flatt, J.P. (1985). Energetics of intermediary metabolism. In *Substrate and Energy Metabolism in Man*, ed. J.S. Garrow & D. Halliday, p. 58–69. London: J. Libbey.

Fleming, M.R. (1985). The thermal physiology of the feathertail glider *Acrobates pygmaeus*. *Australian Journal of Zoology*, **33**, 667–81.

Forbes, G.B. (1984). Energy intake and body weight: a re-examination of two classic studies. *American Journal of Clinical Nutrition*, **34**, 349–50.

Forbes, G.B., Brown, M.R., Welle, S.L. & Lipinski, B.A. (1986). Deliberate over-feeding in women and men: energy cost and composition of gain. *British Journal of Nutrition*, **56**, 1–9.

Foster, D.O. & Frydman, M.L. (1978). Non-shivering thermogenesis in the rat. 2. Measurement of blood flow with microspheres point to brown adipose tissue as the dominant site of the calorigenesis induced by noradrenaline. *Canadian Journal of Physiology and Pharmacology*, **56**, 110.

Fox, R.H. (1973). Heat acclimation and the sweating response. In *Heat Loss from Animals and Man*, ed. J.L. Monteith & L.E. Mount, pp. 277–303. London: Butterworths.

France, J. & Thornley, F.H.M. (1984). *Mathematical Models in Agriculture*. London: Butterworths.

Fuller, H.L., Dale, N.M. & Smith, C.F. (1983). Comparison of heat production of chicken measured by energy balance and by gaseous exchange. *Journal of Nutrition*, **113**, 1403–8.

Fuller, M.F. (1983). Energy and nitrogen balances in young pigs maintained at constant weight with diets of differing protein content. *Journal of Nutrition*, **113**, 15–20.

Fuller, M.F. & Boyne, A.W. (1972). The effect of environmental temperature on the growth and metabolism of pigs given different amounts of food. 2. Energy metabolism. *British Journal of Nutrition*, **28**, 373–84.

Fuller, M.F., Cadenhead, A., Mollison, G. & Seve, G. (1987a). Effects of amount and quality of dietary protein on nitrogen metabolism and heat production of growing pigs. *British Journal of Nutrition*, **58**, 277–85.

Fuller, M.F., Reeds, P.J., Cadenhead, A., Seve, B. & Preston, T. (1987b). Effects of amount and quality of protein on nitrogen metabolism and protein turnover of pigs. *British Journal of Nutrition*, **58**, 287–300.

Furuse, M. & Yokata, H. (1984). Protein and energy utilization in germ-free and conventional chicks given diets containing different amounts of dietary protein. *British Journal of Nutrition*, **51**, 255–64.

Gabrielsen, G.W., Mehlum, F. & Nagy, K.A. (1987). Daily energy expenditure and energy utilization of free-ranging black-legged kittiwakes. *The Condor*, **89**, 126–32.

Gagge, A.P., Burton, A.C., & Bazett, H.C.A. (1941). A practical system of units for the description of the heat exchange of man with his environment. *Science*, **94**, 2445.

Gagge, A.P., Herrington, L.P. & Winslow, C.-E.A. (1937). Thermal interchanges between the human body and its atmospheric environment. *American Journal of Hygiene*, **26**, 84–102.

Gallivan, G.J. & Best, R.C. (1986). The influence of feeding and fasting on the metabolic rate and ventilation of the Amazonian manatee (*Trichechus ininguis*). *Physiological Zoology*, **59**, 552–7.

Garby, L., Lammert, O. & Nielsen, E. (1986). Energy expenditure over 24 hours on low physical activity programmes in human subjects. *Human Nutrition: Clinical Nutrition*, **40C**, 141–150.

Garrow, J.S. (1965). The use and calibration of a small whole body counter for measurement of total body potassium in malnourished children. *West Indian Medical Journal*, **24**, 73–81.

Garrow, J.S. (1978). *Energy Balance and Obesity in Man*, 2nd edn. Amsterdam: Elsevier/North Holland.

Garrow, J.S. (1979). Problems in measuring human energy balance. In *Assessment of Energy Metabolism in Health and Disease*, ed. S. Calvert, pp. 2–5. Columbus, Ohio: Ross Laboratories.

Garrow, J.S. (1981). *Treat Obesity Seriously: A Clinical Manual*. Edinburgh: Churchill Livingstone.

Garrow, J.S. (1985). Resting metabolic rate as a determinant of energy expenditure in man. In *Substrate and Energy Metabolism in Man*, ed. J.S. Garrow & D. Halliday, pp. 102–7. London: J. Libbey.

Garrow, J.S., Durrant, M.L., Mann, S., Stalley, S. & Warwick, P.M. (1978). Factors determining weight loss in obese patients in a metabolic ward. *International Journal of Obesity*, **2**, 441–6.

Garrow, J.S., Murgatroyd, R., Toft, R. & Warwick, P. (1977). A direct calorimeter for clinical use. *Journal of Physiology (London)*, **267**, 15P–16P.

Garrow, J.S., Stalley, S., Diethelm, P., Pittet, P.H., Hesp, R. & Halliday, D. (1979). A new method for measuring the body density of obese adults. *British Journal of Nutrition*, **42**, 173–83.

Garry, R.C., Passmore, R., Warnock, G.M. & Durnin, J.V.G.A. (1955). Expenditure of energy and the consumption of food by miners and clerks, Fife, Scotland. *Medical Research Council Special Report Series*, **No. 289**. London: HMSO.

Gatenby, R.M. (1977). Conduction of heat from sheep to ground. *Agricultural Meteorology*, **18**, 387–400.

Gessaman, J.A. (1973). *Ecological Energetics of Homeotherms*. Utah State University Monograph Series No 20. Logan, Utah: Utah State University Press.

Giaja, J. (1925). Summit metabolism and metabolic quotient. *Annales de Physiologie et Physicochemie*, **1**, 596–602.

Giaja, J. & Gelineo, S. (1933). Sur la resistance au froid de quelques homeotherms. *Archives Internationale Physiologie*, **37**, 20–68.

Gill, M., Thornley, J.H.M., Black, J.L., Oldham, J.D. & Beever, D.E. (1984). Simulation of the metabolism of absorbed energy yielding nutrients in young sheep. *British Journal of Nutrition*, **52**, 621–49.

Girard, H. & Grima, M. (1980). Allometric relation between blood oxygen uptake and body mass in birds. *Comparative Biochemistry and Physiology*, **66A**, 439–45.

Goldspink, G. (1981). The use of muscles during swimming and running from the point of view of energy saving. *Symposium Zoological Society of London*, **48**, 219–38.

Golley, F.B., Ryszkowksi, L. & Sokur, J.T. (1975). The role of small mammals in temperate forests and grasslands. In *Small Mammals, their Productivity and Population Dynamics*, ed. F.B. Golley, K. Petrusewicz & L. Ryszkowski, pp. 223–41. Cambridge: Cambridge University Press.

Gonzalez-Jiminez, E. & Blaxter, K.L. (1962). The metabolism and thermal regulation of calves in the first month of life. *British Journal of Nutrition*, **16**, 199–212.

Graham, A.M. (1982). Assessment of nutritional intake. *Proceedings of the Nutrition Society*, **41**, 343–8.

Graham, N. McC. (1967). The metabolic rate of fasting sheep in relation to total and lean body weight and the estimation of maintenance requirements. *Australian Journal of Agricultural Science*, **18**, 127–36.

Graham, N. McC., Searle, T.W. & Griffiths, D.A. (1974). Basal metabolic rate in lambs and young sheep. *Australian Journal of Agricultural Science*, **25**, 957–71.

Graham, N. McC., Wainman, F.W., Blaxter, K.L. & Armstrong, D.G. (1959). Environmental temperature, energy metabolism and heat regulation in closely clipped sheep. 1. Energy metabolism in closely clipped sheep. *Journal of Agricultural Science (Cambridge)*, **52**, 13–24.

Grande, F. (1980). Energy expenditure of organs and tissues. In *Assessment of Energy Metabolism in Health and Disease*, ed. S. Calvert, pp. 88–92. Columbus, Ohio: Ross Laboratories.

Green, B. (1984). Composition of milk and energetics of growth in marsupials. *Symposium Zoological Society of London*, **51**, 369–87.

Griffiths, M. (1968). *Echidnas*. Oxford: Oxford University Press.

Guernsey, D.L. & Edelman, J.S. (1983). Regulation of thermogenesis by thyroid hormones. In *Molecular Basis of Thyroid Hormone Action*, ed. J. Hoppenheimer & H.H Samuels, pp. 293–324. New York: Academic Press.

Gulick, A. (1922), A study of weight regulation in the adult human body during over-nutrition. *American Journal of Physiology*. **60**, 371–95.

Gump, F.E. (1980). Use of insensible water loss to calculate resting energy expenditure. In *Assessment of Energy Metabolism in Health and Disease*, ed. S. Calvert, pp. 49–53. Columbus, Ohio: Ross Laboratories.

Günther, B. (1975). Dimensional analysis and theory of biological similarity. *Physiological Reviews*, **55**, 659–99.

Gurr, M.I., Mawson, R., Rothwell, M.J. & Stock, M. (1980). Effects of manipulating dietary protein and energy intake on energy balance and thermogenesis in the pig. *Journal of Nutrition*, **110**, 532–42.

Gustafson, F.L. & Benedict, F.G. (1928). The seasonal variation in basal metabolism. *American Journal of Physiology*, **86**, 43–57.

Haecker, T.L. (1920). *Investigations in Beef Production*. University of Minnesota Agricultural Experiment Station, Bulletin No. 193. University of Minnesota.

Hails, C.J. & Bryant, D.M. (1979). Reproductive energetics of a free-living bird. *Journal of Animal Ecology*, **48**, 471–82.

Hainsworth, F.R. & Wolf, L.L. (1975). Wing disc loading: implications and importance for humming bird energetics. *American Naturalist*, **109**, 229–33.

Hall, W.C. & Brody, S. (1934). Growth and development with special reference to domestic animals. 32. The energy cost of walking in cattle and horses of various ages and body weights. *Research Bulletin University of Missouri Agricultural Experiment Station*, **No. 208**.

Halliday, D. & Miller, A.G. (1977). Precise measurement of total body water using trace quantities of deuterium oxide. *Biomedical Mass Spectrometry*, **4**, 82–7.

Hammel, H.T. (1955). Thermal properties of fur. *American Journal of Physiology*, **182**, 369–76.

Handbook of Physics and Chemistry (1985), 65th edn, ed. R.C. Weast. Boca Raton, Florida: CRC Press.

Hardy, J.D. (1949). Heat transfer. In *Physiology of Heat Regulation and the Science of Clothing*, ed. L.H. Newburgh, pp. 78–108. Philadelphia: W.B. Saunders.

Hardy, J.D. (1961). Physiology of temperature regulation. *Physiological Reviews*, **41**, 521–606.

Hardy, J.D. & Dubois, E.F. (1938). Basal Metabolism, radiation, convection and vaporization at temperatures of 22 to 35 °C. *Journal of Nutrition*, **15**, 477–97.

Hardy, J.D. & Milhorat, A.T. (1941). Basal metabolism and heat loss of young women at temperatures from 22 to 35 °C. *Journal of Nutrition*, **21**, 383–404.

Hardy, J.D. & Soderstrom, G.F. (1938). Heat loss from the nude body and peripheral blood flow at temperatures of 22 °C to 35 °C. *Journal of Nutrition*, **16**, 493–510.

Hargrove, J.L. & Gessaman, J.A. (1973). An evaluation of respiratory rate as an indirect monitor of free living metabolism. In *Ecological Energetics of Homeotherms*, Monograph Series University of Utah vol. 20, ed. J.A. Gessaman, pp. 77–85. Logan, Utah: Utah State University Press.

Harlow, H.J. (1981). Metabolic adaptation to prolonged food deprivation by the American badger (*Taxidea taxus*). *Physiological Zoology*, **54**, 276–84.

Hart, J.S. (1962). Seasonal acclimatization in four species of small wild bird. *Physiological Zoology*, **35**, 224–36.

Hart, J.S. (1971). Rodents. In *Comparative Physiology of Temperature Regulation*, vol. 2, ed. G.C. Whittow, pp. 2–149. New York: Academic Press.

Heath, M.E. & Ingram, D.L. (1981). The metabolism of young pigs reared in a hot or cold environment. *Journal of Thermal Biology*, **6**, 19–22.

Heglund, N.C., Cavagna, G.A. & Taylor, C.R. (1982). Energetics and mechanisms of terrestial locomotion. 3. Energy changes of the center of mass as a function of speed and body size in birds and mammals. *Journal of Experimental Biology*, **97**, 22–56.

Heitzman, R.J. (1978). The use of hormones to regulate nutrient utilization in farm animals. *Proceedings of the Nutrition Society*, **37**, 295–9.

Helmholtz, H. von. (1847). *Uber die Erhaltung der Kraft*. Berlin: G. Reiner.

Helmholtz, H. von. (1848). Uber die wärmeenwicklung bei der Muskelaction. *Wissenschaftlichen Abhandlung*, **2**, 745–63.

Hemingway, A. & Stuart, D.G. (1963). Shivering in man and animals. In *Temperature: Its Measurement and Control in Science and Industry*. ed. J.D. Hardy, chapter 36. New York: Reinhold.

Hemmingsen, A.M. (1950). The relation of standard (basal) metabolism to total fresh weight of living organisms. *Report of the Steno Memorial Hospital and Nordinsk Insulin Laboratorium*, **4**, 1–58.

Hemmingsen, A.M. (1960). Energy metabolism as related to body size and respiratory surfaces and its evolution. *Report of the Steno Memorial Hospital and Nordinsk Insulin Laboratorium*, **9**, 6–110.

Hennemann, W.W. 3rd. (1983). Relationship among body mass, metabolic rate and the intrinsic rate of natural increase in mammals. *Oecologia (Berlin)*, **56**, 104–8.

Herreid, C.F. & Kessel, B. (1967). Thermal conductance in birds and mammals. *Comparative Biochemistry and Physiology*, **21**, 405–14.

Herrington, L.P. (1940). The heat regulation of small laboratory animals at various environmental temperatures. *American Journal of Physiology*, **129**, 123–9.

Herrington, L.P., Winslow, C.E.A. & Gagge, A.P. (1937). The relative influence of radiation and convection upon vasomotor temperature regulation. *American Journal of Physiology*, **120**, 133–43.

Hervey, G.R. & Tobin, G. (1983). Luxuskonsumption, diet induced thermogenesis and brown fat. *Clinical Science*, **64**, 7–18.

Hess, G.H. (1838). The evolution of heat in multiple proportions [translated title of paper not seen], *Poggendorf's Annalen der Chemie und Physik*, **47**, 210–?.

Heusner, A. (1956). Mise en évidence d'une variation nycthémérale de la calorification independante du cycle de l'activité chez le rat. *Compte rendu Societé Biologie*, **150**, 1240–51.

Heusner, A.A. (1985). Body size and energy metabolism. *Annual Reviews of Nutrition*, **5**, 267–93.

Hey, E.N. (1969). The relation between environmental temperature and oxygen consumption in the new-born baby. *Journal of Physiology (London)*, **200**, 589–603.

Hey, E.N. & Katz, G. (1969). Evaporative water loss in the new-born baby. *Journal of Physiology (London)*, **200**, 605–19.

Hey, E.N., Katz, G. & O'Connell, B. (1970). The total thermal insulation of the new-born baby. *Journal of Physiology (London)*, **207**, 683–98.

Hey, E.N. & Mount, L.E. (1967). Heat losses from babies in incubators. *Archives of Diseases of Childhood*, **42**, 75–84.

Hill, A.V. (1965). *Trails and Trials in Physiology*. London: Edward Arnold.

Hill, A.V. & Hartree, W. (1920). The four phases of heat production in muscle. *Journal of Physiology (London)*, **54**, 84–128.

Hill, J.R. & Rahimtulla, K.A. (1965). Heat balance and metabolic rate of new-born babies in relation to temperature and the effect of age and weight on metabolic rate. *Journal of Physiology (London)*, **180**, 239–65.

Hill, L. & Campbell, J.A. (1922). Observations on the resting metabolism of children and adults in Switzerland. *British Medical Journal*, **1**, 385–9.

Himms-Hagen, J. (1985). Brown adipose tissue metabolism and thermogenesis. *Annual Reviews of Nutrition*, **5**, 69–94.

HMSO (1973). *SI, the International System of Units*. London: HMSO.

Hocking, P.M. (1987). Nutritional interactions with reproduction in birds. *Proceedings of the Nutrition Society*, **46**, 217–25.

Hoffmann, L. (1958). Zur berechnung der wärmeproduction und zur kontrolle eines gesamtstoffwechselversuche über das gesetz zue erhaltung der energie. *Wissenschafliche Abhandlungen Deutsche Akademie der Landwirtschaftwissenschaften zu Berlin*, **37**, 83–91.

Hoffmann, L., Klein, M. & Schiemann, R (1986). Untersuchungen an ratten zur nahrstoffabhnagigkeit des energieerhaltungsbedarfs. *Animal Nutrition (Berlin)*, **11**, 981–93.

Hoffmann, L., Klippel, W. & Schiemann, R. (1967). Untersuchungen uber den energieumsatz beim pferd unter besonderer berücksichtigung der horizontalbewegung. *Archives für Tierenährung*, **17**, 441–9.

Hoffmann, L., Schiemann, R. & Jentsch, W. (1971). Energetische ververtung der nahrstoffe in futterrationen. In *Energetische Futterbewertung und Energienormen*, ed. R. Schiemann, K. Nehring, L. Hoffmann, W. Jentsch & A. Chudy, pp. 118–67. Berlin: Deutscher Landwirtschaftverlag.

Holmes, C.W. & McLean, N.R. (1974). The effect of low ambient temperatures on the energy metabolism of sows. *Animal Production*, **19**, 1–12.

Holmes, F.L. (1985). *Lavoisier and the Chemistry of Life*. Madison: University of Wisconsin Press.

Homsher, E. (1987). Muscle enthalpy production and its relation to actomyosin ATP-ase. *Annual Reviews of Physiology*, **49**, 673–90.

Horn, H.S. & Rubenstein, D.I. (1984). Behavioural adaptation and life history. In *Behavioural Ecology: An Evolutionary Approach*, 2nd edn, ed. J.R. Krebs & N.B. Davies, pp. 279–98. Oxford: Blackwell.

Houseman, R.A., McDonald, I. & Pennie, K. (1973). The measurement of total body water in living pigs by deuterium oxide dilution and its relation to body composition. *British Journal of Nutrition*, **30**, 149–56.

Hoyt, D.F. & Taylor, C.R. (1981). Gait and the energetics of locomotion in horses. *Nature (London)*, **292**, 239–40.

Hudson, J.W. (1978). Shallow daily torpor: a thermoregulatory adaptation. In *Strategies in Cold: Natural Torpidity and Thermogenesis*, ed. L.C.H. Wang & J.W. Hudson, pp. 66–108. New York: Academic Press.

Hudson, R.J. & Christopherson, R.J. (1985). Maintenance metabolism. In *Bioenergetics of Wild Herbivores*, ed. R.J. Hudson & R.J. White, pp. 121–42. Boca Raton, Florida: CRC Press.

Huggett, A. St G. & Widdas, W.F. (1951). The relation between mammalian foetal weights and conception age. *Journal of Physiology (London)*, **114**, 306–17.

Hulbert, A.J. (1980). The evolution of energy metabolism in mammals. In *Comparative Physiology: Primitive Mammals*, ed. K. Schmidt-Nielsen, L. Bolis & C.R. Taylor, pp. 129–39. Cambridge: Cambridge University Press.

Hull, D. & Segall, M.M. (1965). The contribution of brown adipose tissue to heat production in the new-born rabbit. *Journal of Physiology (London)*, **181**, 449–57.

Hungate, R.E. (1966). *The Rumen and its Microbes*. New York: Academic Press.

Hutchens, J.O. (1975). Heat of combustion, enthalpy and free energy of formation of amino acids and related compounds. In *Handbook of Biochemistry*, pp. B7–9. Boca Raton, Florida: CRC Press.

Hutchinson, J.C.D. & Brown, D.B. (1969). Penetrance of cattle coats by radiation. *Journal of Applied Physiology*, **26**, 454–64.

Huxley, A.F. (1980). *Reflections on Muscle*. Liverpool: Liverpool University Press.

Huxley, A.F. & Niedergerke, R. (1954). Structural changes in muscle during contraction. *Nature (London)*, **173**, 971–3.

Huxley, H.E. (1969). The mechanism of muscular contraction. *Science*, **164**, 1356.

Huxley H.E. & Hanson, J. (1954). Changes in the cross-striations of muscle during contraction and their structural interpretation. *Nature (London)*, **173**, 973–6.

Huxley, J.S. (1932). *Problems of Relative Growth*. London: Methuen.

Huyssen, V. & Lacy, R.C. (1985). Basal metabolic rates in mammals. Taxonomic differences in the allometry of SMR and body mass. *Comparative Biochemistry and Physiology*, **81A**, 741–54.

Hytten, F.E. & Leitch, I. (1971). *The Physiology of Human Pregnancy*, 2nd edn. Oxford: Blackwells.

Ingram, D.L. (1974). Heat loss and its control in pigs. In *Heat Loss from Animals and Man*, ed. J.L. Monteith & L.E. Mount, pp. 233–54. London: Butterworths.

International Critical Tables (1929). vol. 5, ed. E.W. Washburn for the National Research

Council. New York: McGraw Hill.

Irving, L. (1964). Terrestial animals in cold. In *Handbook of Physiology, Section 4 Adaptation to Environment*, pp. 361–77. Washington DC: American Physiological Society.

Jacobsen, N.K. (1980). Differences in thermal properties of white-tailed deer pelage. *Journal of Thermal Biology*, **5**, 151–8.

Jakob, M. & Hawkins, G.A. (1957). *Elements of Heat Transfer*. London: John Wiley & Sons.

Jansky, L. (1961). Body organ cytochrome oxidase and its relation to basal and maximal metabolism. *Nature (London)*, **189**, 921–2.

Jansky, L. (1973). Non-shivering thermogenesis and its thermo-regulatory significance. *Biological reviews*, **48**, 85–132.

Janssen, W.M.M., Terpstra, K., Beeking, F.F.E. & Bisalsky, A.J.N. (1979). *Feeding Values for Poultry*. Netherlands: Spelderholt Institute for Poultry Research.

Jenkins, F.A. (1971). Limb posture and locomotion in the Virginian opossum (*Didelphus marsupialis*) and in other non-cursorial mammals. *Journal of Zoology (London)*, **165**, 303–15.

Jenness, H. & Sloan, R.E. (1970). The composition of milks of various species: a review. *Dairy Science Abstracts*, **32**, 599–612.

Jequier, E. & Schutz, Y. (1983). Long-term measurements of energy expenditure in humans using a respiration chamber. *American Journal of Clinical Nutrition*, **38**, 989–98.

Johnson, R.E. & Kark, R.M. (1947). Environment and food intake in men. *Science*, **105**, 378.

Johnston, I.A. & Goldspink, G.A. (1973). A study of swimming performance of the crucian carp in relation to the effects of exercise and recovery on biochemical changes in the myotomal muscles. *Journal of Fish Biology*, 5, 249–260.

Jones, C.G. (1984). Shelter studies using thermal models of cattle. PhD Thesis. University of Aberdeen.

Jones, C.W. (1981). *Biological Energy Conservation: Oxidative Phosphorylation*, 2nd edn. London: Chapman & Hall.

Joyce, J.P. & Blaxter, K.L. (1964a). Respiration in sheep in cold environments. *Research in Veterinary Science*, **5**, 506–16.

Joyce, J.P. & Blaxter, K.L. (1964b). The effect of air movement, air temperature and infrared radiation on the energy requirements of sheep. *British Journal of Nutrition*, **18**, 5-27.

Joyce, J.P., Blaxter, K.L. & Park, C. (1966). The effect of natural outdoor environments on the energy requirements of sheep. *Research in Veterinary Science*, **7**, 342–59.

Kay, R.N.B. (1979). Seasonal changes of appetite in deer and sheep. *Agricultural Research Council Research Reviews,* **5**, 13-18.

Kaye, G.W.C. & Laby, T.H. (1973). *Tables of Physical and Chemical Constants*, 14th edn. London: Longmans.

Kellner, O. & Köhler, A. (1896). Untersuchungen uber den Stoff-und energieumsatz volljahrigen Ochsen bei Erhaltungsfutter. *Landwirtschaft Versuchsstationen*, **47**, 331–81.

Kendleigh, S.C., Dol'nik, V.R. & Govrilov, V.M. (1977). Avian energetics. In *Graniverous Birds in Ecosystems*, ed J. Pinowksi & S.C. Kendeigh, pp. 127–204. Cambridge: Cambridge University Press.

Kennedy, P.M., Christopherson, R.J. & Milligan, L.P. (1976). The effect of cold exposure of sheep on digestion, rumen turnover time and efficiency of microbial synthesis. *British Journal of Nutrition*, **36**, 213–42.

Kennedy, P.M., Christopherson, R.J. & Milligan, L.P. (1982). Effects of cold exposure on feed protein degradation, microbial synthesis and transfer of plasma urea to the rumen of sheep. *British Journal of Nutrition*, **47**, 521–35.

Kerslake, D. McK. (1972). *The Stress of Hot Environments*. Cambridge: Cambridge University Press.

Kerslake, D. McK. (1983). Effects of climate. In *The Body at Work: Biological Ergonomics*, ed. W.T. Singleton, pp. 235–97. Cambridge: Cambridge University Press.

Keys, A. & Brozek, J. (1953). Body fat in adult man. *Physiological Reviews*, **33**, 245–325.

Keys, A., Brozek, A., Henschel, A., Micckelsen, O. & Taylor, H.L. (1950). *The Biology of Human Starvation*. Minneapolis: University of Minnesota Press.

Kharasch, M.S. (1929). Heat of combustion of organic compounds. *Journal of Research of the National Bureau of Standards*, **2**, 596–627.

Kielanowski, J. (1965). Estimates of the energy cost of protein deposition in growing animals. In *Energy Metabolism*, ed. K.L. Blaxter, pp. 13–20. London: Academic Press.

Kihlstrom, J.E. (1972). Period of gestation and body weight in some placental mammals. *Comparative Biochemistry and Physiology*, **43A**, 673–9.

Kinney, J.M., Weissman, C. & Askanazi, J. (1985). Influence of nutrients on ventilation. *Nutrition Abstracts and Reviews, Ser. A*, **54**, 917–29.

Kirchgessner, M. (1982). Efficiency of utilization of dietary energy by the laying hen in relation to different energy and protein supply. In *Energy Metabolism of Farm Animals*, ed. A. Ekern & B. Sundstol, pp. 266–9. Aas, Norway: Agricultural University of Norway.

Kirkwood, R.N., Cumming, D.C. & Aherne, F.X. (1987). Nutrition and puberty in the female. *Proceedings of the Nutrition Society*, **46**, 177–92.

Kleiber, M. (1932). Body size and metabolism. *Hilgardia*, **6**, 315–53.

Kleiber, M. (1947). Body size and metabolic rate. *Physiological Reviews*, **27**, 511–41.

Kleiber, M. (1961). *The Fire of Life*. New York: Wiley.

Koivisto, V.A. (1986). The physiology of marathon running. *Science Progress (Oxford)*, **70**, 109–27.

Koong, L.J., Nienaber, J.A. & Mersmann, H.J. (1983). Effects of plane of nutrition on organ size and fasting heat production in genetically obese and lean pigs. *Journal of Nutrition*, **111**, 1626–31.

Koong, L.J., Nienaber, J.A., Pekas, J.C. & Yen, J-T. (1982). Effects of plane of nutrition on organ size and fasting heat production in pigs. *Journal of Nutrition*, **112**, 1638–42.

Krebs, H.A. (1950). Body size and tissue respiration. *Biochimica Biophysica Acta*, **4**, 249–69.

Krebs, H.A. (1960). The cause of the specific dynamic action of foodstuffs. *Arzeneimittel-Forschung*, **10**, 369–73.

Krebs, H.A. (1964). The metabolic fate of amino acids. In *Mammalian Protein Metabolism*, vol. 1, ed. H.N. Munro & J.B. Allison, pp. 125–76.

Krogh, A. (1916). *The Respiratory Exchange of Animals and Man*. London: Longmans Green.

Kuhn, G. (1894). Futterungs und Respirationsversuche mit volljahrigen Ochsen über die Fettbildung aus Kohlenhydraten. *Landwirtschaft Versuchsstationen*, **44**, 259–312.

Kuhn, T.S. (1955). Energy conservation as an example of simultaneous discovery. In *Critical Problems in the History of Science*, ed. M. Clagett, pp. 321–56. Madison: University of Wisconsin Press.

Kuhnen, G. (1986). Oxygen and carbon dioxide concentrations in burrows of euthermic and hibernating golden hamsters. *Comparative Biochemistry and Physiology*, **84A**, 517–22.

Kushmerick, M.J. (1983). Energetics of muscular contraction. In *Handbook of Physiology, Section X: Skeletal Muscle*, ed. L.D. Peachey, pp. 189–236. Bethesda: American Physiological Society.

Lavigne, D.M., Innes, S., Worthy, G.A.J., Kovacs, K.M., Schmitz, O.J. & Hickie, J.P. (1986). Metabolic rate of seals and whales. *Canadian Journal of Zoology*, **64**, 279–84.

Lavoisier, A.L. (1777). Experiences sur la decomposition de l'air dans le poulmon, et sur un des principaux usages de la respiration dans l'economie animale. *Archives de l'Academie des Sciences*. Note: there were several editions of this paper – see Holmes (1985) for a full account of them.

Lavoisier, A.L. (1790). Letter to Joseph Black. Published by A. Mielo (1943). Una lettera di A. Lavoisier a J. Black, *Archeion*, **25**, 238–9.

Lavoisier, A.L. & Laplace, P.S. (1783). Memoire sur la chaleur. *Memoires de l'Academie des Sciences*, pp. 404 et seq. (Original not seen: see Holmes 1985).

Lawes, J.B. & Gilbert, J.H. (1861). On the composition of oxen, sheep and pigs and of their increase whilst fattening. *Journal of the Royal Agricultural Society of England*, **21**, 1–92.

Lawrence, P.R. & Pearson, R.A. (1985). Factors affecting the measurement of draught work output and power of oxen. *Journal of Agricultural Science (Cambridge)*, **105**, 703–14.

Leaver, J.D. (1982). *Herbage Intake Handbook*. Hurley, England: British Grassland Society.

Leblanc, J. & Labrie, A. (1981). Glycogen and non-specific adaptation to cold. *Journal of Applied Physiology*, **51**, 1428–36.

LeFebre, E.A. (1964). The use of D_2O^{18} for measuring energy metabolism in *Columbia livia* at rest and in flight. *Auk*, **81**, 403–16.

Leicester, H.M. (1951). Germain Henri Hess and the foundation of thermochemistry. *Journal of Chemical Education*, **28**, 581–3

Leitch, I., Hytten, F.E. & Billewicz, W.W.Z. (1959). The maternal and neonatal weights of some mammals. *Proceedings of the Zoological Society of London*, **133**, 11–28.

Lewis, G.N. & Randall, M. (1923). *Thermodynamics*. New York: McGraw Hill.

Lewis, T. (1930). Observations upon the reactions of the vessels of the human skin to cold. *Heart*, **15**, 177.

Liebig, J. von (1842). *Animal Chemistry or Organic Chemistry in its Application to Physiology and Pathology*, trans. W. Gregory. London: Taylor & Walton.

Lifson, N., Gordon, G.B. & McClintock, R. (1955). Measurement of total carbon dioxide production by means of DO^{18}. *Journal of Applied Physiology*, **7**, 704–10.

Lifson, N. & McClintock, R. (1966). Theory of use of turnover rates of body water for measuring energy and material balances. *Journal of Theoretical Biology*, **122**, 46–74.

Lighthill, M.J. (1975). *Mathematical Bio-fluid dynamics*. Philadelphia: Society of Industrial and Applied Mathematics.

Lindstedt, S.L. & Calder, W.A. 3rd. (1981). Body size, physiological time and longevity of homeothermic animals. *Quarterly Review of Physiology*, **56**, 1-16.

Lindstedt, S.L. (1980). Energetics and water economy of the smallest desert animal. *Physiological Zoology*, **53**, 82–97.

Linzell, J.L. (1972). Milk yield, energy loss in milk and mammary gland weight in different species. *Dairy Science Abstracts*, **34**, 351–60.

Lipmann, F. (1941). Metabolic generation and utilization of phosphate bond energy. *Advances in Enzymology*, **1**, 99–162.

Livesey, G. (1984). The energy equivalent of ATP and the energy value of food proteins and fats. *British Journal of Nutrition*, **51**, 15–28.

Livesey, G. (1985). Mitochondrial uncoupling and the isodynamic equivalents of protein, fat and carbohydrate at the level of biochemical energy provision. *British Journal of Nutrition*, **53**, 381–9.

Livingstone, S.D. (1968). Calculation of mean body temperature. *Canadian Journal of Physiology and Pharmacology*, **46**, 15–17.

Lloyd, B.B. & Zacks, R.M. (1972). The mechanical efficiency of treadmill running against a horizontal impeding force. *Journal of Physiology (London)*, **233**, 355–63.

Lloyd, J.T. (1970). Background to the Joule–Mayer controversy. *Notes and Records of the Royal Society of London*, **25**, 211–25.

Lobley, G.E. & Lovie, J.M. (1979). The synthesis of myosin, actin and the major proteins in rabbit skeletal muscle. *Biochemical Journal*, **182**, 867–74.

Lofti, M. & MacDonald, I.A. (1976). Energy losses associated with oven drying and the preparation of rat carcasses for analysis. *British Journal of Nutrition*, **36**, 305–9.

Lord, R.D. (1960). Litter size and latitude in North American mammals. *American Naturalist*, **64**, 488–99.

Lusk, G. (1915). Animal calorimetry. An investigation into the causes of the specific dynamic action of foodstuffs. *Journal of Biological Chemistry*, **20**, 555–617.

Lusk, G. (1928). *The Elements of the Science of Nutrition*. Philadelphia: Saunders.

McArthur, A.J. (1980). Air movement and heat loss from sheep. 3. Components of insulation in a controlled environment. *Proceedings of the Royal Society of London, Series B*, **209**, 219–37.

McArthur, A.H. (1981). Thermal resistance and sensible heat loss from animals. *Journal of Thermal Biology*, **6**, 43–7.

McArthur, A.J. & Monteith, J.L. (1980a). Air movement and heat loss in sheep. 1. Boundary layer insulation of a model sheep with and without fleece. *Proceedings of the Royal Society of London, Series B*, **209**, 187-208.

McArthur, A.J. & Monteith, J.L. (1980b). Air movement and heat loss in sheep. 2. Thermal insulation of fleece in wind. *Proceedings of the Royal Society of London, Series B*, **209**, 209–17.

MacArthur, R.H. & Wilson, E.O. (1967). *The Theory of Island Biogeography*. Princeton: Princeton University Press.

McBride, G.E. & Christopherson, R.J. (1984). Effects of cold exposure on young growing lambs. *Canadian Journal of Animal Science*, **64**, 403–10.

McCance, R.A. & Mount, L.E. (1960). Severe undernutrition in growing and adult animals. 5. Metabolic rate and body temperature in the pig. *British Journal of Nutrition*, **14**, 509–18.

McCracken, K.J. (1986). Nutritional obesity and body composition. *Proceedings of the Nutrition Society*, **45**, 91–100.

McCracken, K.J. & Caldwell, B.J. (1980). Effect of temperature and energy intake on heat loss and energy retention of early weaned pigs. In *Energy Metabolism*, ed. L.E. Mount, pp. 445–53. London: Butterworths.

McCracken, K.J. & McAllister, A. (1984). Energy metabolism and body composition of young pigs given low protein diets. *British Journal of Nutrition*, **51**, 225–34.

McCracken, K.J. & McNiven, M.A. (1983). Effects of overfeeding by gastric intubation on body composition of adult female rats and on heat production during feeding and fasting. *British Journal of Nutrition*, **49**, 193–202.

McDowell, R.E., Moody, E.G., van Soest, P.J., Lehmann, R.P. & Ford, G.L. (1969). Effect of heat stress on energy and water utilization of lactating cows. *Journal of Dairy Science*, **52**, 188–94.

McGilvery, R.W. & Goldstein, G.W. (1983). *Biochemistry: A Functional Approach*. Philadelphia: Saunders.

McGlashan, M.L. (1979). *Chemical Thermodynamics*. London: Academic Press.

McKay, L.F. & Eastwood, M.A. (1984). A comparison of bacterial fermentation end products in carnivores and primates including man. *Proceedings of the Nutrition Society*, **43**, 35A.

McLean, J.A. (1974). Loss of heat by evaporation. In *Heat Loss from Animals and Man*, ed. J.L. Monteith & L.E. Mount, pp. 19–31. London: Butterworths.

McLean, J.A. & Calvert, D.T. (1972). Influence of air humidity on the partition of heat exchanges of cattle. *Journal of Agricultural Science (Cambridge)*, **78**, 303–7.

McLean, J.A., Downie, A.J., Jones, C., Stombaugh, D.P. & Glasbey, C.A. (1983a). Thermal adjustment of steers *Bos taurus* to abrupt changes in environmental temperatures. *Journal of Agricultural Science (Cambridge)*, **100**, 305–14.

McLean, J.A., Downie, A.J., Watts, P.R. & Glasbey, C.A. (1982). Heat balance of ox steers *Bos taurus* in steady-temperature environments. *Journal of Applied Physiology*, **52**, 324–32.

McLean, J.A., Stombaugh, D.P. & Downie, A.J. (1983b). Body heat storage in steers *Bos taurus* in fluctuating thermal environments. *Journal of Agricultural Research (Cambridge)*, **100**, 315–22.

McLean, J.A. & Tobin, G. (1987). *Animal and Human Calorimetry*. Cambridge: Cambridge University Press.

McLean, J.A., Whitmore, W.T., Young, B.A. & Weingardt, R. (1984). Body heat storage, metabolism and respiration of cows abruptly exposed and acclimatized to cold and 18 °C environments. *Canadian Journal of Animal Science*, **64**, 641–53.

MacLeod, M.G., Jewitt, T.R. & Andersen, J.E.M. (1987). Energy utilization and physical activity in intubated and self-fed male domestic fowl. *Proceedings of the Nutrition Society*, **46**, 148A.

McMahon, T.A. (1973). Size and shape in biology. *Science*, **179**, 1201–4.

McNab, B.K. (1969). The economics and temperature regulation in neotropical bats. *Comparative Biochemistry and Physiology*, **31**, 227–68.

McNab, B.K. (1980). Food habits, energetics and the population biology of mammals. *American Naturalist*, **116**, 106–24.

McNab, B.K. (1986). The influence of food habits on the energetics of eutherian mammals. *Ecological Monographs*, **56**, 1–19.

McNeil, G., Rivers, J.P.W., Payne, P.R., de Britto, J.J. & Abel, R. (1987). Basal metabolic rate of Indian men: no evidence of adaptation to a low plane of nutrition. *Human Nutrition: Clinical Nutrition*, **41C**, 473–83.

Magnus, G. (1847). Über die in Blute Enthalten gase, sauerstoffe, stickstoff und Kohlensaure. *Annalen der Physik und Chemie*, **40**, 583–606.

Marr, J.W. (1971). Individual dietary surveys: purposes and methods. *World Reviews of Nutrition and Dietetics*, **13**, 105–32.

Marston, H.R. (1948). Energy transactions in sheep. *Australian Journal of Scientific Research, B*, **1**, 93–129.

Martin, A.K. (1978). The metabolism of aromatic compounds in ruminants. In *The Hannah Research Institute, 1928–78*, ed. J.H. Moore & J.A.F. Rook, pp. 148–63. Ayr: Hannah Research Institute.

Martin, R.D. (1984). Scaling effects and adaptation strategies in mammalian lactation. *Symposium Zoological Society of London*, **51**, 87–117.

Mason, E.D. & Benedict, F.G. (1934). The effect of sleep on human basal metabolism with particular reference to South Indian women. *American Journal of Physiology*, **108**, 377–83.

Mathieu, O., Krauer, R., Hoppeler, H., Gehr, P., Alexander, R. McN., Taylor, C.R. & Weibel, E.R. (1981). Design of the mammalian respiratory system: scaling mitochondrial volume in skeletal muscle to body mass. *Respiratory Physiology*, **44**, 13–28.

May, R.M. (1981). *Theoretical Ecology: Principles and Applications*. Oxford: Blackwell.

May, R.M. & Rubenstein, D.I. (1984). Reproductive strategies. In *Reproduction in Mammals. 4. Reproductive Fitness*, 2nd edn, ed. C.R. Austin & R.V. Short, pp. 1–23. Cambridge: Cambridge University Press.

Mayer, J.R. (1842), Bemerkungen über die krafte der unbelebten Natur. *Liebig's Annalen der Chemie und Pharmacie*, **42**, 233–40.

Mayes, R.W.`& Lamb, C.S. (1983). The possible use of n-alkanes as indigestible faecal markers. *Proceedings of the Nutrition Society*, **43**, 39A.

Maynard, L.A. (1944). The Atwater system of calculating the caloric value of diets. *Journal of Nutrition*, **28**, 443–52.

Meeh, K. (1879). Oberflachenmessungen des menschlichen korpers. *Zeitschift für Biologie*, **15**, 39–52.

Mellor, D.J. & Cockburn, F. (1986). A comparison of energy metabolism in the new-born, infant piglet and lamb. *Quarterly Journal of Experimental Physiology*, **71**, 361–79.

Mendelsohn, E. (1964). *Heat and Life. The Development of the Theory of Heat and Life*. Cambridge, Mass: Harvard University Press.

Merrill, A.L. & Watt, B.K. (1955). *Energy Value of Foods – Basis and Derivation*. USDA Handbook No. 74. Washington DC: US Department of Agriculture.

Meyerhof, O. & Schulz, W. (1935). Über die energieverhaltnisse bei der enzymatischen milchsauerbildung und der synthese der phophagen. *Biochemische Zeitschrift*, **281**, 292–305.

Miller, A.T. & Blyth, C.S. (1953). Lean body mass as a metabolic reference standard. *Journal of Applied Physiology*, **5**, 311–16.

Miller, D.S., Mumford, P. & Stock, M.J. (1967a). Gluttony. 2. Thermogenesis in over-eating man. *American Journal of Clinical Nutrition*, **20**, 1223–9.

Miller, D.S. & Payne, P.R. (1962). Weight maintenance and food intake. *Journal of Nutrition*, **78**, 255–61.

Miller, T.L. & Wolin, M.J. (1986). Methanogens in human and animal digestive tracts. *Systematic Applied Microbiology*, **7**, 223–9.

Milligan, L.P. & McBridge, B.W. (1985). Energy cost of ion pumping by animal tissues. *Journal of Nutrition*, **115**, 1374–82.

Milligan, L.P. & Summers, M. (1986). The biological basis of maintenance and its relevance to assessing responses to nutrients. *Proceedings of the Nutrition Society*, **45**, 185–93.

Millward, D.J., Garlick, P.J. & Reeds, P.J. (1976). The energy cost of growth. *Proceedings of the Nutrition Society*, **35**, 339–49.

Ministry of Agriculture, Fisheries & Food (1976). *Nutrient Allowances and Composition of Feedingstuffs for Ruminants*. Advisory paper No. 11, 2nd edn. London: HMSO.

Mitchell, D. (1974). Convective heat transfer from man and other animals. In *Heat Loss from Animals and Man*, ed. J.L. Monteith & L.E. Mount, pp. 59–76. London: Butterworths.

Mitchell, D., Wyndham, C.H., Atkins, A.R., Vermeullen, H.S., Hofmeyr, H.S., Strydom, N.B. & Hodgson, T. (1948). Direct measurement of the thermal responses of nude resting men in dry environments. *Archives für gesamte Physiologie des Menschen un der Tiere*, **303**, 324–43.

Mitchell, H.H. (1962). *Comparative Nutrition of Man and Domestic Animals*, Vols. 1 & 2. New York: Academic Press.

Mitchell, M.A. (1985). Measurement of forced convective heat transfer in birds: a forced wind tunnel calorimeter. *Journal of Thermal Biology*, **10**, 87–95.

Moe, P.W. & Tyrrell, H.F. (1980). Methane production in dairy cows. In *Energy Metabolism*, ed. L.E. Mount, pp. 59–62. London: Butterworths.

Møllgaard, H. & Lund, A. (1929). Om grundtraekkene af malkekvaegets ernaeringslaehre. 131 Beretning Førsøgslaboratoriet. Copenhagen: Kommission hos Aug. Bang.

Monteith, J.L. (1972). *Survey of Instruments for Micrometeorology*. Oxford: Blackwell.

Monteith, J.L. (1973). *Principles of Environmental Physics*. London: Edward Arnold.

Montgomery, G.G. & Sunquist, M.E. (1978). Habitat selection and use by two-toed and three-toed sloths. In *Ecology of Arboreal Folivores*, ed. C.G. Montgomery, pp. 329–59. Washington DC: Smithsonian Institution.

Morgan, C.A., Whittemore, C.T. & Cockburn, J.H.S. (1984). The effect of level and source of protein, fibre and fat in the diet on the energy value of compound feeds. *Animal Feed Science and Technology*, **11**, 11–34.

Morowitz, H.J. (1978). *Foundations of Bioenergetics*. New York: Academic Press.

Moulton, C.R. (1923). Age and chemical development in mammals. *Journal of Biological Chemistry*, **57**, 79–97.

Moulton, C.R., Trowbridge, P.F. & Haigh, L.D. (1922) Studies in animal nutrition. 3. Changes in chemical composition on different planes of nutrition. *University of Missouri Agricultural Experiment Station*, Research Bulletin No. 55.

Mount, L.E. (1967). The heat loss of new born pigs to the floor. *Research in Veterinary Science*, **8**, 175–86.

Mount, L.E. (1974). The concept of thermoneutrality. In *Heat Loss from Animals and Man*, ed. J.L. Monteith & L.E. Mount, pp. 425–39.

Mount, L.E. (1977). The use of heat transfer coefficients in estimating sensible heat loss from the pig. *Animal Production*, **25**, 271–80.

Mount, L.E. (1978). Heat transfer between animal and environment. *Proceedings of the Nutrition Society*, **37**, 5–12.

Mount, L.E. (1979). *Adaptation to the Thermal Environment*. London: Edward Arnold.

Mount, L.E. & Rowell, J.G. (1960). Body size, body temperature and age in relation to the metabolic rate of the pig in the first five weeks after birth. *Journal of Physiology* (*London*), **154**, 408–16.

Mount, L.E. & Stephens, D.B. (1970). The relation between body size and maximum and minimum metabolic rates in the new-born pig. *Journal of Physiology* (*London*), **207**, 417–27.

Murgatroyd, P.R., Davies, H.L., Goldberg, G.R., Cole, T.J. & Prentice, A.R. (1985). Continuous measurement of energy expenditure by direct and indirect calorimetry over 12 days in man. *Abstract: XIII Congress of Nutrition*, ed. T.G. Taylor & N.K. Jenkins, London; John Libbey.

Murray, J.A. (1922). The chemical composition of animal bodies. *Journal of Agricultural Science (Cambridge)*, **12**, 103–110.

Nagy, K.A. (1987). Field metabolic rate and food requirement scaling in mammals and birds. *Ecological Monographs*, **57**, 112–28.

Nagy, K.A. & Costa, D.P. (1980). CO_2 production in animals: an analysis of potential errors in the doubly labelled water method. *American Journal of Physiology*, **238R**, 466–73.

Nagy, K.A. & Martin, R.W. (1985). Field metabolic rate, water flux, food consumption and time budget of Koalas *Phascolaretus cinereus*. *Australian Journal of Zoology*, **33**, 655–65.

Nagy, K.A. & Montgomery, G.G. (1980). Field metabolic rate, water flux and food consumption in 3-toed sloths. *Journal of Mammalogy*, **61**, 564–73.

Nash, K. (1970). *Elements of Chemical Thermodynamics*. Reading, Massachusetts: Addison-Wesley.

National Research Council (1981). *Effect of Environment on Nutrient Requirements of Domestic Animals*. Washington, DC: National Academic Press.

Needham, D.M. (1971). *Machina Carnis: The Biochemistry of Muscular Contraction in its Historical Development*. Cambridge: Cambridge University Press.

Needham, J. (1931). *Chemical Embryology*. London: Macmillan.

Nehring, K. Beyer, M. & Hoffmann, B. (1970). *Futtermitteltabellenwerk*. Berlin: Deutsch Landwirtschaftverlag.

Neumann, R.O. (1902). Experimentelle beitrage zur lehre von dem taglichen nährungsbedarf des menschen unter besonderer berucksichtigung der notwendigen eiweissmenge. *Archives für Hygeine und Bacteriologie* **45**, 1–87.

Newburgh, L.H. (1949). *Physiology of Heat Regulation and the Science of Clothing*. Philadelphia: Saunders.

Newsholme, E.A. (1982). The inter-relationship between metabolic regulation, weight control and obesity, *Proceedings of the Nutrition Society*, **41**, 183–91.

Newsholme, E.A. & Crabtree, B. (1976). Substrate cycles in metabolic regulation and in heat generation. *Biochemical Society Symposia*, **41**, 61–109.

Nicholl, M.E. & Thompson, S.D. (1987). Basal metabolic rates and energetics of reproduction in therian mammals: Marsupials and placentals compared. *Symposium of the Zoological Society of London*, **57**, 7–27.

Nicholls, D.G. & Locke, R. (1983). Thermogenic mechanisms in brown fat. *Physiological Reviews*, **64**, 1–64.

Nielsen, M. (1938). Die regulation der körpertemperatur bei muskelarbeit. *Scandinavian Archiv für Physiologie*, **79**, 193–217.

Noble, R.C. (1986). Lipid metabolism in the chick embryo. *Proceedings of the Nutrition Society*, **45**, 17–25.

Noblet, J. & Etienne, M. (1987). Metabolic utilization and maintenance requirements of lactating sows. *Journal of Animal Science*, **64**, 774–81.

Oftedal, O.T. (1984). Milk composition, milk yield and energy output at peak lactation: A comparative review. *Symposium of the Zoological Society of London*, **51**, 33–85.

Oftedal, O.T., Boness, D.J. & Tedman, R.A. (1987). The behaviour, physiology and anatomy of lactation in the Pinnipedia. In *Current Mammalogy*, ed. H.H. Genoways, pp. 175–245. New York: Plenum Press.

Ogle, C. (1934). Climatic effects on the growth of the mouse. *American Journal of Physiology*, **107**, 635–40.

Oikawa, S. & Itazaway, Y. (1984). Allometric relationship between tissue respiration and body mass in the carp. *Comparative Biochemistry and Physiology*, 77A, 445–18.

Olsson, K.-E. & Saltin, B. (1970). Variation in total body water with muscle glycogen in man. *Acta Physiologica Scandinavia*, **80**, 11–18.

Ørskov, E.R. & Allen, D.M. (1966a). Utilization of salts of volatile fatty acids by growing sheep. 1. Acetate, propionate and butyrate as sources of energy for young growing lambs. *British Journal of Nutrition*, **20**, 295–17.

Ørskov, E.R. & Allen, D.M. (1966b). Utilization of salts of volatile fatty acids by growing sheep. 2. Effects of frequency of feeding on the utilization of acetate and propionate by young lambs. *British Journal of Nutrition*, **20**, 509–17.

Ørskov, E.R., Grubb, D.A., Smith, J.S., Webster, A.J.F. & Corrigall, W. (1979). Efficiency of utilization of volatile fatty acids for maintenance and energy retention by sheep. *British Journal of Nutrition*, **41**, 541–51.

Orst, B.S., Nagy, K.A. & Ricklefs, R.E. (1987). Energy utilization by Wilson's storm petrel (*Oceanitis oceanicus*). *Physiological Zoology*, **60**, 200–10.

Oscai, L.B., Brown, M.M. & Miller, W.C. (1984). Effect of dietary fat on food intake, growth and body composition in rats. *Growth*, **48**, 415–24,

Osuji, P.O. Gordon, J.G. & Webster, A.J.F. (1975). Energy exchanges associated with eating and rumination in sheep given grass diets of different physical forms. *British Journal of Nutrition*, **34**, 59–71.

Owen, E., Kavle, Owen, R.S., Polansky, M., Caprio, S., Mozzoli, M.A., Kendrick, Z.V., Bushman, M.C. & Boden, G. (1986). A reappraisal of caloric requirements in healthy women. *American Journal of Clinical Nutrition*, **44**, 1–19.

Paechtner, J. (1931). Der Gaswechsel. In *Handbuch der Ernährung und des Stoffwechsels der Landwirtschaftlichen Nutztiere*, Band 2, ed. E. Mangold. Berlin: Springer-Verlag.

Pagan, J.D. & Hintz, H.F. (1986). Equine energetics. 2. Energy expenditure in horses during sub-maximal exercise. *Journal of Animal Science*, **63**, 822–30.

Parks, J.R. (1982). *A Theory of Feeding and Growth*. Berlin: Springer-Verlag.

Partridge, G.G., Fuller, M.F. & Pullar, J.D. (1983). Energy and nitrogen metabolism of lactating rabbits. *British Journal of Nutrition*, **49**, 507–16.

Partridge, G.G., Lobley, G.E. & Fordyce, R.A. (1986). Energy and nitrogen metabolism of lactating rabbits during pregnancy, lactation and concurrent pregnancy and lactation. *British Journal of Nutrition*, **56**, 199–207.

Passmore, R. (1967). Energy balance in man. *Proceedings of the Nutrition Society*, **26**, 97–101.

Passmore, R. & Durnin, J.V.G.A. (1967). *Energy, Work and Leisure*. London: Heinemann Educational Books.

Passmore, R., Meiklejohn, A.P., Dewar, A.P. & Throw, R.K. (1955). Energy utilization in overfed thin young men. *British Journal of Nutrition*, **9**, 20–7.

Paul, A.A. & Southgate, D.A.T. (1978) *McCance & Widdowson's the Composition of Foods*, 4th edn. London: HMSO.

Payne, P.R. & Wheeler, E.F. (1968). Comparative nutrition in pregnancy and lactation. *Proceedings of the Nutrition Society*, **27**, 129–38.

Pearson, O.P. (1950). The metabolism of humming birds. *Condor*, **52**, 145–52.

Pedley, T.J. (1977). *Scale Effects in Animal Locomotion*. London: Academic Press.

Pembrey, M.S. (1898). Animal heat. In *Textbook of Physiology*, vol. 1, ed. E.A. Schafer, p. 838. London: Hodder & Stoughton.

Pennycuick, C.J. (1969). The mechanics of bird migration. *Ibis*, **111**, 525–56.

Pennycuick, C.J. (1972). Soaring behaviour and performance of some East African birds observed from a motor glider. *Ibis*, **114**, 178–217.

Pennycuick, C.J. (1975a). On the running of the gnu (*Connochaetis taurinus*) and other animals. *Journal of Experimental Biology*, **63**, 775–99.

Pennycuick, C.J. (1975b). Mechanics of flight. In *Avian Biology*, vol. 5, ed. D.S. Farmer & J.R. King, pp. 1–73. New York: Academic Press.

Perry, S.V. (1985). The biochemistry and physiology of the muscle cell. *Proceedings of the Nutrition Society*, **44**, 235–43.

Peters, R.H. (1983). *The Ecological Implications of Body Size*. Cambridge: Cambridge University Press.

Petrusewicz, K. & MacFadyen, A. (1970). *Productivity of Terrestial Animals: Principle and Methods*. Oxford: Blackwell.

Phillips, J.G., Butler, P.J. & Sharp, P.J. (1985). *Physiological Strategies in Avian Biology*. Glasgow: Blackie.

Pinshaw, B., Fedak, M.A. & Schmidt-Nielsen, K. (1977). Terrestial locomotion in penguins: It costs more to waddle. *Science*, **195**, 592–4.

Pitts, G.C. & Bullard, T.R. (1968). *Body Composition of Animals and Man*. Washington, DC: National Academy of Sciences.

Poczopko, P. (1971). Metabolic levels in adult homeotherms. *Acta Theriologica*, **16**, 1–21.

Pospisilova, D. & Janksy, L. (1976). Effects of various adaptive temperatures on oxidative capacity of brown adipose tissue. *Physiologia Bohemoslovaca*, **25**, 519–22.

Prange, H.D. & Schmidt-Nielsen, K. (1970). The metabolic cost of swimming in ducks. *Journal of Experimental Biology*, **53**, 763–7.

Prentice, A.M., Coward, W.A., Davies, H.L., Murgatroyd, P.R., Black, A.E., Goldberg, G.R., Ashford, J., Sawyer, M. & Whitehead, R.G. (1985). Unexpectedly low levels of energy expenditure in healthy women. *Lancet*, **i**, 1419–22.

Prentice, A.M. & Whitehead, R.G. (1987). The energetics of human reproduction. *Symposium of the Zoological Society of London*, **57**, 275–304.

Priestley, J. (1774). *Experiments and Observations on Different Kinds of Air*. London: J. Johnson.

Prothero, J.W. (1979). Maximal oxygen consumption in various animals and plants. *Comparative Biochemistry and Physiology*, **64A**, 463–6.

Pugh, L.G.C.E. (1971). The influence of wind resistance in running and walking and the efficiency of work against horizontal and vertical forces. *Journal of Physiology (London)*, **48**, 518–22.

Quinn, T.J. & Martin, J.E. (1985). A radiometric determination of the Stegan-Boltzmann constant and thermodynamic temperature between −40 and +100 °C. *Philosophical Transactions of the Royal Society of London*, **316**, 85–189.

Rahn, H., Paganelli, C.V. & Ar, A. (1975). Relation of avian egg weight to body weight. *Auk*, **92**, 750–65.

Rand, R.P., Burton, A.C. & Ing, T. (1965). The tail of the rat in temperature regulation and acclimatization. *Canadian Journal of Physiology and Pharmacology*, **43**, 257–67.

Rashevsky, N. (1960). *Mathematical Biophysics*. New York: Dover.

Rayner, J.M.V. (1979). A vortex theory of animal flight. *Journal of Fluid Mechanics*, **91**, 697–763.

Rayner, J.M.V. (1982). Avian flight energetics. *Annual Review of Physiology*, **44**, 109–19.

Reeds, P.J. & Fuller, M.F. (1983). Nutrient intake and protein turnover. *Proceedings of the Nutrition Society*, **42**, 463–71.

Reeds, P.J., Fuller, M.F. & Nicholson, B.A. (1985). Metabolic basis of energy expenditure with particular reference to protein. In *Substrate and Energy Metabolism in Man*, ed. J.S. Garrow & D. Halliday, pp. 46–57. London: J. Libbey.

Reeds, P.J. & Harris, C.I. (1981). Protein turnover in animals: man in his context. In *Nitrogen Metabolism in Man*, ed. J.C. Waterlow & J.M.L. Stephen, pp. 391–408. Barking, Essex: Applied Science Publishers.

Reeds, P.J., Wahle, K.W.J. & Haggarty, P. (1982). Energy costs of protein and fat synthesis. *Proceedings of the Nutrition Society*, **41**, 155–9.

Richards, F.J. (1959). A flexible growth function for empirical use. *Journal of Experimental Botany*, **10**, 290–300.

Richards, S.A. (1970). The biology and comparative physiology of thermal panting. *Biological Reviews*, **45**, 223–64.

Richards, S.A. (1976). Evaporative water loss in domestic fowls and its partition in relation to ambient temperature. *Journal of Agricultural Science (Cambridge)*, **87**, 527–32.

Richards, S.A. (1977). The influence of loss of plumage on temperature regulation in laying birds. *Journal of Agricultural Science (Cambridge)*, **89**, 393–8.

Riclefs, R.E. (1974). Energetics of reproduction in birds. In *Avian Energetics*, ed. R.A. Paynter Jr., pp. 152–297. Cambridge, Massachusetts: Nuttall Ornithological Club.

Robbins, C.T. & Robbins, B.L. (1979). Fetal and neonatal growth patterns and maternal reproductive effort in ungulates and sub-ungulates. *American Naturalist*, **114**, 101–16.

Robinson, J.J. (1977). The influence of maternal nutrition on ovine foetal growth. *Proceedings of the Nutrition Society*, **36**, 9–16.

Robinson, J.J. (1986). Changes in body composition during pregnancy and lactation. *Proceedings of the Nutrition Society*, **45**, 71–80.

Robinson, J.J., McDonald, I., Fraser, C. & Crofts, R.M.J. (1977). Studies on reproduction in prolific ewes. 1. Growth of the products of conception. *Journal of Agricultural Science (Cambridge)*, **88**, 539–52.

Robinson, J.J., McDonald, I., Fraser, C. & Gordon, J.G. (1980). Studies on reproduction in prolific ewes. 6. The efficiency of energy utilization for conceptus growth. *Journal of Agricultural Science (Cambridge)*, **94**, 331–8.

Robinson, W.R., Peters, R.H. & Zimmermann, J. (1983). The effects of body size and temperature on metabolic rate of organisms. *Canadian Journal of Zoology*, **61**, 281–8.

Romanoff, A.L. (1967). *Biochemistry of the Avian Embryo*. New York: Wiley.

Rook, J.A.F. & Thomas, P.C. (1983). *Nutritional Physiology of Farm Animals*. London: Longmans.

Root, H.K. & Benedict, F.G. (1926). Insensible perspiration, its relation to human physiology and pathology. *Archives of Internal Medicine*, **38**, 1–34.

Rosenmann, M., Morrison, P.R. & Feist, D. (1975). Seasonal changes in metabolic capacity of red-backed voles. *Physiological Zoology*, **48**, 303–10.

Rothell, N.J. & Stock, M.J. (1979). A role for brown adipose tissue in diet induced thermogenesis. *Nature (London)*, **281**, 31–5.

Rothwell, N.H. & Stock, M.J. (1982). Effects of feeding a palatable cafeteria diet on energy balance in young and adult lean (+/?) Zucker rats. *British Journal of Nutrition*, **47**, 461–71.

Rothwell, N.H. & Stock, M.J. (1983). Diet induced thermogenesis. In *Mammalian Thermogenesis*, ed. L. Girardier & M.J. Stock, pp. 208–33. London: Chapman & Hall.

Rothwell, N.J. & Stock, M.J. (1985). Biological distribution and significance of brown adipose tissue. *Comparative Biochemistry and Physiology*, **82A**, 745–51.

Rothwell, N.J. & Stock, M.J. (1986). Influence of environmental temperature on energy balance, diet induced thermogenesis and brown fat activity in 'cafeteria fed' rats. *British Journal of Nutrition*, **56**, 123–9.

Roy, J.H.B., Huffman, C.F. & Reineke, E.P. (1957). The basal metabolism of the new-born calf. *British Journal of Nutrition*, **11**, 373–81.

Rubenstein, D.I. (1985) Evolutionary ecology of mammalian life histories and social organisation. (Quoted by May & Rubenstein (1984): original not seen).

Rubner, M. (1883). Über die einfluss der körpergrosse auf stoff und kraftwechsel. *Zeitschrift fur Biologie*, **19**, 535–62.

Rubner, M. (1885). Calorimetrische untersuchungen. *Zeitschrift fur Biologie*, **42**, 261–75.

Rubner, M. (1902). *Die Gesetz des Energieverbrauchs bei der Ernährung*. Leipzig: Franz Deuticke.

Sacher, G.A. & Staffeldt, E.F. (1974). Relation of gestation time to brain weight for placental mammals: implications for the growth of vertebrate young. *American Naturalist*, **108**, 593–615.

Schiemann, R. (1958). *Kritische Betrachtungen über die Entwicklung der Starkewertlehre Oskar Kellners*, Abh. Deutsche Akademie Landwirtschaftswissenschaften No 31. Berlin: Akademieverlag.

Schiemann, R., Bock, H.D., Keller, J., Hoffman, L., Krawielitzki, K. & Klein, M. (1983). Methodische untersuchungen zum eiweisumsatz und zur bioenergetic des eiweisansatzes bei wachsenden tieren. *Archives für Tierernährung*, **33**, 57–74.

Schiemann, R., Hoffman, L. & Nehring, K. (1961). Die ververtung reiner nahrstoff. 2. Versuche mit schweinen. *Archive für Tierernährung*, **11**, 265–83.

Schiemann, R., Nehring, K., Hoffman, L., Jentsch, W., & Chudy, A. (1971). *Energetische*

Futterbewertung und Energienormen. Berlin: Deutsche Landwirtschaftverlag.

Schmidt-Nielsen, K. (1964). *Desert Animals. Physiological Problems of Heat and Water*. Oxford: Clarendon Press.

Schmidt-Nielsen, K. (1984). *Scaling: Why Animal Size is so Important*. Cambridge: Cambridge University Press.

Schmidt-Nielsen, K. (1985). *Animal Physiology: Adaptation and Environment*. Cambridge: Cambridge University Press.

Schmidt-Nielsen, K., Crawford, E.C. & Hammel, H.T. (1981a). Respiratory water loss in the camel. *Proceedings of the Royal Society of London, Series B*, **211**, 291–303.

Schmidt-Nielsen, K., Schroter, R.C & Shkolnik, A. (1981b). Desaturation of exhaled air in camels. *Proceedings of the Royal Society of London, Series B*, **211**, 305–19.

Schoenheimer, R. (1942). *The Dynamic State of Body Constituents*. Cambridge, Massachusetts: Harvard University Press.

Schoenheimer R. & Rittenberg, D. (1940). The study of intermediary metabolism of animals with the aid of isotopes. *Physiological Reviews*, **20**, 218–48.

Schofield, C. (1985). An annotated bibliography of source material for basal metabolic rate data. *Human Nutrition: Clinical Nutrition*, **39C**, Suppl. **1**., 42–91.

Schofield, W.N. (1985). Predicting basal metabolic rate, new standards and review of previous work. *Human Nutrition: Clinical Nutrition*, **39C**, Suppl. **1**., 5–41.

Scholander, P.F., Hock, R., Walters V. & Irving, L. (1950). Heat regulation in some arctic and tropical mammals and birds. *Biological Bulletin*, **99**, 237–58.

Scholes, T.A. & Hinkle, P.C. (1984). Energetics of ATP-driven electron transfer from cytochrome C to fumarate and from succinate to NAD in submitochondrial particles. *Biochemistry*, **23**, 3341–45.

Schulz, A.R. (1978). Simulation of energy metabolism in the simple stomached animal. *British Journal of Nutrition*, **39**, 235–54.

Schutz, Y. (1984). Terminology: factors and constants in studies on energy metabolism in humans. In *Human Energy Expenditure*, Report No. 5 of Action Project on Nutrition in the European Community, ed. A.J.H. van Es, pp. 169–81. Wageningen: Euro-Nut.

Sclafani, A.A. & Springer, D. (1976). Dietary obesity in adult rats: similarities to hypothalamic and human obesity syndromes. *Physiology Behaviour*, **17**, 461–71.

Searle, T.W. (1970). Prediction of body composition of sheep from tritiated water space and body weight – test of published equations. *Journal of Agricultural Science (Cambridge)*, **75**, 497–500.

Searle, T.W., Graham, N. McC., & O'Callaghan, M. (1972). Growth in sheep. 1. The chemical composition of the body. *Journal of Agricultural Science (Cambridge)*, **79**, 371–82.

Seguin, A. & Lavoisier, A.L. (1793). *Premier Memoire sur la Respiration des Animaux*. Paris: Memoire de l'Academie des Sciences.

Sellers, E.A. & Scott, J.W. (1954). Electrical activity of skeletal muscle of normal and acclimatized rats on exposure to cold. *American Journal of Physiology*, **177**, 372–6.

Shetty, P.S. (1984). Adaptive change in basal metabolic rate and lean body mass in chronic undernutrition. *Human Nutrition: Clinical Nutrition*, **38C**, 443–52.

Shkolnik, A. (1980). Energy metabolism in hedgehogs: 'primitive' strategies? In *Comparative Physiology: Primitive Mammals*, ed. K. Schmidt-Nielsen, L. Bolis & C.R. Taylor, pp. 148–54. Cambridge: Cambridge University Press.

Short, R.V. (1984). Species differences in reproductive mechanisms. In *Reproductive Strategies. 4. Reproductive Fitness*, ed. C.R. Austin & R.V. Short, pp. 24–61. Cambridge: Cambridge University Press.

Simpson, A.M., Webster, A.J.F., Smith, J.S. & Simpson, C.A. (1978). Energy and nitrogen metabolism of red deer in cold environments: a comparison with cattle and sheep. *Comparative Biochemistry and Physiology*, **60A**, 251–6.

Simon, E. (1988). Glossary of terms for thermal physiology. *European Journal of Physiology*, **410**, 567–87.

Siri, W.E. (1961). Body volume measured by gas dilution. In *Techniques for Measuring Body Composition*, ed. J. Brozek & A. Henschel, p. 223. Washington DC: National Academy of Sciences.

Sleeth, C.K. & van Liere, E.J. (1937). The effect of environmental temperature on the emptying time of the stomach. *American Journal of Physiology*, **118**, 272–5.

Snellen, J.W., Mitchell, D. & Wyndham, C.H. (1970). Heat of evaporation of sweat. *Journal of Applied Physiology*, **29**, 40–4.

Southgate, D.A.T. & Durnin, J.V.G.A. (1970). Calorie conversion factors. An experimental re-assessment of the factors used in the calculation of the energy value of human diets. *British Journal of Nutrition*, **24**, 517–35.

Southwood, T.R.E. (1981). Bionomic strategies and population parameters. In *Theoretical Ecology*, 2nd edn, ed. R.M. May. pp. 30–52. Oxford: Blackwell.

Spray, C.M. & Widdowson, E.M. (1950). The effect of growth and development on the composition of mammals. *British Journal of Nutrition*, **4**, 332–53.

Spurr, G.B., Hutt, B.K. & Horvath, S.M. (1954). Responses of dogs to hypothermia. *American Journal of Physiology*, **179**, 139–45.

Steiniger, J. (1983). *Untersuchungen zum tagesenergieumsatz und zur Nährungsinduzierten Thermogenese bei Adipositas*. Dissertation, Akademie Wissenschaft, DDR.

Stokes, R.H. (1967). Thermodynamics of aqueous urea solutions. *Australian Journal of Chemistry*, **20**, 2087–100.

Sturtevant, J.M. (1945). Calorimetry. In *Physical Methods of Organic Chemistry*, vol. 1, ed. A. Weissenberg, pp. 311–434. New York: Interscience.

Swain, S. & Farrell, D.J. (1975). Effects of different temperature regimens on body composition and carry-over effects on energy metabolism of growing chicken. *Poultry Science*, **54**, 513–7.

Swan, H. (1972). Comparative metabolism: surface versus mass solved by the hibernators. In *Bioenergetics*, ed R.E. Smith, pp. 25–31. USA: Federation of American Societies for Experimental Biology.

Swift, R.W. (1931). The effects of low environmental temperature on metabolism. 2. The influence of shivering, subcutaneous fat and skin temperature on heat production. *Journal of Nutrition*, **5**, 227–49.

Swift, R.W. & French, C.E. (1954). *Energy Metabolism and Nutrition*. New York: Scarecrow Press.

Swinbank, W.C. (1963). Long wave radiation from clear skies. *Quarterly Journal of the Royal Meteorological Society*, **89**, 339–51.

Sykes, A.H. (1977). Nutrition–environment interactions in poultry. In *Nutrition and the Climatic Environment*, ed. W. Haresign & H. Swan, pp. 17-29.

Tadesse, K. & Eastwood, M.A. (1978). Metabolism of dietary fibre components in man assessed by breath hydrogen and methane. *British Journal of Nutrition*, **40**, 393–6.

Taylor, C.R. & Heglund, N.C. (1982). Energetics and mechanics of terrestial locomotion. *Annual Reviews of Physiology*, **44**, 97–107.

Taylor, C.R., Heglund, N.C. & Maloiy, G.M.O. (1982). Energetics and mechanics of terrestial locomotion. 1. Metabolic energy consumption as a function of speed and body size in birds and mammals. *Journal of Experimental Biology*, **97**, 1–21.

Taylor, C.R., Heglund, N.C., McMahon, T.A. & Looney, T.R. (1980b). Energetic cost of generating muscular force during running: a comparison of large and small animals. *Journal of Experimental Biology*, **86**, 9-18.

Taylor, C.R., Maloiy, G.M.O., Weibel, E.R., Langman, V.A., Zamau, J.M.Z., Seeherman, H.J. & Heglund, N.C. (1980a). Design of the mammalian respiratory system. 3. Scaling maximum aerobic capacity to body mass: wild and domestic animals. *Respiratory Physiology*, **44**, 25–37.

Taylor, C.R., Schmidt-Nielsen, K. & Raab, J.L. (1970). Scaling of energetic cost of running to body size in mammals. *American Journal of Physiology*, **219**, 1104–7.

Taylor, P.M. (1960). Oxygen consumption in new-born rats. *Journal of Physiology (London)*, **154**, 153–68.

Taylor, St. C.S. (1965). A relation between mature weight and time taken to mature in mammals. *Animal Production*, **7**, 203–20.

Thomson, D.J. & Cammell, S.B. (1979). The utilization of chopped and pelleted lucerne *Medicago sativa* by growing lambs. *British Journal of Nutrition*, **41**, 297–310.

Titus, H.W. & Yull, M.A. (1928). The growth of Rhode Island Reds and the effects of feeding skim milk on the constants of their growth curves. *Journal of Agricultural Research*, **36**, 515–40.

Tomasi, T.E. (1985). Basal metabolic rates and thermoregulatory activities in four small mammals. *Canadian Journal of Zoology*, **63**, 2534–7.

Toutain, P.L., Toutain, C., Webster, A.J.F. & McDonald, J.D. (1977). Sleep and activity, age and fatness and the energy expenditure of confined sheep. *British Journal of Nutrition*, **38**, 445–54.

Tsuzuki, T., Harper, D.O. and Hunt, H. (1958). Heats of combustion. 7. The heats of combustion of some amino acids. *Journal of Physical Chemistry Reference Data*, **62**, 1594.

Tucker, V.A. (1968). Respiratory exchange and evaporative water loss in the flying budgerigar. *Journal of Experimental Biology*, **48**, 67–87.

Tucker, V.A. (1973). Bird metabolism during flight: evaluation of a theory. *Journal of Experimental Biology*, **58**, 589–709.

Tyler, P. & Calder, P. (1985). *Fish Energetics*. New York: Croom Helm.

Tyndale-Biscoe, H. (1973). *The Life of Marsupials*. London: Edward Arnold.

Tracy, C.R., Hammond, K.A., Lechleitner, R.A., Smith, W.J. 2nd, Thompson, D.B., Whicker, A.D. and Williamson, S.C. (1983). Estimating clear day radiation: an evaluation of three models. *Journal of Thermal Biology*, **8**, 247–51.

Tyrrell, H.F., Reynolds, P.J. & Moe, P.W. (1976). Effect of basal ration consumed upon utilization of acetic acid for lipogenesis by mature cattle. In *Energy Metabolism of Farm Animals*. ed. M. Vermorel, pp. 57–60. Clermont-Ferrand: G. de Bussac.

Underwood, C.R. & Ward, E.J. (1966). The solar radiation area of man. *Ergonomics*, **9**, 155–68.

United States National Academy of Sciences (1969). *US–Canadian Tables of Feed Composition*, Publication No. 1684, National Research Council. Washington, DC: National Research Council.

Utter, J.M. & LeFebre, E.A. (1970). Energy expenditure for free flight by the purple martin (*Progne subis*). *Comparative Biochemistry and Physiology*, **35**, 713–19.

Van Es, A.J.H. (1984). *Human Energy Expenditure*. Concerted Action Project on Nutrition in the European Community Rept. No. 5. Wageningen: Euro Nut.

Van Es, A.J.H., De Groot, L. & Voght, J.E. (1986). Energy balances of eight volunteers fed on diets supplemented with either lactitol or saccharose. *British Journal of Nutrition*, **56**, 545–54.

Van Es, A.J.H., Vogt, J.E., Deurenberg, P., van Raay, J. & van der Beek, E. (1982). Energy requirements of man. In *Energy Metabolism of Farm Animals*, ed. A. Ekern & F. Sundstol, pp. 148–51. Aas, Norway: Agricultural University of Norway.

Van Kampen, M. (1976). Activity and energy consumption in laying hens. 1. The energy cost of nesting activity and oviposition. *Journal of Agricultural Science (Cambridge)*, **86**, 471–3.

Van Staveren, W.A. & Durenberg, P. (1984). Validation of a repeated 24-hours recall method of assessing food consumption in a selected group of women. In *Human Energy Expenditure*, Concerted Action Project on Nutrition in the European Community Rept. No. 5, ed. A.J.H. van Es. Wageningen: Euro Nut.

Vermorel, M., Remond, B., Vernet, J. & Liamadis, D. (1982). Utilisation of body reserves by high producing cows in early lactation. In *Energy Metabolism of Farm Animals*, ed. A. Ekern & B. Sundstol, pp. 18–21. Aas, Norway: Agricultural University of Norway.

Verstegen, W.M.A., Brasscamp, E.W. & van der Hel, W. (1978). Growing and fattening of pigs in relation to temperature of housing and feeding level. *Canadian Journal of Animal Science*, **58**, 1–13.

Vickery, W.L. & Milar, J.S. (1984). The energetics of huddling by endotherms. *Oikos*, **43**, 88–93.

Vleck, D. (1979). Energy cost of burrowing by the pocket gopher. *Physiological Zoology*, **52**, 122–36.

Vogel, P. (1980). Metabolic levels and biological strategies in shrews. In *Comparative Physiology: Primitive Mammals*, ed. K. Schmidt-Nielsen, L. Boris & C.R. Taylor. Cambridge: Cambridge University Press.

Vohra, P., Wilson, D.O. & Stopes, T.D. (1975). Meeting the energy needs of poultry. *Proceedings of the Nutrition Society*, **34**, 13-19.

Voit, E. (1901). Über die grosse des energiebedarfes der tiere im hungerzustande. *Zeitschrift für Biologie*, **41**, 120–71.

Wainman, F.W., Dewey, P.J.S. & Boyne, A.W. (1981). *Compound Feeding Stuffs for Ruminants*. Edinburgh: Department of Agriculture and Fisheries for Scotland.

Wainman, F.W., Dewey, P.J.S. & Boyne, A.W. (1984). *Feeding stuffs Evaluation Unit, 4th Report*. Edinburgh: Department of Agriculture and Fisheries for Scotland.

Wallis, R.L. (1979). Responses to low temperatures in small marsupial animals. *Journal of Thermal Biology*, **4**, 105–11.

Walsberg, G.E. (1983). Avian ecological energetics. In *Avian Biology*, vol. 7, ed. D.S. Farmer & J.R. King, pp. 161–220. London: Academic Press.

Walser, E.S. (1977). Maternal behaviour in mammals. *Symposium of the Zoological Society of London*, **41**, 313–31.

Ware, D.M. (1978). Bioenergetics of pelargic fish. *Journal of the Fisheries Board of Canada*, **35**, 220–8.

Warner, A.C. (1981). Rate of passage of digesta through the gut of mammals and birds. *Nutrition Abstracts and Reviews, B*, **51**, 789–820.

Waterlow, J.C. (1986). Notes on the new international estimates of energy requirements. *Proceedings of the Nutrition Society*, **45**, 351–60.

Waterlow, J.C., Garlick, P.J. & Millward, D.J. (1978). *Protein Turnover in Mammalian Tissues and in the Whole Body*. Amsterdam: Elsevier North Holland.

Wathes, C.M. & Clark, J.A. (1981). Sensible heat transfer from the fowl: boundary layer resistance of a model fowl. *British Poultry Science*, **22**, 175–83.

Weast, R.C. (1985). *CRC Handbook of Chemistry and Physics*, 65th edn. Boca Raton, Florida: CRC Press.

Weathers, W.W. (1981). Physiological thermoregulation in heat-stressed birds: consequences of body size. *Physiological Zoology*, **54**, 345–61.

Weathers, W.W., Buttemer, W.A., Hayworth, A.M. & Nagy, K.A. (1984). An evaluation of time budget estimates of daily energy expenditure in birds. *Auk*, **101**, 459–72.

Weaver, M.E. & Ingram, D.L. (1969). Morphological changes in swine associated with environmental temperature. *Ecology*, **50**, 710–13.

Webb, P. (1980). The measurement of the energy exchange in man: an analysis. *American Journal of Clinical Nutrition*, **33**, 1299–310.

Webb, P. (1984). *Human Calorimeters*. New York: Praeger.

Webster, A.J.F. (1971). Prediction of heat losses from cattle exposed to cold outdoor environments. *Journal of Applied Physiology*, **30**, 684–90.

Webster, A.J.F. (1972), Act of eating and its relation to the heat increment of feed in ruminants. *Proceedings of the International Sympoisum on Environmental Biology*, pp. 42–8. USA: Federation of American Societies for Experimental Biology.

Webster, A.J.F. (1974). Heat loss from cattle with particular emphasis on the effect of cold. In *Heat Loss from Animals and Man*, ed. J.L. Monteith & L.E. Mount, pp. 205–23. London: Butterworths.

Webster, A.J.F. (1978). Measurement and prediction of methane production, fermentation heat and metabolism in the tissues of the ruminant gut. In *Ruminant Digestion and Feed Evaluation*, ed. D. Osbourn, D.E. Beever & D.J. Thomson, pp. 8.1–8.10. London: Agricultural Research Council.

Webster, A.J.F. & Blaxter, K.L. (1966). The thermal regulation of two breeds of sheep exposed to air temperatures below freezing point. *Research in Veterinary Science*, **7**, 466–79.

Webster, A.J.F., Osuji, P.O., White, F. & Ingram, J.F. (1975). The influence of food intake on portal blood flow and heat production in the digestive tract. *British Journal of Nutrition*, **34**, 125–39.

Webster, J.D., Welsh, G., Pacy, P.O. & Garrow, J.S. (1986). Description of a human direct calorimeter with a note on the cost of clerical work. *British Journal of Nutrition*, **55**, 1–6.

Weekes, T.E.C. & Webster, A.J.F. (1975). Metabolism of propionate in the tissues of the sheep gut. *British Journal of Nutrition*, **33**, 425–38.

Weir, J.B.deV. (1949). New methods of calculating metabolic rate with special reference to protein metabolism. *Journal of Physiology (London)*, **109**, 1–9.

Weis-Fogh, T. (1977). Dimensional analysis of hovering flight. In *Scale Effects in Animal Locomotion*, ed. T.J. Pedley, pp. 405–20. London: Academic Press.

Welch, W.R. (1984). Temperature and humidity of expired air: interspecific comparisons and significance for loss of respiratory heat and water from endotherms. *Physiological Zoology*, **57**, 366–75.

Wenger, C.B. (1972). Heat of vaporization of sweat: thermodynamic considerations. *Journal of Applied Physiology*, **32**, 456–9.

Westra, R. & Christopherson, R.J. (1976). Effects of cold on digestibility, retention time of digesta, reticulum motility and thyroid hormones in sheep. *Canadian Journal of Animal Science*, **56**, 699–708.

White, R.G. & Yousef, M.K. (1978). Energy expenditure in reindeer walking on roads and on tundra. *Canadian Journal of Zoology*, **56**, 215–23.

Whitelaw, F.G. (1974). Measurement of energy expenditure in the grazing ruminant. *Proceedings of the Nutrition Society*, **33**, 163–72.

Wickler, S.J. (1980). Maximal thermogenic capacity and body temperature of white-footed mice (*Peromysus*) in summer and winter. *Physiological Zoology*, **53**, 338–46.

Wiley, F.H. & Newburgh, L.H. (1931). The doubtful nature of luxuskonsumption. *Journal of Clinical Investigation*, **10**, 733–44.

Wilkie, D.R. (1968). Heat, work and phosphorylcreatine breakdown. *Journal of Physiology (London)*, **195**, 157–63.

Will, L.C. & McCay, C.M. (1943). Ageing, basal metabolism and retarded growth. *Archives of Biochemistry*, **2**, 481–5.

Williams, P.E.V., Pagliani, I., Innes, G.M., Pennie, K., Harris, C.I. & Garthwaite, P. (1987). Effects of a beta-agonist (clenbuterol) on growth, carcass composition, protein and energy metabolism of veal calves. *British Journal of Nutrition*, **57**, 417-28.

Winberg, G.G. (1960). Rate of metabolism and food requirements of fish. *Fisheries Research Board, Translation Services Report No. 194*.

Winchester, C.F. (1943). The energy cost of standing in horses. *Science*, **97**, 24.

Windmueller, H.G. & Spaeth, A.E. (1966). Perfusion *in situ* with tritium oxide to measure hepatic lipogenesis and lipid secretion. *Journal of Biological Chemistry*, **241**, 2891–9.

Worthy, A.J. & Lavigne, D.M. (1987). Mass loss, metabolic rate and energy utilization by harp seal pups during the post weaning fast. *Physiological Zoology*, **60**, 352–64.

Young, B.A. (1966). Energy expenditure and respiratory activity of sheep during feeding. *Australian Journal of Agricultural Research*, **17**, 355–62.

Young, T. (1807). *A Course of Lectures on Natural Philosophy*. London.

Zeuthen, E. (1947). Body size and metabolic rate in the animal kingdom. *Laboratory Carlsberg, Chemical Series*, **26**, 17-165.

Zeuthen, E. (1953). Oxygen uptake related to body size in organisms. *Quarterly Review of Biology*, **28**, 1–12.

Zuntz, N. & Hagemann, O. (1898). Untersuchungen über den stoffechsel des pferdes bei ruhe und arbeit. *Landwirtschaft Jahrbuch*, **27(3)**, 440.

Zuntz, N. & Schumberg, W.A.E.F. (1901). *Studien zu einer Physiologie des Marsches*. Berlin: Hirschwald.

INDEX OF ANIMAL SPECIES

Mammals and birds mentioned in the text are here indexed according to their taxonomic classes and orders. Broad classifications – such as 'herbivora' or 'ungulates' are not indexed.

SUBJECT INDEX